T0339772

ENERGY TRANSFORMATION TOWARDS SUSTAINABILITY

ENERGY TRANSFORMATION TOWARDS SUSTAINABILITY

Edited by

MANUELA TVARONAVIČIENĖ
Professor, Department of Business Technologies and Entrepreneurship
Faculty of Business Management
Vilnius Gediminas Technical University
Vilnius, Lithuania

BEATA ŚLUSARCZYK
Associate Professor, Faculty of Management
Institute of Logistics and International Management
Czestochowa University of Technology
Czestochowa, Poland

ELSEVIER

Elsevier
Radarweg 29, PO Box 211, 1000 AE Amsterdam, Netherlands
The Boulevard, Langford Lane, Kidlington, Oxford OX5 1GB, United Kingdom
50 Hampshire Street, 5th Floor, Cambridge, MA 02139, United States

Notices
Knowledge and best practice in this field are constantly changing. As new research and experience broaden
our understanding, changes in research methods, professional practices, or medical treatment may become
necessary.

Practitioners and researchers must always rely on their own experience and knowledge in evaluating and
using any information, methods, compounds, or experiments described herein. In using such information or
methods they should be mindful of their own safety and the safety of others, including parties for whom
they have a professional responsibility.

To the fullest extent of the law, neither the Publisher nor the authors, contributors, or editors, assume any
liability for any injury and/or damage to persons or property as a matter of products liability, negligence or
otherwise, or from any use or operation of any methods, products, instructions, or ideas contained in the
material herein.

Library of Congress Cataloging-in-Publication Data
A catalog record for this book is available from the Library of Congress

British Library Cataloguing-in-Publication Data
A catalogue record for this book is available from the British Library

ISBN: 978-0-12-817688-7

For information on all Elsevier publications visit our website at
https://www.elsevier.com/books-and-journals

Publisher: Brian Romer
Acquisition Editor: Graham Nisbet
Editorial Project Manager: Ali Afzal-Khan
Production Project Manager: Selvaraj Raviraj
Cover Designer: Miles Hitchen

Typeset by TNQ Technologies

Contents

Contributors

Juozas Baublys Independent expert

Gatis Bazbauers Institute of Energy Systems and Environment, Riga Technical University, Riga, Latvia

Jacek Brożyna Rzeszow University of Technology, The Faculty of Management, Department of Quantitative Methods, Rzeszów, Poland

Dainius Genys Energy Security Research Center at Vytautas Magnus University, Kaunas, Lithuania

Ruth C. Hughes Sidhu School of Business and Leadership, Wilkes University, Wilkes-Barre, PA, United States

Martina Iorio Università degli Studi Roma Tre, Rome, Italy

Tadas Jakstas Energy Security Expert at NATO Energy Security Centre of Excellence, NATO Civilian Expert on Energy Resilience and Crisis Management

Dileta Jatautaitė General Jonas Žemaitis Military Academy of Lithuania, Vilnius, Lithuania

Justinas Juozaitis Research Centre, General Jonas Žemaitis Military Academy of the Republic of Lithuania, Vilnius, Lithuania

Girts Karnitis Faculty of Computing, University of Latvia, Riga, Latvia

Edvins Karnitis Faculty of Computing, University of Latvia, Riga, Latvia

Sebastian Kot North-West University, Faculty of Economic and Management Sciences, South Africa and Faculty of Management, Czestochowa University of Technology, Czestochowa, Poland

Grzegorz Mentel Department of Quantitative Methods, Rzeszow University of Technology

Salvatore Monni Università degli Studi Roma Tre, Rome, Italy; Roma Tre University, Rome, Italy

Joanna Nakonieczny Department of Finance, Banking and Accountancy, Rzeszow University of Technology

Gunnar Prause Department of Business Administration, Tallinn University of Technology, Tallinn, Estonia

Jurgita Raudeliūnienė Vilnius Gediminas Technical University, Vilnius, Lithuania

Ugis Sarma Institute of Energy Systems and Environment, Riga Technical University, Riga, Latvia; JSC Latvenergo, Riga, Latvia

Agnė Šimelytė Department of Economics Engineering, Faculty of Business Management, Vilnius Gediminas Technical University, Vilnius, Lithuania

Beata Ślusarczyk Faculty of Management, Czestochowa University of Technology, Czestochowa, Poland; North-West University, Faculty of Economic and Management Sciences, South Africa

Rasa Smaliukiene Vilnius Gediminas Technical University, Vilnius, Lithuania; General Jonas Žemaitis Military Academy of Lithuania, Vilnius, Lithuania

Marleen A. Troy Department of Environmental Engineering and Earth Sciences, Wilkes University, Wilkes-Barre, PA, United States

Tarmo Tuisk Department of Business Administration, Tallinn University of Technology, Tallinn, Estonia

Manuela Tvaronavičienė Vilnius Gediminas Technical University, Vilnius, Lithuania

Preface

Dear readers,

We offer you a collective monograph on energy consumption issues, written by an international team of authors. Each author provided his/her approach and shared expertise in different, but very much related areas. Together, all the pieces of this book comprise a rich mosaic reflecting complexity of the energy stewardship attempts.

This collective monograph sheds light on such questions as the relationship between economic development and energy demand, consumption of energy and pollution, impact of large infrastructure projects on sustainability, income inequality and energy consumption, energy security concerns, and the best international practices of energy stewardship.

On behalf of this international authors' team, I wish you enjoyable reading.

Let us all, in one way, or another, contribute to our sustainable future!

With kind regards,

Prof. Manuela Tvaronavičienė
Editor
Vilnius Gediminas Technical University
Lithuania

Acknowledgments

Part of the research presented in this book was supported by the European Research Council under the European Union's Horizon 2020 research and innovation program, Marie Sklodowska-Curie Research and Innovation Staff Exchanges ES H2020-MSCA-RISE-2014 CLUSDEVMED (2015-2019) grant number 645730730.

1

Global energy consumption peculiarities and energy sources: role of renewables

Manuela Tvaronavičienė[1], Juozas Baublys[2],
Jurgita Raudeliūnienė[1], Dileta Jatautaitė[3]

[1]Vilnius Gediminas Technical University, Vilnius, Lithuania; [2]Independent expert; [3]General Jonas Žemaitis Military Academy of Lithuania, Vilnius, Lithuania

OUTLINE

Introduction

The World Energy Council predicts the pattern of economy growth, and if it happens, the demand for global energy will increase by 45%—60% (optimistic variant) or by 35% (pessimistic variant in comparison with 2010) by 2030. By 2030 in the European Union (EU), the total energy demand is planned to increase by 15%—20%. According to the International Energy Agency, in 2030 the world primary energy resources will increase to approximately 334 million barrels per day in oil equivalent, i.e., ~1.5 times more than in 2000 when 205 million barrels were consumed. Currently, 41% of the world's primary energy consumption is by the three largest countries (United States, China, and Russia), which possess only 38% of the world's primary energy sources. Although energy resources are diminishing, the demand for them is increasing. In order to reduce the negative impact of production on the environment in the world, more energy-efficient technologies have to be implemented in all areas, and the use of renewable energy sources has to be encouraged.

Increasing energy efficiency is one of the most important directions for sustainable economic development in all states, economic and social structures, ensuring fulfillment of the expectations of human existence.

Limited resources, especially in the future, and their highly uneven geographic distribution, cause a problem of energy security. Ensuring energy security is one of the most important national interests of any state because the national security depends on energy security.

While examining energy security and energy efficiency as interrelated processes, it is necessary to assess global trends in the world and in the EU, the expected shifts by implementing the latest technologies, and global efforts by neutralizing threats of climate change. Decisions should be based on a systematic analysis of potentially possible energy scenarios, covering the diversity of fossil fuels and renewable energy sources, diversification of energy supply and energy availability for consumers at affordable prices.

Modern energy is one of the main means of human survival in the world. Energy policy is an area of economic and political activity of the state, involving all energy sectors related to the existing major resources of the Earth and various types of energy production and use. Economic activity in the energy field is about exploration, production, storage, supply, rational distribution, sale, and distribution of energy materials and products (oil, natural and other gases, coal, fossil fuel, and renewable energy resources).

At present, the world is focusing on shale gas production. It is expected that the gas extracted from shale will significantly reduce natural gas consumption. By 2009, the United States had extracted a large amount of shale gas and consumed 650 billion m^3 gas in total, bypassing Russia by more than 40 billion m^3.

In 1992, at the United Nations Conference "Environment and Development" a program on environmental protection and international development promotion for the 21st century was endorsed. According to it, the world energy strategy should be based not only on technical and economic criteria but also on comprehensive, cohesive activity of society. The development of long-term energy policies of individual countries has to be based on the modeling of a potential future, from medium- to long-term scenarios, with broad-based insight into the development of potential political, technological, and environmental factors. The most significant feature of current energy supply systems is the dominance of fossil fuel in the global energy balance, and hence, the dependence of many countries on energy imports is increasing. Industrial development, mechanization, and automation of production processes, population growth and increasing mobility, the pursuit of better working conditions and greater expectations of comfort and other factors have contributed to the ever-increasing energy needs.

For several decades, there has been a continuous debate on energy status at various levels, potential future generations' problems due to decreasing storage of traditional fossil fuel resources in the crust of the Earth, the dramatic and reckless and overwhelming effects of the use of major energy sources, and the dramatic increase in greenhouse gas emissions in the atmosphere and the high impact of these gases' on climate change and global warming. In

1968 the Rome Club united efforts of 30 experienced scientists, prominent politicians, and entrepreneurs to identify trends in the development of the world economy and to assess the potential dangers. By applying the special method of dynamic modeling (Forrester, 1971), world population figures, the amount of food needed, the amount of natural resources consumed, as well as the volume of industrial production and the extent of environmental pollution were analyzed. A decade ago, after assessing the trends in these factors and their interconnections, a very pessimistic future of the world was projected—if the population continues to grow, consume natural resources, and increase industrial production at the same rate as it has at the beginning of the 21st century, pollution would reach disastrous levels with inevitable ecological crisis (Meadows et al., 1972). The main conclusion of the prognoses was that it time to review the priorities of societal development and to abandon reckless economic growth as the only and most important priority; hence, more attention and funding should be devoted to regulating consumption of all major resources and environmental pollution.

The research results of the Rome Club, which gained great attention and considerable criticism, provided the impetus for an active discussion of scientists, diverse organizations, and the public on a rational world development scenario. In 1987 at the United Nations General Assembly, a special report, entitled "Our Common Future" was presented by the International Environmental and Development Commission, formulating the concept of sustainable development, such that sustainable development allows society to meet its current needs without reducing the opportunities for future generations to meet their own needs (World Commission on Environment and Development, 1987).

The principles of sustainable development were internationally approved in 1992 in the Rio de Janeiro World Summit Declaration. The Sustainable Development Action implementation program, "Agenda 21," was later subsequently adjusted, but other specific commitments for priority areas were approved at the 2002 World Summit of Heads of State and Government in Johannesburg, South Africa. All countries were obliged to prepare and start implementing national sustainable development strategies. In 2006, the Council of the EU approved the updated EU Sustainable Development Strategy and mandated member states to ensure consistency between national sustainable development strategies.

Climate change stabilization is one of the main priorities of the EU energy policy. The common and ambitious goals of sustainable development are addressed by the member states in the EU directives and communications, individual country strategies and others, but real progress in meeting the global goals is still lower than expected.

World energy scenarios

The world economy has increased 3.3 times over 4 decades. Global growth rates are slightly slowing down: from 1971 to 1980 growth rates averaged 3.8%, then fell to 3.0% between 1980 and 1990, in 1990—2000 subsequently fell to 2.8%, and in 2000—11 it was 2.6% per annum. However, the overall downward trend in global economic growth was driven by slower economic growth in the Organization for Economic Co-operation and Development (OECD)—based developed countries, generating gross domestic product (GDP) due to the effects of the global economic crisis of 2000—11, which grew by an average of 1.6% per year. Meanwhile, economic growth in developing countries has accelerated over the past 3 decades: GDP growth rates reached 2.4% between 1980 and 1990; in the period 1990—2000 it was 3.1%; and it was 6.1% in 2000–11 per annum. China's generated GDP in 2000–11 grew by an average of 9.9% per year and increased by 2.8 times in 11 years, whereas the global economic growth index for this period was 32%, while in the United States it was only 19% (IEA, 2012, 2013a, 2013b). According to the World Situation Prospects (2018), 2017 global economic growth reached 3.0%, i.e., a sharp increase to 2.4% in 2016; and since 2011 there has been the largest recorded global growth. Labor market indicators are continuing to improve in many countries, and around two-thirds of the global market in 2017 grew faster than in previous years. It is expected that global growth will remain stable at 3.0% in 2018 and 2019 (United Nations, 2018).

The role of developing countries in the global economy has grown rapidly with the share of GDP generated by it in national currencies in terms of purchasing power parity, rising from 34.6% in 2000 to 46.1% in 2011. The role of countries in the Asian region, and in particular China, grew even faster, with the country's share of GDP in the global economy rising from 7.4% in 2000 to 14.6% in 2011. The role of the US economy in this period fell significantly from 23.1% in 2000 to 18.8% in 2011. Moreover, there is an interesting fact that the economies of the top-level countries retain their existing positions. Comparing the 20 top 1980s countries' GDP, 17 members are on the same list, which means that only three new entrants integrated over the entire period. The nominal GDP of the 10 largest countries (United States, China, Japan, Germany, United Kingdom, India, France, Brazil, Italy, and Canada) accounts for about 67% of the world economy, and the top 20 for almost 81%. The remaining 172 countries together account for less than one-fifth of the global economy (Fig. 1.1) (World Economic Outlook Database, 2018).

In spite of all the fears and calls to change energy consumption trends and the considerable efforts made to replace fossil fuels with renewable energy sources, the total amount of energy consumed worldwide, including oil and its products, natural gas and coal, only increased. Thus, worldwide primary energy consumption in 1971–80 grew by an average of 3.0%; in 1980 and 1990 it reached 2.0%; in 1990–2000 it was 1.4%; and in 2000–11 it was 2.4% per annum. In the period 2000–11, rapid economic growth in developing countries resulted not only in an increase in energy consumption in these countries but also in a global trend in energy demand growth. Exceptionally fast (on average 8.6% per year) primary energy consumption increased in China. In 2009, as a result of the economic crisis, global primary energy

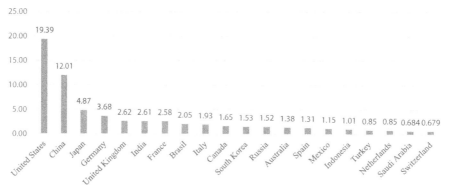

FIGURE 1.1   Nominal GDP in 2017, trillion U.S. dollars.

consumption was 1% lower than in 2008. In 2011 developed world countries consumed the same amount of primary energy as in 2000, 5.3 billion tons (toe). Meanwhile, primary energy consumption in developing countries increased by 66.1%, i.e., from €4.4 billion toe in 2000 to 7.4 billion toe in 2011. Therefore, China according to primary energy consumption indicators in 2009 overtook the United States. In 2011 China consumed 2.7 billion toe primary energy, which increased by 2.5 times over 11 years. The global primary energy growth index for this period was 29.8% (Fig. 1.2).

Primary energy consumption worldwide increased by 10% compared to 2010 and 2017. Thus, in 2017 mainly primary energy consumption was in regions such as Asia Pacific (Australia, Bangladesh, China, China Hong Kong SAR, India, Indonesia, Japan, Malaysia, New Zealand, Pakistan, Philippines, Singapore, South Korea, Sri Lanka, Taiwan, Thailand, Vietnam, and others) (5743.6), North America (United States, Canada, Mexico) (2772.8), Europe (1969.5), Commonwealth of Independent States (CIS) (Azerbaijan, Belarus, Kazakhstan, Russian Federation, Turkmenistan, Ukraine, Uzbekistan, other CIS) (978.0), Middle East (Iran, Iraq, Israel, Kuwait, Oman, Qatar, Saudi Arabia, United Arab Emirates, and others) (897.2), and at least South and Central America (Argentina, Brazil, Chile, Colombia, Ecuador, Peru, Trinidad and Tobago, Venezuela, and other South and Central America countries) (700.6) and Africa (Algeria, Egypt, Morocco, South Africa, and others) (449.5) (Fig. 1.3) (BP Statistical Review of World Energy, 2018).

Energy resources, in particular oil products, used for automotive transport and the widespread penetration of electricity in all areas of human activity, were and still remain one of the most important and essential components of mankind's needs for progress and improvement of living conditions. Natural gas (5.7%), nuclear (3.9%), oil (3.5%), and other (includes

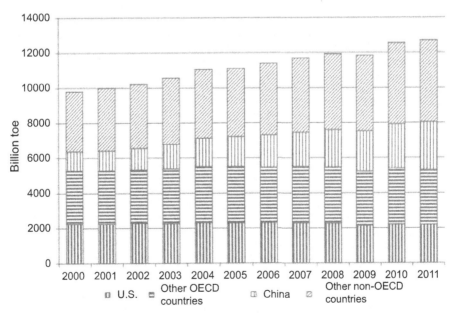

FIGURE 1.2 Global primary energy consumption in 2000–11 (IEA, 2006, 2009, 2012, 2013a, 2013b).

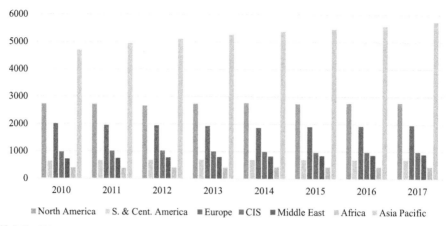

FIGURE 1.3   Primary energy consumption in 2010—17 world (commercially traded fuels, including modern renewables used to generate electricity; Mtoe oil equivalent).

geothermal, solar, wind, tide/wave/ocean, heat, and other) consumption (1.5%) increased compared to 2010 and 2017, while the consumption of coal (10.6%) and biofuel waste (4%) decreased significantly. Oil, natural gas, and coal dominated the world primary energy balance in 2010—2017 (Fig. 1.4) (IEA 2014a; 2014b, 2015, 2016, 2017, 2018).

Nuclear fuel consumption increased from 5.7% to 9.6% in 2010—17, as well as increased geothermal, solar, wind, and tide/wave/ocean heat from 0.9% to 2.4% consumption (Fig. 1.4) (IEA 2013b, 2014a, 2015, 2016, 2017, 2018).

The development of the world's energy is influenced by many factors, the whole of which can be driven by both the different directions of growth of energy needs and the ways of their satisfaction. In 2013 highly acknowledged experts from the International Energy Agency prepared a thorough study on the directions of global energy development by 2035, which outlined three global scenarios: (1) a scenario for existing policies, (2) a scenario for a new policy, and (3) a 450 scenario. A new policy scenario can be considered as a major one that is based

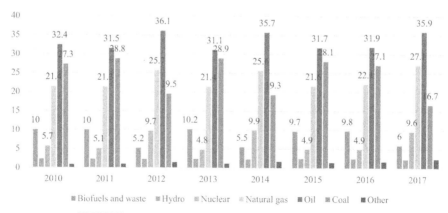

FIGURE 1.4   Global energy consumption structure (2010—17).

on the assumption that energy policy, existing and future technologies will be streamlined and aimed by means of weakening the links between the economy, energy needs, and the growth of greenhouse gas emissions from fuel combustion (IEA, 2013b; Miškinis et al., 2014).

The choice of these scenarios largely depended on two main conclusions drawn from long-term climate change research: (1) it is necessary to slow down the average temperature rise in the world so that it does not exceed 2°C compared to the preindustrial period; (2) this strategic goal can only be achieved if the concentration of greenhouse gases, which is estimated by the number of carbon molecules per million molecules in the atmosphere, does not exceed 450. On the other hand, in recent years much discussion has been conducted on the effects of greenhouse-related factors—whether it is necessary to ensure the rise in temperature below 2°C, what the impact of greenhouse gases is on climate change, or if the impact of the sun on climate change is properly assessed, what the impact of climate change on the world community, individual lifestyles, ecosystems, and water resources is.

All analyzed scenarios show a continuous further increase in energy demands. According to the current policy scenario, it is foreseen that in 2035 the world's primary energy consumption will reach 44%; in the new policy scenario, it will reach 33%; and in the 450 scenario it will be 14% more than in 2011, when the growth of the needs will be determined by the growth of the world economy and population; it is also prognosed that GDP is projected to grow at an average of 3.6% per year and by 2035 will increase 2.3 times, and the world's population will increase to 8.7 billion.

Significant changes in the world's energy balance will be driven by very different growth rates of fossil fuels and renewable energy sources. Carbon, oil, and natural gas in accordance with the 2035 current policy scenario will be 40% consumed, and due to the new policy scenario, there will be 24% more than in 2011.

If scenario 450 succeeds, fossil fuel consumption by 2020 would increase by 4%, but then it would start to decrease by 2035 and this fuel would be consumed 11% less than in 2011. Contrariwise, in the 2035 scenario 450 renewable energy would be 2.3 times consumed, 77% in the new policy scenario, and 58% more in the current policy scenario in comparison with 2011 when the share of renewable energy in the global primary energy balance increased from 13.2% in 2011 and up to 26% in 450 scenario, then up to 18% in the new policy scenario and only up to 15% in the current policy scenario. All scenarios foresee the growth of nuclear fuel needs in 2035. In the current policy scenario 51% will be consumed, while in the new policy scenario 66%, and in 450 (scenario) two or three times more than in 2011. The purpose of the World Energy Future Scenario was to attempt to assess the potential consequences of the accident at the Fukushima Daiichi nuclear power plant for the development of nuclear energy, but at present the scope and pace of development remain uncertain (IEA, 2013b).

Policy actions that tend to stop the growth of energy needs or to expand the supply of energy from new sources will be critical in the medium to long term. Global needs can grow significantly due to fuel consumption in the transport sector of developing countries. It is expected that despite significant increases in the automobile economy, the growth in demand may be driven by a rapid increase in car sales. In order for alternative engine technologies to become economically viable and saturated, including low-oil products, electric cars, etc., the market requires a longer period.

Growth in energy demand can significantly increase the amount of greenhouse gas emissions to the atmosphere and have a decisive impact on climate change. If current trends persist, by 2035 greenhouse gases will increase to 43.1 Gt or 38%. Following the assumptions of the new policy scenario, greenhouse gas emissions will increase to 37.2 Gt or 19%. Such emissions to the atmosphere in the long term will lead to a dangerous global temperature rise. In order to stabilize climate change, additional energy-saving measures should be implemented as electricity generation from coal-fired power plants, more renewable energy sources, and other measures envisaged in the 450 scenario that would allow in 2035 to reduce greenhouse gas emissions to 21.6 Gt or up 69% from 2011 level.

The prospects of latest global energy development by 2035, developed by the global gas and oil company BP (BP, 2014), estimates the growth of energy demand and the structure indicators of primary energy resources as well as greenhouse gas emission quantities that are close to the new policy scenario prepared by experts from the International Energy Agency.

In the long run, by 2050 the main candidates in the world can become primary energy sources that are considered as renewable sources of energy and control nuclear and thermonuclear reactions energy. The World Energy Council warned 22 participants of the World Energy Congress about prevailing myths that hamper the efforts of governments, the energy industry, and civil society to create a sustainable energy future (World Energy Council, 2013), and outlined the potential prospects for future change:

- *The growth of energy demand in the world will slow down and remain the same.* In fact, the demand for energy will continue to rise and by 2050 it will increase and many cases it will mainly depend on economic growth in developing countries;
- *The peak of oil extraction is inevitable because of lack of fossil fuel resources.* Oil shortage is not encountered yet. Improved extraction is executed from existing deposits, new resource pools and new technologies are being discovered that allow to extract unconventional oil and gas;
- *By increasing energy demand, it will be possible to completely satisfy needs from new clean energy sources.* In spite of noticeable increases in the relative contribution of renewable energy sources that are supposed to reach from 20% to 30% in 2050, fossil fuel in 2050 is believed to increase from 10 to 16 billion toe;
- *By 2050 total greenhouse gas emissions can be reduced by 50%.* Even in the best case scenario, global greenhouse gas emissions by 2050 are expected to increase almost twice as much as the objective of 450 ppm;
- *Current business and market models are appropriate.* Current markets and business models are unable to cope with the problems that are raised by growing the share of renewable energy in the overall primary energy structure, decentralized systems and the growing flow of information;
- *Contemporary programs will provide universal access to energy over the next 10–15 years.* Common access to energy will not yet be the present reality. After implementing existing measures in the 2050 world, between 530 and 880 million people will still live without electricity.

The World Energy Council experts in their official statement expressed an alarming position about the future of energy and call for effective action to transform global energy systems, otherwise there may be serious risks pursuing energy security, energy supplies, and environmental sustainability.

Trends in energy consumption in the EU

In the EU, overall energy consumption over 2 decades (1990–2012) was very uneven. In 1990 the EU consumed 1.67 Mtoe and the consumption rose to 1.83 Mtoe in 2000; in addition, in 2012 it dropped to 1.68 Mtoe. Overall, in the period 2006–12 energy consumption fell to 8% in the EU.

Primary energy consumption in Europe (Austria, Belgium, Czech Republic, Finland, France, Germany, Greece, Hungary, Italy, Netherlands, Norway, Poland, Portugal, Romania, Spain, Sweden, Switzerland, Turkey, United Kingdom, and other European countries) in comparison 2010 to 2017 decreased by 3.5%. In 2017, in Europe primary energy was mostly used by countries such as Germany (335.1), France (237.9), the United Kingdom (191.3), and Turkey (157.7) (Fig. 1.5) (BP Statistical Review of World Energy, 2018).

According to the gross inland energy consumption in Europe 1995–2016 (Mtoe), it has been observed that consumption in 1995 and 2016 decreased by 2%. The largest decrease was recorded for solid fuels (34%), petroleum and products (13.3%), electricity (11.1%), and nuclear (4.7%), whilst waste and nonrenewables (61.7%), renewables (60.9%), and gas (12.2%) increased (Fig. 1.6) (European Commission, 2018).

In the face of increasing dependence on energy imports and decreasing domestic resources, and in order to curb climate change and overcome the economic crisis, on October 25, 2012, the EU adopted their 2012/27/EU Directive on energy efficiency. The purpose of this directive is to ensure that in 2020 compared to 2007 primary energy consumption would

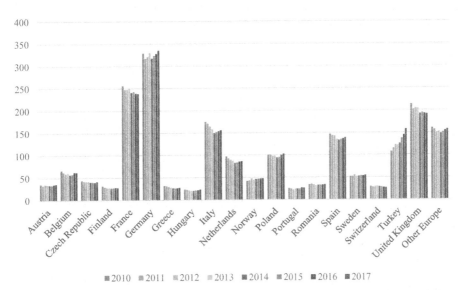

FIGURE 1.5 Primary energy consumption in Europe 2010–17 (commercially traded fuels, including modern renewables used to generate electricity; Mtoe oil equivalent).

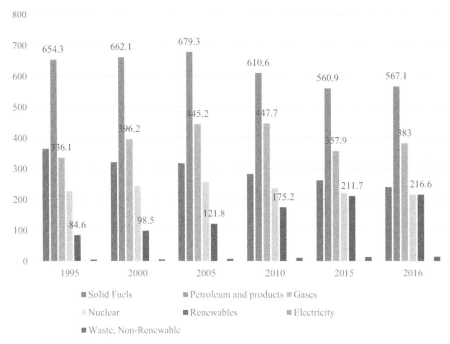

FIGURE 1.6   Gross inland energy consumption in Europe 1995—2016 (Mtoe).

be 20% lower according to Directive (2012)/27/EU forecasts. Based on national indicative energy efficiency targets for individual countries, "by 2020 Member States are aiming at savings of only about 16.4%" (Communication from the Commission to the European Parliament and the Council, 2013). As the economy grows, the needs for final energy, which provides the right conditions for the development of industries, are growing at a somewhat slower pace, as well as the needs of all primary energy sources. The growth of primary energy needs may be slowed down or partly compensated by the implementation of energy efficiency improvement measures in the energy transformation, supply, and consumption chains.

The primary energy intensity indicator describes the overall energy efficiency of all energy sources, from extraction, transformation, transmission, and distribution to end-user installations. The change in primary energy intensity reflects real changes in energy efficiency in each country. When comparing primary energy intensity indicators, two very important aspects need to be taken into account: the size of this indicator is determined not by the amount of energy consumed but by the value of national GDP; the primary energy intensity indicator is highly dependent on the specific features of the country's energy sector, the structure of the energy transformation sector, the losses in energy transformation plants, the import/export ratio of electricity, and the energy resources consumed for nonenergy use.

Purchasing power parity indicators are used to measure the economic level of developed and developing countries. These indicators are based on an assessment of the availability of equal quantities of goods and services in different countries. This principle should undoubtedly be applied to the determination of energy intensity indicators, i.e., GDP generated in all countries from national currencies should be converted into the single international currency (in euros or US dollars at constant prices) using the purchasing power parity method. GDP of developed countries in terms of exchange rate and purchasing power indicators differs little. Meanwhile, in the developing countries, GDP converted to purchasing power parity indicators is 1.5–2.5 times higher than exchange rate indicators. Therefore, the primary energy intensity indicators for developing countries using these two methods can be very different.

Energy resources and energy sources

Primary energy resources are energy sources and supplies that are obtained in nature and used for energy purposes, i.e., electricity energy and heat production. All primary energy resources are divided into two main groups: nonrenewable (finite) and renewable. Nonrenewable (finite) resources are fossils from the depths of the Earth (coal, oil, natural gas [including shale gas], peat, nuclear fuel). Renewable energy resources in nature are constantly being refreshed: hydropower, solar and wind energy, sea tides and low tides, wood and other plants. There is also another group of energy resources such as waste. It is hot water obtained from cooling systems of technological processes, used steam, heat spreading from combustion products burnt in different furnaces, and other flammable waste. Energy resources are measured in terms of energy potential in a ton of oil equivalent (toe) when one toe = 11.63 MWh (Baublys et al., 2011; Baublys, 2014; Nagavičius, 2014).

Global energy consumption is constantly changing. World energy consumes mainly fossil resources. As already mentioned, their stocks are limited and will sooner or later expire. Alternative technologies of energy production are developing rapidly in the world. Consequentially, in the long-term fossil fuels will be replaced by renewable energy.

Despite the fact that the documents provided by the World Energy Council (2013) state that fossil fuels have a very high level of exploration, fossil fuel prices will rise in the future due to difficult accessibility to resources. Easily accessible resources are already mostly drained by nonstop exploitation in the world (Vilemas, 2014).

Various science, production, and business organizations in many countries predict global development of energy consumption. The most widely known global scenarios include: Royal Dutch Shell (Shell)—Dutch and British Oil and Gas Production and Processing Company scenarios; Stockholm Environment Institute (Global Scenario Group); Millennium Project; World Business Council for Sustainable Development; International Energy Agency (IEA); European Renewable Energy Council (EREC); and others.

The EU countries most often rely on the scenarios of Shell and the International Energy Agency when examining energy prospects in their countries. The European Renewable Energy Council on May 27, 2004, published and on December 15, 2005, approved two more accurate world energy consumption scenarios.

The first scenario is called realistic, i.e., Advanced International Policy (AIP) scenario and the second scenario is the Dynamic Current Policy (DCP). According to the AIP scenario, renewable energy is projected to increase by 2040, which could comprise almost 50% of all global energy consumption.

Energy-intensive economies similarly require a lot of energy in the EU. It is expected that primary energy consumption in the EU will grow by 1%–2% annually and will reach 1950

Mtoe by 2030. The fastest and greatest energy consumption is occurring in households, the service sector, and transport. In industry, due to more efficient technologies energy consumption is rising more slowly.

The EU's ability to procure energy from its own internal resources is very limited. In Europe oil is mainly extracted in the North Sea. It is estimated that oil will last up to 25 years, but if the current amount of oil extraction continues at the same current rate, it will last only 8 years. The EU oil accounts for 60% of total energy consumption. Therefore, the EU imports oil from a lot of countries: ~ 50% from OPEC (Iran, Kuwait, Saudi Arabia, Libya, Nigeria, Algeria, and Venezuela), 33% from Russia, 16% from Norway, and others. Prices of oil extracted in the North Sea are from two to seven times higher than in other countries.

It is projected that by 2030–40 at least 75% of oil reserves increase will belong to OPEC countries: Kuwait, Angola, Saudi Arabia, Iraq, and Nigeria.

The resources of natural gas produced in the North Sea are also low. They encompass only 2% of global stocks.

The EU is also facing difficulties in the use of solid fuels (coal, peat, and lignite or brown coal). In the EU the cost of coal production is 4–5 times higher than globally, with the exception of Britain. Most coal consumption is imported from South Africa, Colombia, Indonesia, and other countries. In addition, strict environmental requirements are imposed on coal consumption.

Indicators of the use of energy resources in production. Energy processes are distributed into power, thermal, electrochemical, electrophysical, and lighting.

The efficiency factor (coefficient) or fuel usage factor (coefficient) of any technological process is expressed as:

$$\eta = \eta_k \cdot \eta_{e.g} \cdot \eta_{e.tr} \cdot \eta_{m.p} \cdot \eta_{d.m}; \tag{1.1}$$

here η_k —fuel extraction, transportation and recycling efficiency factor (coefficient); $\eta_{e.g}$ — efficiency factor of energy production; $\eta_{e.tr}$ —energy transport efficiency factor; $\eta_{m.p}$ —the efficiency factor of the transmission mechanism; $\eta_{d.m}$ —the efficiency factor of the machine or the technological aggregate expressed as:

$$\eta_{d.m} = \frac{Q_n + Q_a}{Q_k + Q_{egz} + Q_{kt}}; \tag{1.2}$$

here Q_n — an efficiently consumed heat in the technological process; Q_a — heat of waste energy resources; Q_k — heat produced by burning fuel; Q_{egz} — heat of exothermic (heat release) reactions; Q_{kt} — heat of other energy suppliers.

The energy efficiency factor (coefficient) of the technological process changes over time (usually increases). This indicator can be increased by optimizing the use of waste energy resources. For this purpose, capital investments tend to increase.

Indicators of various energy sources used in any production branch include the level of energy supply (toe/per person; kWh/per person or kW/per person), the capacity of the plant for the production personnel, energy intensity, intensity of the main production funds, energy intensity of production, thermoelectric coefficient, electric fuel factor, electrification factor of power processes (Table 1.1).

TABLE 1.1 Energy supply indicators.

Name of the indicator	Formula	Explanation of variables
Level of energy supply, toe/per person	$\beta_{el.en} = \dfrac{\sum B}{N_{vid}}$	$\sum B$ — Annual requirements of fuel and all types of energy converted into oil equivalent; N_{vid} — the average number of personnel per year.
Energy of work supply, kWh/per person or kW/per person	$\eta_{el.en}^{en} = \dfrac{\sum W}{N_{vid}}$ or $\beta_{el.en}^{g} = \dfrac{\sum P_{i_r}}{N_{vid}}$	$\sum W$ — Cost of annual electricity consumption, kWh; $\sum P_{i_r}$ — the total installed power of all receivers.
Power of equipment for production personnel, kW/per person	$\beta_{d.m} = \dfrac{\sum P_m}{N_{vid}}$	$\sum P_m$ — Total installed power of all types of motors used in shift, kW.
Energy intensity, toe/million EUR	$\beta_{n.en} = \dfrac{\sum Q}{GDP}$	$\sum Q$ — Total cost of energy consumption, toe; GDP is the gross domestic product of a country (in analyzing the energy intensity of individual branches of industry, the energy consumption is divided by the gross value added generated by those industries), million EUR.
Intensity of main production funds, toe/thousand EUR	$\beta_{f.en} = \dfrac{\sum B}{K_{vid}}$	K_{vid} — Average annual value of the main production funds.
Production energy intensity, toe/output per unit	$\beta_{pr} = \dfrac{\sum B}{V}$	V — The annual production output in terms of money or units.
Thermoelectric coefficient, toe/thousand kWh	$\beta_{š.e} = \dfrac{Q}{\sum W}$	$\sum W$ — Annual cost of electricity consumption, kWh.
Electric fuel factor, thousand kWh/toe	$\beta_{e.k} = \dfrac{\sum W}{B}$	B — Annual cost of fuel, toe.
Power process electrification coefficient	$k_{el} = \dfrac{W_e}{W_e + W_{mech}}$	W_e — Electricity consumption of electric transmission, kWh; W_{mech} — Electricity consumption (kWh) of motors with mechanical transmission.

Renewable energy. Renewable energy resources (RES) are biofuels, hydropower, wind energy, geothermal and aerothermal energy, solar energy, bioenergy (energy crops, straw, biogas, landfill gas, municipal waste), and so on. RES are used in three areas: electricity production, heat and biofuel production.

Renewable energy is directly linked to sustainable development strategy. It improves security of energy supply, and it is characterized by almost clean technologies. The use of renewable energy provides opportunities to electrify and provide heat to remote areas. A positive feature of RES is their global accessibility, especially of the sun and the wind.

The potential of renewable energy in the world is inexhaustible. Worldwide RES in 2013 accounted for about 16% of total energy consumption. In 2012, the total installed power of the power plants was 1373 GW (the power of all nuclear power plants in the world was about 300 GW). Among them there are: hydropower capacity (990 GW), wind power capacity (283 GW), and solar PV capacity (100 GW). Around 19.5% of global electricity was generated from RES (REN21, 2012, 2013, 2015, 2017).

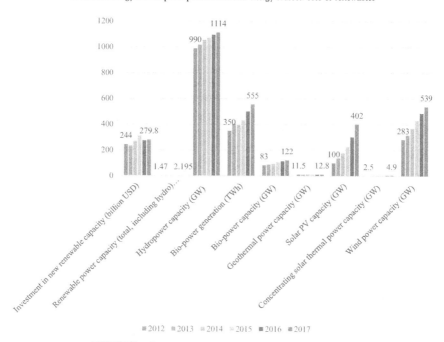

FIGURE 1.7   Renewable energy indicators (2012–17).

Comparing 2012 to 2017, investment in new renewable capacity (billion USD) increased by 13%. During this period, the largest increase was observed in solar PV capacity (75%), with a capacity of 49% and wind power capacity (47%), with the lowest increase in hydropower capacity (11%) and geothermal power capacity (10%). RES utilization is rapidly growing in North and South America, Europe, and Asia. Rapid growth is also found in South Korea, Australia, and France. The leaders in this case are China, the EU, the United States, Brazil, Canada, Russia, and India (Fig. 1.7) (REN21, 2012, 2013, 2015, 2017).

The countries of the world began to absorb RES in the first decade of the 21st century. In countries (pioneers) such as Denmark, Germany, Austria, the United States, Spain, and later in China, the expansion of RES has become a key part of public energy policy, not for economic reasons but because of their understanding of global warming and environmental pollution problems.

There is a great deal of global focus on RES efficiency. RES is a good and perhaps the only real long-term alternative to fossil fuels. However, the widespread use of RES for energy production presents some specific problems. For example, one of the problems is that solar and wind energy is volatile. To compensate for the volatility of power generation systems that are using solar and wind power, two methods are applied: power reservation and energy storage. The power reserve requires additional sources, the power of which can be adjusted by the total instantaneous power of the generators using RES. If RES systems cannot function or if their overall power is not sufficient, reserve energy sources can be activated.

With the expansion of RES technologies, the second method of compensating for energy volatility has become very important—the use of accumulators. Low-power energy storage is commonly used in stand-alone power generation systems. Powerful renewable energy generators often use high-capacity aggregators: high-pressure compressed-air storage underground cavities, hydroelectric power plants, gravity energy storage devices, and others.

The main reasons for the rapid use of renewable resources are as follows: the implementation of RES systems can contribute to reducing environmental pollution; the attractiveness of wind and solar energy systems is increased by the fact that they produce energy without using any fuel; the development of RES systems positively impacts on the economy.

The structure and, in particular, the level of production of renewable energy (AE) varies widely across the EU. On April 23, 2009, European Parliament and Council approved the 2009/28/EC directive on the promotion of the use of energy from renewable sources and set specific aims for each country, which after being realized by 2020, the common share of renewable energy in total final energy consumption should reach at least 20%. In the area of RES production in the EU, Germany is the main leader, then Spain, Italy, France, Sweden, Finland, and Austria. In Italy, about 30% of total RES belongs to geothermal energy. Wind energy is most widely used in Germany, Spain, and Denmark.

It is forecasted that by 2035 in both developed and developing countries, the use of renewable energy sources will grow faster than nuclear energy (also without generating CO_2). Among the countries of the OECD, where energy demand is growing very little, RES will be used mainly in new power plants. In Germany, one of the leading leaders in RES use in the EU, it is estimated that in 2020 at least 40% of electricity will be produced by using RES (23.4% in 2013). Poland, Sweden, and Estonia also envisage extensive use of RES (Vilemas, 2014).

There are many opportunities for expansion with wind power. However, the rapid growth of their generating capacity within the structure may cause a number of problems with total production because of dependence on highly variable wind speed, the management of these power plants in the power system, the lack of required reserve power, electricity quality requirements, and more issues.

Hydropower. Hydropower is one of the energy industries that uses natural water resources. Hydropower is the cheapest, cleanest, and least polluting renewable energies. This energy is found in running or falling water. Water energy can be used in mechanics or power generation.

The amount of energy from this water is called hydropower (Deksnys et al., 2008).

The water flowing through the river is calculated by:

$$W = 9,81\,QHt\eta, \text{ kWh;} \tag{1.3}$$

here Q — river water debit, m^3/s; H — water drop height, m; η — utility factor, t — time, h.
Hydropower (hydraulic) power is calculated:

$$P = \frac{W}{t} = 9,81QH\eta, \text{ kW;} \tag{1.4}$$

The annual hydropower output is calculated as:

$$W_{years} = 8760 \cdot P = 85936QH, \text{ kWh.} \tag{1.5}$$

Hydropower resources in the territory are divided into potential (theoretical), technical, and economic. Potential (theoretical) hydropower resources are the sum of power or energy of all river stretches in the area under investigation. Exceptional stretches of rivers that are technically capable of using water make up technical resources for hydropower. The share of technical hydropower resources is called economic hydropower resources, which are those already in use or potentially could be quite feasible and economically expedient to be used.

From historical sources it is known that water-lifting wheels have been used in Babylon, Egypt, India, and other countries for over 5000 years of ground irrigation. From the second half of the 19th century, hydropower, in the case of power generators, has been most widely used for the production of electricity in hydroelectric power plants (HE), which are divided into large and small power plants. Small HEs are basically different from large HEs. While constructing large HEs, an individual project is designed for each of them. Unified hydraulic and energy equipment, standard series production structures are used for the construction of small HEs. The water dam is the largest, most complex, and expensive of all construction elements in the construction of large HEs.

Prior to the construction of any water economic object, an integrated river use scheme should be established, taking into account the needs of different industries and the various negative impacts on the environment.

Hydropower is closely linked to water protection from river wastage and pollution. Before discharging water waste into rivers and domestic waters, the latter ones should be well cleaned. Hydropower is considered to be a clean means of electricity production.

The price of electricity produced from hydroelectric power plants is considerably low compared to other types of power plants. Water is a renewable source of energy, which helps to diminish the need for fossil fuels such as coal, oil, and natural gas in producing electricity. The advantage of hydroelectric power plants is that they can be switched on or off quickly depending on the change in electricity demand.

The biggest disadvantage of HE is that nature is mostly damaged. Accumulated water bodies land permanently on large areas of agricultural land. HE dams very often disrupt fish migration to spawning grounds.

In the world, the largest shares of electricity generated by hydroelectric power plants are in Norway (99%), Iceland (83%), Canada (70%), and Austria (67%). The world's largest HEs are presented in Table 1.2.

There is no unified definition of small hydroelectric power plants (SHEs) in the EU. Belgium, Greece, Iceland, Spain, and Portugal consider SHEs as those with installed capacity less than 10 MW; in Italy, a SHE is up to 3 MW; in France, it is up to 5 MW. However, the European Commission, European Small Hydroelectricity Association, International Union of Electricity Producers and Suppliers, and a number of the EU countries consider SHEs with an installed capacity below 10 MW. The EU countries such as Spain, France, Sweden, and Germany have the most developed SHEs network. Their SHE installed capacity reaches over 80% of installed SHE power in the EU. Most of the SHEs in the EU were built 40–60 years ago and require renovation.

Hydropower utilization technologies. Water energy in hydroelectric installations is mainly used for industrial purposes. When the main unit is a water wheel (hydraulic turbine), the potential energy of water is converted into mechanical rotational energy.

TABLE 1.2 The largest HEs in the world.

Name	Country	Year of construction	Power, MW
Three gorges	China	2006–11	22,500
Itaipu	Brazil–Paraguay	1984–91–2003	14,000
Guri	Venezuela	1986	10,200
Grand Kulee	United States	1942–80	6809
Sayano-Shushenskaya	Russia	1983	6721
Robert-Bourassa	Canada	1981	5616
Churchill Falls	Canada	1971	5429
Bratsk HE	Russia	1961–67	4515
Jasireta	Argentina–Paraguay	1998	4050
Zhiguli/Kuibyshev	Russia	1957	2357.5
Iron gate	Romania–Serbia	1970	2280
Aswan	Egypt	1970	2100

Depending on the power of the hydrotechnical unit, the flow rate and volume of the water flow, the hydraulic turbines (wheels) can be installed in two ways: reactive turbines (they are completely immersed in the running water of the river, water pressing on the turbine blades forces the wheel to rotate); impulse turbines (these turbines are used in hydropower plants where the potential energy of water pressure is converted into high-pressure kinetic energy).

The HE is a complex of compound hydrotechnical structures and equipment. The scheme of the HE, type of hydrotechnical structures and construction depend on the pressure of the HE, the maximum flow rate of the water, the design of the hydroelectric installations (especially the type of hydroturbine, size, and number of turbines), the geological and topographical status of the hidroknot place, and the hydrological conditions.

The HE construction ground is selected so that the river valley width at that location would be sufficient for all the necessary buildings. There should be comfortable, short water channels, HE building and flood water passages. It is necessary to foresee and to construct passage routes to the construction grounds, to construct power lines to the construction ground, and then to power lines from HE to the common electricity network, as well as to the telecommunications network lines. The administrative buildings and some other ones should be selected close to the location grounds.

Hydrotechnical constructions comprise about half of the total cost of HE construction. The selection of their type depends on the main economic indicators of the HE and the prime cost of electricity. For the reason that the dam is located on the other side of the dam, the water pool accumulates water that is above the lower pool. Then the difference in height H_{st} between upper and lower pools appears.

The power of the HE depends on the effectiveness of the height H, the water flow Q, the operating mode, and water level in the pond regulations. Since these parameters are not constant, the power of the HE and electricity production change.

The instantaneous electric power in kW generated by the HE is determined in the same way as the hydropower of the river:

$$P = 9,81 \cdot Q \cdot H \cdot \eta; \qquad (1.6)$$

here H is the dynamic pressure (useful height of the water reservoir), m; Q — water flow, m^3/s; η — total efficiency factor of HE equipment.

HE dynamic water pressure H equal to static pressure, excluding pressure loss in pressure pipe and water outlet (suction line):

$$H = H_{st} - \Delta H_{sl} - \Delta H_{sr}; \qquad (1.7)$$

here H_{st} is the height of the water reservoir (between the upper and lower basins), m; H_{sl} — dynamic pressure distances in the pressure pipe, m; H_{sr} — dynamic pressure distances in the water outlet (suction pipe), m.

Dynamic pressure distances form from water friction and pipe walls. The exhaust pipe is made wider to reduce dynamic losses.

In addition to the basic parameters (H and Q), the power of the power plant greatly influences the hydropower efficiency factor η_T. It is calculated as the multiplication of the utility of hydraulic turbine, the mechanical transmission η_p, and the electric generator η_g productivity:

$$\eta = \eta_T \cdot \eta_p \cdot \eta_g. \qquad (1.8)$$

The amount of electricity generated by the HE during the time interval is calculated as follows:

$$W_{HE} = P_v \cdot T; \qquad (1.9)$$

here P_v — average HE power in kW over time internal T; T is the time the HE has worked on the calculated average power P_v, h.

A very important parameter of HE is its power utilization coefficient:

$$k_p = W_{HE}/W_v; \qquad (1.10)$$

here W_{HE} — the amount of electricity actually generated by the HE per year, kWh; W_v is the amount of produced electricity if the power plant would operate at nominal installed capacity throughout the year, kWh.

The value of this coefficient is always less than one; it is also different for each HE and different for each year (depending on climate conditions) (Kytra, 2006).

Wind energy. Wind energy is a branch of science and technology that embraces wind energy theory and practice. Wind kinetic energy is transformed into mechanical. Wind energy is an inexhaustible energy resource on the planet and is used in various technological processes, including the propulsion of ships. Sail ships already sailed in Egypt and China in the fifth millennium BC. Remnants of drum-type windmills built in Egypt in 2–1 c. BC have been found. Persians began using wing windmills in the seventh century. Wind engines were used in paper-making technologies, wool latex, sawmills, blacksmithing, oil mills, water pumps, and elsewhere. The first time the windmill was mentioned in Europe was in 833 (Kytra, 2006).

The main equipment of each windmill consisted of a windbreaker, a mechanical power transmission mechanism, and milling machines or other work machines and many additional mechanisms (hoes, grit, rollers, grain peeling and control machines, bag lifting machines). Machines not involved in grain processing (disk saws, turning machines, planing machines, various agricultural machines that were driven by belt drive or cardan) were also often used.

In the late 19th and early 20th centuries after electric machines came into use, wind energy started to be used for power generation with low-power wind turbines.

The main parameters of wind power are: instantaneous wind speed, average annual wind speed, probability of occurrence of a certain wind speed, wind power, and wind energy.

Most the EU countries have a standard height of 10 m above ground level. Although wind speed and airflow power are not constant and vary over time, it has been observed that speed change has some regularities.

Wind speeds have changed over the years. The average wind speeds in winter months are higher than in summer. The regularities of wind speed change determine the time dependence of wind power plant performance. Also, the change in wind speed depends on height. As the altitude increases, the wind speed gradually increases. Since the wind power windmills are higher than the standard wind speed measurement height (10 m), it is necessary to recalculate the wind speed v_x at the height of the vane according to the empirical formula (Adomavičius, 2013):

$$v_x = v_{10} \cdot \left(\frac{h}{10}\right)^{0,142}; \qquad (1.11)$$

here v_{10} — wind speed at 10 m, m/s; h_x — height of wind turbine shaft of wind power plant, m.

The wind speed of accidental change makes it difficult to estimate wind energy in a particular area over time (per month or per year). Wind power is an instantaneous wind parameter.

The power of moving air in any area depends on wind speed and air density:

$$P = 0,5 \cdot \rho \cdot v^3; \qquad (1.12)$$

Here P — wind power per unit area, W/m^2; p — air density, kg/m^3; v — wind speed, m/s.

The instantaneous wind power is most dependent on its speed. Wind power per unit area is called the power density, which is the power of the unit of airflow passing through the plane perpendicular to the wind direction (m^2).

At constant wind velocity v and constant air density p, the air energy W passing through area S is calculated over time t as follows:

$$W = 0,5 \cdot \rho \cdot v^3 \cdot S \cdot t, \text{ Wh.} \qquad (1.13)$$

The air density, depending on air temperature and atmospheric pressure, is calculated as follows:

$$\rho = \rho_0 \frac{p \cdot T_0}{p_0 \cdot T}; \qquad (1.14)$$

here $\rho_0 = 1,204$ kg/m^3 − relative air density under normal climatic conditions; p − atmospheric pressure, kPa; p_0 − atmospheric pressure under normal climatic conditions (101.3 kPa); T − air temperature, K; T_0 − normal air temperature (293 K).

Air temperature does not have a significant effect on wind power. Wind speed and direction can be predicted for a few days with some accuracy. Such forecasts are very important in power systems, where the total power of wind power plants is high.

Wind speed variation is a random process that can be described by the statistical mathematics method using the Weibull function:

$$f(v) = \frac{k}{A} \left(\frac{v}{A}\right)^{k-1} \exp\left[-\left(\frac{v}{A}\right)^k\right]; \qquad (1.15)$$

Here v − wind speed, m/s; k − parameter of the graph form; A − scale parameter.

Parameters k and A are determined from experimentally generated wind speed duration histogram. The wind velocity histogram and the speed distribution according to Weibull function (1.15) can be used to calculate wind power at each point on the histogram by multiplying the wind speed discrete value by the wind turbine power factor.

The highest wind speeds are found in oceans and seas at medium geographical latitudes and are of high seasonal nature. In high atmospheric layers, winds are homogeneous. However, the wind is about 1 km high and is heavily affected by local conditions on the surface. Surface roughness, various buildings, trees, and so forth reduce wind speed. Different obstacles create turbulent (swirling) air movement. Vivid turbulent air movements occur during a storm, when strong wind gusts, swirls, are subject to strong wind direction and wind speeds. The higher the surface roughness, the lower the wind speed.

The change of wind speed also depends on the degree of openness of the obstacle, which is the ratio of the open area of the obstacle to the obstacle in the whole area. The turbulent wind flow zone, when the wind blows into a small obstacle, many times exceeds the volume occupied by the obstacle. At higher altitudes, wind speed is less affected by obstacles on the ground. The curve of the wind speed change dependence on the height is called the wind speed cut. The curve of wind speed dependence on height is an exponential function that is drawn on a logarithmic scale and becomes a straight line. The surface roughness height is when the height is obtained by crossing a straight line in the ordinate and when the wind speed is zero.

The roughness of the area is determined by the roughness classes:

Class 0 − Surface of water (lake, sea);
Class 1 − Open Landscape;

Class 2 — Ground surface with small tree bands and farmsteads;

Grade 3 — Agricultural region with protective strips, groves, and settlements.

The direction of the wind is constantly changing. The change of wind direction in each area is reflected in the so-called wind rose. This is the wind direction change data for each specific area in a compass-scale diagram divided into 8, 12, or 16 angles by world countries, North - South - East - West, as well as intermediate directions. Wind speed is the most important parameter for wind power use. For this purpose, wind atlases are created in all geographical latitudes.

Wind turbines. Modern wind turbines used for electricity generation are of two types: with a horizontal axis and with a vertical axis. In European countries, most modern wind turbines are used with a horizontal axis because they are more efficient than turbines with a vertical axis. However, the latter turbines also have the advantage of simplifying the installation and operation of the electric generator, gear, and other mechanisms at the bottom of the tower. There is no need to orient the blades in the direction of the wind, because they work equally well when blowing winds of any direction, but in the wind power market only the high-power horizontal axis turbines remain.

Modern turbines are divided into three groups according to technological solutions and due to the most important features: turbines according to the power adjustment method; speed (constant speed, two constant speeds, and variable speed); and in compliance with the use of mechanical drive (with gear and without gear).

Modern high-power turbines are characterized by three specific wind speed points: the minimum wind speed at which the turbine starts to rotate; stop speed — the maximum speed at which the turbine stops (when the wind speed is 25 m/s); and the speed of destruction (stored up to 41 m/s).

The main elements of the turbine type with the horizontal axis are: rotor, cabin, tower, foundation. The wind turbine rotor consists of a blade and a rotor hub. The rotor blades are subject to high demands on sufficient mechanical resistance to loads and climatic effects, and aerodynamic efficiency (to maximize wind power).

Wind turbine cabs are equipped with the following main equipment: main shaft, mechanical drive, disc brake, electric generator, converter, switching machine, control cabinet, turbine direction control scheme. The wind direction meter signals the change in wind direction. The electric motor, after receiving the signal, rotates the turbine cab at an appropriate angle through the intermediate mechanical drive.

High-power wind turbines use AC synchronous or asynchronous electric generators. Low-power (<20 kW) turbines also use DC generators.

Asynchronous generators are suitable for all sizes of wind turbines. However, they are not effective with low wind loads. Synchronous electric generators are used in wind turbines with a stable rotor speed, irrespective of wind speed random variations (fluctuations). Asynchronous generators are suitable for turbines whose speed depends on wind speed.

Turbines with a power of more than 1.5 kW are predominantly adapted to work at variable wind speeds. Such turbines have several advantages over fixed speed turbines: start working when wind speeds are less than 3 m/s, work effectively at wind speeds of different speeds. The power they give is stable. However, such turbines should be equipped with semiconductor inverters. Therefore, these wind power plants are connected to the electricity grid through inverters.

Depending on the wind speed, the windmill is rotated at a different frequency. The latter rotates a synchronous generator within the frequency range of 15–72 Hz. The iron smooths out the alternating current, and the inverter continues to convert by replacing the alternating current of the network (should be 50 Hz). The inverter also adjusts the output voltage from the generator to the current network voltage. Everything is controlled by a power plant computer that can replace the 0.02 s inverter.

Speed boxes are not used in the last type of power plants. So, operation has become simpler for there is no need to change the oil and grease every year. Only the bearings of the fans are applied during the operation. There are not so many parts that wear as there used to be in the gearboxes. This could be achieved by the use of permanent magnets instead of electromagnets in the wind power plants.

In 2011 the European Organization for Transmission of Electricity (ENTSO-E) published new rules for all power generators connected to public electricity networks. The rules are based on the European Commission Regulation (EC) 714/2009, which came into force in 2011 as articles 6 and 10 and as Council Directive EC/72/09. Under these rules, wind and solar power plants are divided into four types: A, B, C, and D (Table 1.3).

Type A includes all small generators from 800 W to 30 kW. They only have the requirement of power management by frequency. These generators are connected to low, sometimes medium voltage networks. The 30 kVA power limit is important so that up to that power the EU power plants are called small power plants, where generators can be single phase. The small power plants are connected only to the low-voltage (up to 1000 V) electricity network.

Type B generators are subject to reactive power control and damage support requirements. They are connected to medium and only occasionally to low-voltage networks. Some of these plants use asynchronous generators and speed boxes. Therefore, this type of power plant is classified as a type A power plant. Wind power plants with the same power and synchronous generators and all power electronic inverters of the same capacity are classified as type B power plants.

C-type wind power plants and their parks are already subject to stricter requirements. They have to participate in the system reserve and start self-reliance if a complete system

TABLE 1.3 Classification of power plants and their parks according to ENTSO-E.

Synchronization area	Power threshold for type B generators	C-type generator power threshold	Power type D generator threshold
Continental Europe	1,0 MW	50 MW	75 MW
Nordic states	1,5 MW	10 MW	30 MW
Great Britain	1,0 MW	10 MW	30 MW
Irish	0,1 MW	5 MW	10 MW
Baltic states	0,5 MW	10 MW	15 MW

crash occurs. The generators of these power plants are connected to medium and to 110 kV networks.

Type D wind farms and other power plants within a common system are subject to the same system static and dynamic control requirements. In parks there must be wind power in the primary and secondary reserves, limiting the generated power during the excess energy, as well as increasing or decreasing the generated power at the speed of the requirement for other thermal power plants operating in the system. Similarly, type D power plants must be able to predict future generation of electricity and merge into electricity trading. The generators of these power plants are connected to 110 kV and to voltage networks.

Also, in the list of requirements for type D wind power plants, it is stated that during a total electrical system accident, such power plants have to start independently and start producing electricity for a limited set of local users.

The requirement for participation in the reserve is most easily met by wind farms by switching off some of the power plants during normal system operation.

In the future, it is planned to use hot semiconductors in the construction of wind power plants. Such power plants will be lighter and have smaller foundations. Electricity management will become even more intelligent. Each wind power plant will be controlled by the wind speed directly measured against it. Electricity generated by wind power plants will be accumulated in car battery storage and other ways.

In large wind turbines, the tower is fitted with a steel tube that taps downwards. The tower has to withstand wind pressure at 50 m/s. The foundation is supposed to withstand the weight of several hundred tons of turbine. Therefore, the requirements for the resistance of the foundation are high. The tower is fixed to the foundation with screws. The foundation is made of steel reinforced concrete.

The experience of many countries has shown that relatively inexpensive grants are good for using renewable energy sources. Indeed, in the year 2000 the installed capacity of wind

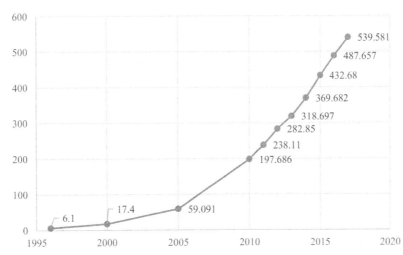

FIGURE 1.8   Global cumulative installed wind capacity in 1996–2017 (MW) (Global Wind Energy Council, 2006, 2016, 2018).

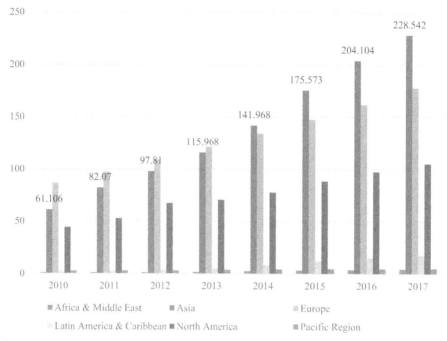

FIGURE 1.9   Global installed wind power capacity (MW) — regional distribution in 2010—17 (Global Wind Energy Council, 2012, 2016, 2018).

power plants was 17,400 MW; in 2010 it reached almost 200,000 MW (200 GW); and in 2013 it exceeded 300.000 MW (300 GW), of which 6.5 GW came from offshore wind farms. Worldwide growth of wind power capacity significantly changed between 1996 and 2017, correspondingly from 6.1 to 539.581 MW (Fig. 1.8).

While examining the 2017 period, it was detected that most installed wind power capacity is in such regions as Asia (228,542), Europe (178,096), North America (105,321), and least in the Pacific (5,193), Africa and Middle East (4,538). Comparing 2010 to 2017, the biggest change in installed wind power capacity occurred in Latin America and the Caribbean (89%), Africa and the Middle East (77%), Asia (73%), and North America (58%) (Fig. 1.9) (Global Wind Energy Council, 2012, 2016, 2018).

In the EU, the highest change in installed power capacity (GW) in 2005 and 2017 was in areas such as gas (99%), wind (76%), and fuel oil (76%). At the end of 2013, the total installed capacity of wind power plants amounted to 117,000 MW (117 GW) and produced 257 TWh (7.8%) of electricity. In 2012, Denmark's wind power plants produced 27% and Spain's produced 21.1% of total electricity consumption in these countries. Throughout the entire 2000—12 period in the EU, the combined capacity of wind and solar power plants far exceeded the capacity of other types of newly installed power plants. In Denmark, it is forecast that in 2020 50% of the total electricity consumption will be generated in wind power plants. The total of the EU installed capacity for different types of power plants in 2017 will basically comprise gas (188 GW), wind (169 GW), and coal (148 GW) (Fig. 1.10) (Statista, 2018).

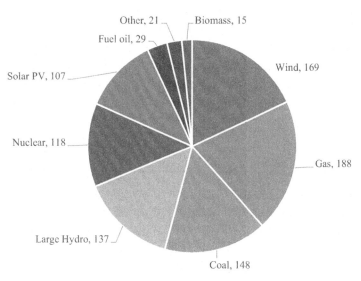

Other, 21 — ┌─ Biomass, 15

Fuel oil, 29 —

Solar PV, 107 —

Wind, 169

Nuclear, 118 —

Gas, 188

Large Hydro, 137 —

Coal, 148

■ Wind ■ Gas ■ Coal ■ Large Hydro ■ Nuclear ■ Solar PV ■ Fuel oil ■ Other ■ Biomass

FIGURE 1.10 Installed power capacity in the EU (EU-28) in 2017 by generation type (GW).

Wind power parks, whether they are on land or offshore, work on electricity interconnections.

Each scheme of wind turbine connection to the local electricity grid depends on the type of generator installed in the wind turbine and the speed mode of the turbine.

Wind power parks are sometimes used to build one power transformer for several turbines.

Most wind turbines operate in variable mode when the wind speed changes. The frequency of the electricity they generate also varies over a wide range. Therefore, these wind power plants are connected to the electricity grid through inverters.

Various types of voltage inverters are used. These devices are quite expensive.

However, the wind turbine connection through the inverters has several important advantages: (1) the wind turbine can operate at variable speed by increasing its efficiency; (2) it reduces turbine rotation speeds at low wind speeds and thus reduces noise; (3) it makes it possible to regulate reactive power in the network.

The use of wind energy is a key factor in addressing environmental issues such as reducing the greenhouse effect (CO_2) and reducing acid rain. Wind power plants, compared with thermal plants, prevent sulfur and nitrogen compounds, various dusts, and slag and volatile ash. Most importantly, however, the abandonment of imported fossil fuels is accompanied by economic and political independence from the importing countries.

Although wind power plants have many advantages over other power plants, some environmental problems that are not common to other sources of electricity need to be taken into account when designing large wind farms.

Wind turbines with turbines operating at a height of 100 m and above produce a visual impact on a few kilometers' radius. Visual effects include moving reflections from long rotor blades.

Another major problem with wind power plants is turbine noise. In low-power power plants, noise is generated by mechanical drives and electric generators, and higher power with long rotor blades causes aerodynamic noise.

The noise level of wind power plants by the turbines themselves is between 90 and 100 dB. Moving away from the power plant, disturbing noise significantly reduces. At 40 m, this noise level is only 40 dB, and at a distance of 500 m, only 25—30 dB is heard.

In the EU, there are norms that allow the construction of wind farms no closer than 150 m from residential buildings. Power plants with a capacity greater than 1 MW are allowed to be built at least 200—300 m from the house.

Wind power parks occupy a large area of land. However, on cultivated land they occupy small areas compared to other power plants, as earthworks are allowed to work next to them.

An important environmental problem is the impact of birds on wind turbines. However, research in some European countries shows that wind turbines do not pose a serious threat to birds. In Denmark, where there are about 500 wind farms, birds are killed 40 times less on the farms than on roads.

The sun is the most powerful source of renewable energy on Earth. The theoretical annual global solar energy potential is estimated to be 900,000,000 TWh and is about 60 times the theoretical annual global wind power potential, about 2200 times the theoretical annual geothermal potential, around 4500 times for biomass and about 36,000 times for biomass. annual theoretical annual global potential of hydropower. In spite of this, solar power potential for electricity and heat generation is still the lowest. This is not a coincidence: the solar energy itself is scattered, weakly concentrated, and its parameters stochastically vary widely, depending on the time of day and year.

The statistics of countries with widespread use of wind power plants shows that the commercial wind power plant generates 50 times more electricity during its life cycle than it was used in its production and construction.

The economic viability of wind power plants is determined by the costs of electricity generation, which depend on various factors: from the wind regime at the site of the power plant; from capital investment to power plants; from capital interest; wind power efficiency and wind power plant operating costs.

The comparative cost of wind power in the EU is diminishing year by year. The rapid reduction of capital costs for power plant production over the last decades has had a significant impact on the reduction of wind energy comparative costs.

The European Wind Energy Association and the Greenpeace document (EWEA 2003a,b) predict a further rapid growth of the wind industry, a decline in comparative investment, electricity, in wind power plants, the decline in prices and the rapid growth of jobs in the wind industry. According to the level of production costs, the electricity generated by wind power plants, will be competitive for energy produced in other types of power plants, including nuclear.

Solar energy. The Sun is the most powerful source of renewable energy on Earth. The theoretical annual global solar energy potential is estimated to be 900,000,000 TWh and is about 60 times the theoretical annual global wind power potential, about 2200 times the theoretical annual geothermal potential, around 4500 times for biomass, and about 36,000 times for theoretical annual global potential of hydropower. In spite of this, solar power potential for electricity and heat generation is still lowest. This is not a coincidence—the solar energy itself is scattered, weakly concentrated, and its parameters stochastically vary widely, depending on the time of day and year.

The solar radiation density that reaches the Earth's atmosphere is called the sun constant at $1367 \, W/m^2$. However, the Earth absorbs only 51% of solar energy. The other part is absorbed by atmospheric air and clouds (19%) and is also reflected from the Earth's surface, air, and clouds (30%).

In the photosphere, from the Earth's visible surface of the sun an electromagnetic wave spectrum is formed in the range of $0.1-3 \, \mu m$. Emitted into the atmosphere the spectrum of sunlight is divided into three parts: ultraviolet, visible to the human eye, and infrared. The range of visible light on the surface of the Earth is from 0.28 to $0.76 \, \mu m$ and can hold 49% of the total spectrum. Infrared radiation also occupies 49% of all spectrum beams, while ultraviolet rays are only 2%.

There are many scattered dark spots at the top of the photosphere. These are relatively cold regions with a temperature of only 3800 K (the surface temperature of the sun is 6000 K). These spots are due to the interaction of the magnetic fields. The diameter of these spots is up to 50,000 km.

The spectrum of the sun reaching the Earth's surface (AM1.5) differs in size and in the shape of a curve from the spectrum of radiation above the atmosphere (AM0). Spectral curves are distorted by ozone (O_3), oxygen (O_2), water vapor (H_2O), and carbon dioxide (CO_2), which does not absorb light well. Ozone is more influential in ultraviolet and visible light ranges, oxygen—at visible and infrared junctions, water vapor, and carbon dioxide—in the infrared zone (bright depressions on the AM15 curve in the infrared range).

The photosphere is surrounded by a very thin and translucent layer of the sun's atmosphere, consisting of two parts: a chromosome and a solar crown. The chromosome is visible only during the sun's eclipse, as it is lit by fire around the sun.

The layer above the chromosome is called the sun crown. It stretches millions of kilometers further into space and is visible only during the full solar eclipse.

On the surface of the Earth, there is a strict energy (heat) balance entering the surface and emitting into space. If this beam balance were damaged, the temperature would increase or decrease on Earth.

Solar energy, which enters the atmosphere, influences all climatic processes on the Earth's surface: 46% is affected by land and ocean surface heating; water movement cycle − 23%; wind and ocean currents − 1%; and plant photosynthesis − 0.023%.

Solar energy resources on the Earth's surface are 1800 times higher than the energy consumed on the planet.

The power of solar radiation on the Earth's surface is constantly changing not only because of the sun's time of day and time of the year but also because of the deterioration of climatic conditions (atmospheric pollution or cloudiness) (Šuri et al., 2005). At each point of the Earth's ball, the amount of solar energy differs not only in its maximum or average value but also in its variation over the year. Closer to the Equator, the intensity of solar radiation is higher and it changes the least compared to other Earth latitudes. The average annual amount of solar energy in different parts of the planet varies from 2200 kWh/m^2 to 400 kWh/m^2 per year except for the southern and northern poles.

Solar energy is mainly used to produce electricity using solar cells and water and indoor heating with solar collectors.

The main elements of the technological scheme for converting solar energy into electricity are semiconductor solar cells (batteries).

The use of solar cells is very wide. Photovoltaic systems are considered to be renewable energy sources and are particularly useful in regions where power transmission is not possible, such as artificial Earth satellites, space stations, remote radiotelephones, or water pumping facilities. Solar cells are becoming more and more widespread in homes, military polygons, and so on.

Solar cells are made up of two semiconductor materials that are intimately interconnected. An electric field appears in the n-p junction zone. Worldwide, solar cells are manufactured in different types. However, from the outset, when the use of solar energy for power generation started on an industrial scale, polycrystalline and monocrystalline silicon elements with various material admixtures were mostly used.

The solar cell consists of two different semiconductor materials. Both of these materials, taken separately, are electrically neutral. By illuminating an element of sunlight, it contains ionizing silicon atoms. The latter, exposed to the potential difference at the p-n junction, polarizes to two opposite load areas.

In the n layer at the upper electrode free electrons accumulate, and in the p layer atoms accumulate by losing the electrons. When the upper (p) and lower (n) layers are connected to the outer chain, free electrons travel to the p-silicon layer and recombine there with the holes by releasing the energy of their charge. The current starts to flow in the external connected circuit.

The main electrical characteristics of a solar cell are characterized by ampervoltmetric characteristics.

The point A, in the ampervoltmetric characteristic, I, corresponds to the current and voltage values of the element at which the maximum power P_m in the element is formed. The maximum power of the element P_m is called the peak power, whose unit of measurement (watt) is denoted by the index p (W_p).

The current strength of the element depends very much on the intensity of the light energy. The intensity of light on the solar cell voltage has little effect.

Solar cells are assembled into so-called modules. Modules are assembled into panels according to the required voltage and current values and the latter into solar cell fields.

To increase panel power, more solar cells are required. However, at the same time the construction gauge increases.

Solar power plants can work autonomously as intended, i.e., supplying electricity only to consumers, and can be integrated with local networks.

Higher-power solar power plants occupy large areas. Therefore, in settlements they are usually mounted on roofs or walls of houses. Solar cells are also often located in unused areas of land.

Self-contained solar power systems require energy storage (accumulators) during the night time and on very cloudy days. If a solar power plant is connected to local networks, energy storage is not necessary.

The maximum amount of solar energy in any plane falls when the sun's rays are perpendicular to it. As a result, solar panels should, as the circumstances allow, be oriented in such a way that the level of solar energy they produce is most efficient. However, this is not always the case.

Most solar energy enters the panel of elements when it leans at a certain angle to the horizon and faces south. It is always important to try to direct the items to the sun at the optimum angle set for that particular location. The optimum angles are different in each geographic latitude and time of year.

Tracking systems are used to optimally orient them to the sun for solar panels at an optimum angle to the horizon both at the daytime and at different times of the year.

One of the key indicators of the economic assessment of solar power plants is the energy recovery time of solar cell modules. It is the time that the power plant has to work to produce as much electricity as it consumes in the production of power modules and other required materials.

The energy recovery time depends on the solar cell module manufacturing technology and the solar cell operating conditions.

The European Solar Power Industry Association predicts a rapid decline in solar electricity–generated energy prices in the 21st century (Lynch 2009, May 7–8, Athens, Greece). As the prices of solar power plants fall, they will grow rapidly both in the world and in the EU.

Heat energy production. Solar heating systems consist of the following elements: solar collectors, hot water tanks, additional heating (electric or other fuel) equipment, pumps, piping, and control units. Assembling of heating systems depends upon the purpose of the use: water for hygiene purposes or heating premises or both.

A solar collector is the main element of a solar thermal unit. Solar heating systems use four types of solar collectors: flat, vacuum, concentrated, and air heating.

The main elements of the solar flat collector are the absorber, the tube with the intermediate heat carrier—the water, the transparent cover, the thermal insulation layer, and the housing.

In the collector absorber, solar energy is converted into heat energy, which is why it is made of a very heat-resistant tin (copper or aluminum). In the absorber, when the temperature rises above the environment, it starts to emit energy into the infrared range. The ratio of radiated and absorbed energy is called the degree of emission.

So that the absorber is more efficient, i.e., reducing energy loss, the surface of the absorber is coated with a selective coating.

In order to limit heat emission, the absorber with the tubes of the heat carrier is covered with a transparent coating that retains the heat radiation of the absorber.

The transparent lid is made of glass without iron. This glass spreads 91% of light.

The heat accumulated in the absorber is transferred to the circulating copper tubes for the liquid. Copper tubes are soldered to the absorber or are fitted with ducts by mounting channels where the heat carrier flows.

The collector frame is made of galvanized aluminum or reinforced glass fabric.

To reduce the heat loss of the absorber, a thermal insulation layer is laid underneath it.

Flat indoor air heating solar collectors work in a similar way to solar heating solar panels.

Combined solar panels are highly efficient because they heat air indoors and water at the same time. These collectors differ from the above-mentioned water heater collectors by the fact that the absorber and the heat carrier tubes are blown with an air fan. Blown air through the tubes is drawn from the collector to the buildings.

The design of vacuum solar collectors is more complex than those of flat panels. The tube with the fluid carrying the heat and the absorber is inserted into the glass tube from where air is withdrawn. The vacuum in the tube is a good heat insulator, which eliminates the heat loss of the absorber. The tube is made of glass that passes through the sun's short wavelengths but does not pass the black surface (absorber) infrared radius.

Two liquid tube shapes are used in the direct circuit diagrams: the U-shaped tube and the tubes adhered to each other. The middle tube is supplied with a liquid mixture consisting of water and glycol space, and the fluid, which is located between the inner tube and the outer tube, returns back to that space.

Sun rays raise and evaporate the liquid in the tube. The vapor settles on the condenser walls and, when cooled, the condenser rinses, condenses, and flows back into the tube. Thus, there is heat exchange between the fluid mixture flowing in the tube and the water (heat carrier) flushing the condenser.

Solar collectors in concentrating solar cells fall into a convex focusing mirror. The reflected radiation from the mirror focuses on the tube where the heat carrier and absorber flow. The disadvantage of concentrating solar collectors is that they only receive direct sunlight but do not accept diffused light.

Concentrated solar collectors are mainly used in solar thermal power plants that require high steam parameters, as well as in other production technologies that require steam.

The parameters of solar collectors describing their quality are: efficiency, conversion factor, and heat loss coefficient.

The efficiency of the collector is determined by the ratio between the energy of the collector surfaces. Loss of solar light from the heating system and the collector itself reduces the efficiency of the collector.

Loss of light is produced by reflecting light from the collector glass cover and absorber, which can reach about 20% of the total light flow. The amount of damage does not depend on the temperature of the collector or the ambient temperature.

The heat loss in the manifold depends on the design of the collector itself, the properties of the materials used in it, and the quality of the production.

The maximum efficiency of the collector is achieved when the intensity of the solar light reaches $1000 \, W/m^2$ and when the heat carrier circulates so intensely that the average temperature of the collector equals the ambient temperature. As the difference between collector and ambient temperatures increases, heat loss increases and efficiency decreases.

The collector conversion coefficient k_k or maximum collector efficiency is when the internal temperature $[(T_{in} + T_{out})/2]$ of the heat carrier collector equals the ambient temperature.

The heat loss coefficient k_k is the ratio of the average relative loss (W/m^2) of the collector area to the temperature difference between the collector and the environment (Table 1.4).

The amount of thermal energy Q produced by the solar collector is calculated as follows:

$$Q = I \cdot \eta \cdot S_k, \, Wh \qquad (1.16)$$

TABLE 1.4 Range of possible values for the basic parameters of solar collectors.

Type of collector	Conversion rate, k_k	Heat loss coefficient, k_n, W/m², °C	Operating temperature range, °C
Open	0.82–0.97	10.0–30.0	To 40
Flat	0.66–0.83	2.0–5.3	20–80
Vacuum	0.62–0.84	0.7–2.0	50–120
Tank	0.56	2.4	20–70
Air collector	0.75–0.90	8.0–30.0	20–50

here Q is the heat energy, Wh;

I — Sunlight intensity, W/m²;

η — collector efficiency coefficient at specific solar light intensity and temperature difference between collector surface and environment;

S — the surface area of the solar collector, m².

An important criterion for the efficient operation of solar collectors is their orientation in a certain direction and at the right angle. It is best to install solar panels on the roof (flat or pitched, or on the ground).

The work of solar collectors should be constantly monitored. This action is performed by collector controls. Sensors and controllers are connected to these electronic systems. The system constantly monitors the data provided by the sensors and regulates the solar collector system at work.

A very important device in a solar heating system is a heat accumulator—a storage tank for hot water, which is a metal cylindrical container with double walls and good thermal insulation between the walls. Depending on the size of the heating system, heat energy storage tanks (water tanks) range from 300 to 1200 L.

Some of the excellent prospects for using solar panels are evident in the technology of drying agricultural produce. Solar collectors are efficient and competitive with other water heating technologies. Their utility factor is 90%. However, the disadvantage of water heating systems with solar collectors is that, during the cold season, when solar energy intensity drops sharply, their performance drops. It is proposed to use hybrid water heating systems to eliminate this deficiency. In this hybrid system, the main function of the solar collector is to heat cold water supplied from the water supply. In the time of the year when the intensity of solar radiation decreases (in winter, autumn), water is stopped being heated by the typical heat exchanger, which is usually located in the heating unit of each block of flats or water boiler heated by traditional fuel.

In such a hybrid heating system, solar collector efficiency is chosen so that the heat exchanger from a district heating system or boiler should not be used during a sunny season. Apparently, the energy needed to heat the premises or prepare hot water for hygienic purposes can be used more rationally.

Bioenergy. The bioenergy resources are divided into three groups according to the area of use: solid biomass (wood, agricultural crop waste—straw and other crops) for electricity and heat production; biogas (waste and sewage, landfill); and biofuels (Šateikis, 2006).

Biomass resources are enormous and widely distributed throughout the world. In nature, biomass occurs (growing) during the photosynthesis process. While acting on the visible spectrum of solar energy, plants absorb CO_2 from the air that converts water into carbohydrates during photosynthesis and through chlorophyll in leaves. At that moment oxygen, O_2, is released into the air.

$$\text{Total photosynthesis equation}: 6CO_2 + 12H_2O \xrightarrow[\text{chlorophyll}]{\text{light}} C_6H_{12}O_6 + 6H_2O + 6O_2$$

Solid biomass by burning releases energy that plants accumulate while growing up.

The process of decomposing biomass is the reverse process, i.e., biomass, with O_2, releases carbon dioxide CO_2, water H_2O, and mass-accumulated energy.

Biomass by status (consistency) is divided into two groups: dry biomass (wood, paper waste, agri-crop biomass, etc.); wet biomass (seaweed, organic liquid waste, livestock manure, etc.). According to origin, biomass is divided into plant and animal.

Before storage and use, biofuels are first dried and processed according to technological requirements. Biomass is dried outdoors by using solar energy and hot chimney gas in furnaces, in steam dryers, or in special drying plants.

The following conditions should be met when increasing the use of plant biomass: there should be no effect on the consumption of food in the country; it should not reduce the diversity of agriculture and forests by avoiding monoculture; and there should be protection of the environment without damaging soil nutrients and water balance without causing soil erosion. It is important to select rational energy crop cultivation technologies with minimal environmental impact.

There is high untapped potential of biofuel production both in high power and small cogeneration plants producing electricity and heat.

Biomass energy is produced in two ways: thermochemical and biochemical. The thermochemical method is biomass combustion or gasification or pyrolysis.

Biomass burning is the oldest used technology to extract heat. The combustion process takes place in three phases:

In phase 1, heat (up to 1300°C) is released during the chemical reaction (with biomass to oxygen), i.e., a so-called exothermic reaction occurs.

In phase 2, there is an endothermic reaction that consumes the released heat when pyrolysis occurs, i.e., decomposition of the substance without air (up to 600°C) that produces flammable gas. During phase 3 (up to 150°C), biomass drying takes place—water evaporation.

According to the combustion method and the biomass supply to the furnace, the following biomass combustion systems exist:

1. A grating systems where flammable biomass is supplied on moving gratings;
2. A bottom biomass feed system, where the combustible biomass is supplied by a screw conveyor to the lower part of the furnace, i.e., below the burning mass zone (this method of burning is suitable for firing chips);

3. An inflating system, where biomass is crushed and air is injected into the combustion chamber (suitable for wood sawdust or wood dust usage);
4. A flammable fuel system, with high airflow from the bottom of the combustion zone, raising fuel (sawdust, chips, wood dust), which is constantly flown in the combustion chamber.

The highest temperature is reached in the combustion chamber and is transferred to a heat exchanger where the heat energy is transferred to the heat carriers. Gas enters the chimney at much lower temperatures but it becomes heavily contaminated. Therefore, before entering the exhaust chimney, the gas is cleaned and discharged.

During combustion, the released gas still exposes high temperatures. Therefore, before the chimney enters the gas, part of the heat is released into the economizer, which preheats water before entering the water and steam drum or it is used for other needs. Hence, saving heat energy increases the efficiency of the entire water heating system.

When operated with biomass, fuel becomes insufficient in its energy quality. Therefore, it is mixed with coal, gas, oil products, or other waste.

Gas contains many impurities that pollute the environment when burning biomass. Therefore, they it is cleaned before entering the environment. The main pollutants in the gas are NO_x, CO, and solid particles (ash). There are also many other chemical elements and their compounds.

Gas cleaning technology consists of two stages. In the first stage, the NO_x and CO fractions are purified. In the second stage, the particles are cleaned.

Gasification of biomass. Instead of natural gas or other conventional fuels (made of oil), other technologies, using gas produced from solid fuel (biomass or coal), are also possible. Biomass gasification is a thermochemical conversion technology. Gasification of solids takes place in four phases: drying, oxidation (combustion), distillation (pyrolysis), and reduction (solid fraction decomposition − gasification).

During biomass oxidation (combustion), heat is released and biomass organic molecules are transformed into carbon dioxide (CO_2) and water vapor (H_2O). At that point, inorganic biomass components (ash) also fall out and pyrolysis also releases some resin. During the next phase, heavy molecules break down at high temperatures into lighter organic molecules and carbon monoxide (CO).

The final gasification product produces flammable gases comprising carbon monoxide, hydrogen, nitrogen, and water vapor. These gases are used as fuel in boiler houses for electricity and heat generation, internal combustion engines, or gas turbines.

Solid biomass gasification is carried out in metal tanks (reactors). The reactors are divided into heat and power. Gases produced in heat reactors are used for combustion in boilers, dryers, and furnaces. Power reactor gas is used as fuel in internal combustion engines.

Biomass in gasification reactors is mixed with air in various ways:

1. Parallel flows (fuel and air concentrated in the same direction);
2. Counter flow (fuel from above, air from bottom);
3. Cross-flow (fuel supplied from above, air flow from side to side of fuel flow);
4. When the fuel is running in the reactor (air is supplied from the side below the supply fuel).

Low power units use combined biomass energy conversion technologies when the gasification reactor is connected to a direct gas combustion chamber. Such a connection between the gasification reactor and the combustion chamber increases the overall efficiency of the equipment.

Technology for the production of industrial waste and landfill gas. When decomposing dead organic matter, it decomposes and releases biogas (methane ,CH_4, and carbon dioxide, CO_2), water vapor, and solid waste (compost) into the environment.

In nature, the rotting process takes place with the help of microorganisms. This microbiological process is called methanogenesis. The process of methanogenesis takes place without oxygen. This rotting process is caused by anaerobic microorganisms that live and reproduce in an oxygen-free environment. The process of methanogenesis by technical means is called biomethanization. In this way, any organic material (waste from urban, industrial, and agricultural farms, wastewater, energy crops, etc.) is suitable for gas production. The biomethane gas consists of 50%–70% CH_4 and CO_2.

Anaerobic (oxygen-free) gas production process takes many forms: one-step, two-stage, waste landfill, and hybrid. Anaerobic digestion of biomass takes place in metal reservoirs, so-called reactors, which are supposed to be leakproof from ambient air (oxygen), as anaerobic bacteria are sensitive to oxygen. The reactors must be well insulated.

The one-step process is continuous, i.e., without the main process stop or interruption, except when the waste from the reactor is disposed and another portion of biomass is loaded. Two-stage gas extraction takes place in two sequentially connected reactors. A combined process is also used to produce biogas, where anaerobic process (without oxygen) occurs in one reactor (A), and an aerobic process (gas release in an atmosphere containing free oxygen) occurs in another reactor (B).

Methanogenesis processes in reactors take place at a certain temperature: processes that occur at 30–35°C are called mesophilic, and at 50–60°C, are called thermophilic.

Prior to the supply of the biomass to the reactor, it is first sorted, processed, crushed, and mixed. The gases produced in the reactors are cleaned of various impurities before entering the low-pressure tanks.

The comparative yield of biogas depends on the type of biomass and the time and temperature of the presence of the reactor (Kytra, 2006).

In landfills, where various population waste (debris) is accumulated, various chemical processes take place, whereby pollutants that penetrate into the contaminated ground and groundwater, as well as flammable gases, pollute the environment and damage human health.

Before installing a modern landfill, the bottom of the landfill is a material that keeps water and gas from entering the ground. When the landfill is fully completed, it is covered so that no gases and odors can penetrate into the environment. In such an arranged landfill, vertical wells are placed from above and drilled to collectors, which collect the separated gases. The composition of the gas extracted from the garbage dump differs little from the biogas produced in the reactors by methanogenesis (Kytra, 2006).

Landfill gas, as well as biogas produced in reactors, is used as natural gas in internal combustion engines, turbines, and heating equipment. If the gas accumulated in the landfill is not supplied to consumers, it must be incinerated in order to avoid explosion.

Biofuels for road transport. Biofuels are one of the renewable energy fuels produced from a wide range of biomass and are used primarily in road transport with internal combustion engines.

The adopted EU Directive 2003/30/EC forecasts that by 2020 biofuel consumption in the transport balance should reach up to 20%, if biofuels and other renewable fuels for transport are in use.

Cars can use bioethanol, biodiesel, and blends with mineral oil, vegetable oil, and bio-oil. Bioethanol is produced from sugar-containing raw materials (sugar beet, sugarcane), starch-containing raw materials (potatoes, cereals, wood, straw, and lignocellulose), and municipal waste.

Biodiesel is a vegetable oil ether, rapeseed ethyl, or methyl ester produced from vegetable oil and alcohol (ethanol or methanol). Vegetable oil is used in specially prepared engines.

Despite some of the advantages of using first-generation biofuels, there are some problems too. In particular, the production of such fuels poses a threat to food production and biodiversity. The production of these fuels is more expensive than petroleum products. At the same time, the production of biofuels throughout the cycle produces a significant contribution to greenhouse gas emissions. In the future, it is planned to produce second-, third-, and fourth-generation biofuels.

Second-generation biofuels are fuels that use lignocellulosic biomass as a raw material (wood and its waste, agricultural waste—straw, maize sticks, and energy crops).

Second-generation biofuels can be produced by using two technological processes: a biochemical process whereby enzymes and other microorganisms break down the cellulose and hemicellulose components into sugar that is fermented into ethanol; and a thermochemical process that produces gas (a mixture of CO and CO_2) from biomass by pyrolysis or gasification.

Comparing the first- and second-generation biofuels, the latter could be produced significantly more from biomass grown on the same land area compared to plants used for first-generation biofuels.

Third-generation biofuels are expected to be produced from improved genetically modified crops.

Geothermal energy. Fourth-generation biofuel production technologies will comprise the cultivation of genetically modified plants that absorb large amounts of CO_2 from the atmosphere accumulated in plants. Thereafter, the biomass from the plant biomass will be efficiently produced from biofuels by biochemical processes. However, using fourth-generation biofuel production technologies is the hardest part of carbon capture and conservation.

The source of geothermal energy is in the depths of the Earth. That is the heat energy and mantle heat of the radioactive elements (thorium — Th 232, uranium — U 238, 235, potassium — K 40, and radium — Ra 202, 204). The 2900-km-thick layer above the outer core of the Earth is called the mantle. It consists of a mixture of rocks and magma and accounts for 82% of the Earth's ball volume. Magma is a high temperature (>1200°C) viscous liquid with a wide range of gases. Above the mantle the Earth's crust has a thickness of 20—65 km. The crust comprises about 16% of the Earth's total ball volume. The earth's crust is broken into pieces called tectonic plates. Lithosphere consists of the Earth's crust and part of the upper mantle layer. In the depths of the oceans, the lithosphere has a thickness of up to 80 km and in continental regions up to 200 km.

A small amount of geothermal energy is accumulated in the upper layers of the Earth when heating from the sun. However, the largest sources of geothermal energy are found deep in the Earth without visible signs on the surface. This deep energy in some parts of the Earth opens up in the form of geysers, hot currents, or volcanoes.

Since there is no balance between the heats that are generated in space above the Earth's surface, it is thought that the Earth is slowly cooling down. Earth's most intense geothermal energy locations are located at the edges of tectonic plates.

The thermal energy resources at a depth of about 10 km are divided into three groups: hydrogeothermic, petrogeothermic, and shallow geothermic.

In the hydrogeothermic energy sources, hot water is accumulated at depths of 5−10 km of the Earth's crust. In geological fractures, geysers (sprays) emit vapors (above 100°C) or hot water (up to 100°C). Geyser water is poorly mineralized. There are various gases in the water vapor. The water fountains in the geysers rise to a height of 30−60 m, the water vapor clouds rise to 300 m. One outburst throws up to several hundred thousand liters of water. Geysers stop throwing hot water or water vapor when the magma fireplace in the area cools down and does not fall into the underground recesses. There are over 1000 geysers in the globe.

Thermal energy from hydrogeothermic fields is extracted through geological wells.

The shallow geothermal energy sources are at a depth of 1−100 m and are between 10 and 12°C (except in winter). Such ground temperature for water or indoor heating is accumulated only with the help of heat pumps.

The advantage of geothermal energy is that it is relatively clean and safe. The production of this type of energy does not require special and changing climatic conditions and its production can be continuous. Geothermal energy recovery plants can counterbalance seismic stability, especially in petrogeothermic sources when water is injected into hot rocks. It should also be noted that these sources pollute the environment by discharging carbon dioxide, nitrogen oxide, and other gases into the environment.

In various technological processes geothermal energy was first used in Italy in the 19th century. The largest users of this energy in the world are the United States, Italy, Mexico, Iceland, and the Philippines.

According to the temperature of the extracted water, these energy utilization technologies are divided into three groups: high temperature (more than 150°C), medium temperature (100−150°C), and low temperature (30−100°C).

High-temperature hot water at temperatures above 150°C can be used to produce electricity.

The first turbine using direct dry geothermal steam was built in Larderello, Italy, at the beginning of the 20th century (1904).

When hot geothermal water with a temperature higher than 150°C is obtained from the well, a different technology for generating electricity is used. At a geothermal temperature of 100−150°C, whose parameters are not suitable for direct electricity generation, an intermediate so-called binary circuit with a circulating liquid heat carrier with a boiling point lower than the boiling point of water is used.

Geothermal water in the intermediate circuit gives heat through the heat exchanger to the heat carrier circulating in the second closed circuit.

Such a technological structure is also used in electricity generation technology when geothermal water is contaminated with various mineral impurities.

Geothermal water at temperatures between 50 and 100°C can be used directly for indoor heating, agricultural production technology, and hygiene.

The use of geothermal energy below 50°C is not directly effective for either water or indoor heating. In order to increase the geothermal energy efficiency, heat pumps are used to accumulate it, and heat is taken from the environment, i.e., from the ground, water, or air.

The use of low-temperature geothermal heat with heat pumps is quite feasible with the help of various technologies based on the same operating principle.

When using geothermal energy and compared to conventional organic fuels, much less emissions are released into the environment, especially sulfur oxide, SO_2, CO_2, and nitrogen oxides. The advantage of using geothermal energy compared to other renewable energy sources is that energy is supplied nonstop all 24 h and does not depend on the season.

Hydrogen in the power industry. Hydragenium H is the first chemical element of period 1A group 1A of the periodic element system, which is a divalent molecule (H_2). Hydrogen was discovered by English scientist Cavendish Henry in 1766.

Hydrogen is an odorless, colorless, and tasteless gas. Hydrogen in nature is in the form of three isotopes. The first two are called stable isotopes: (1) Light H Hydrogen (Gr. *protos* — first) the lightest and most common hydrogen isotope. The protos is made up of a nucleus (proton) and one electron. (2) Heavy hydrogen D (deuterium, Gr. *deuteros* — second). Deuteron's core, deutron, consists of one proton and one neuron. The molecule consists of two atoms D_2. The third hydrogen isotope T (tritium, Gr. *tritos* — third) is a radioactive hydrogen isotope having a mass number of 3. The tritium nucleus consists of one proton and two neutrons, a divalent gas used in thermonuclear fusion as an isotopic indicator.

Although hydrogen is among the most common chemical elements in the universe, there is no free hydrogen on Earth. It is found in organic compounds and water. At room temperature, hydrogen consists of ortho hydrogen (75%) (Gr. *orthos* — true), paravandenyl (25%) (Gr. *para* — to), and its molecule consists of two atoms. Hydrogen binds to almost all chemical elements except inert gases and precious metals, and blends and explodes with air at a ratio of 2:1. At temperatures above 550°C, hydrogen forms water (H_2O) when combining with oxygen.

Hydrogen is the most common chemical element in space—its plasma is about 75% of the universe. In the chemical industry, hydrogen is used in the production of hydrochloric acid, ammonia, methanol, ethanol, and liquid fuel. It is also used for the extraction of metals from their fluorides or oxides, the hydrogenation of hydrocarbons, the purification of petroleum products, the welding and the cutting of metals (the flame temperature of the H_2 and O_2 mixture is about 2000°C). Hydrogen is used as a reducer in fuel cells to produce electricity. The resulting liquid hydrogen-oxygen mixture is used as a rocket fuel.

In 1952 in the United States and in 1953 in the Soviet Union hydrogen bombs were blown, which were many times more powerful than nuclear bombs. For the first time, uncontrolled thermonuclear fusion was triggered. A very large amount of energy was released when the deuterium and tritium nuclei and at very high temperatures ($>10^7$ K) exposed. The specific thermonuclear energy (energy per unit mass of reactive nuclei or one nucleon) is several times higher than the nuclear energy obtained by sharing uranium nuclei.

The goal of modern science is to create a controlled thermonuclear fusion technology that will enable to gain enormous amounts of energy on earth.

Hydrogen is seen as a potential fuel for future transport, as it is an alternative to polluting the environment in fossil fuels. Hydrogen is considered to be a potential energy carrier, which is expected to play a key role in the energy sector in the near future (Rifkin, 2012).

The use of hydrogen for energy purposes is one of the most important and least-polluting environments of future energy, and some countries such as the United States, Japan, and some EU countries have set their strategic energy plans largely to shift to hydrogen-based energy by 2050 (Rifkin, 2012).

In each country it is necessary to address a number of technological, legal, educational, and social challenges in the transition to a hydrogen-based energy system that uses renewable energy sources as primary energy sources and hydrogen as the main energy carrier.

Coordination between scientists and industrial operators at national and international levels is necessary due to problems that have arisen. For this purpose, the EU has established a European Hydrogen and Fuel Cell Technology Platform.

The use of hydrogen as an energy source will make it possible to radically change the areas of energy production, transmission, and consumption. Special attention is paid to the development of completely clean transport systems. There are already a number of car factories that have developed car demos that use hydrogen as fuel. When it comes to running such cars, it is necessary to solve problems of hydrogen infrastructure of filling stations.

Hydrogen production. A variety of technologies have been developed to extract hydrogen from various materials. It can be obtained by: (1) water electrolysis; (2) the process of reforming from natural gas; (3) using any energy from renewable sources (wind, solar, etc.); (4) using nuclear power (high-temperature radiators); (5) by microorganisms; and (6) plasma decomposition at high-temperature plasma and others.

All of the listed hydrogen production technologies, with the exception of the first—water electrolysis—and second—reforming—are still only at the fundamental research stage, and their energy efficiency is rather low and not yet economically viable in terms of commercial objectives.

The electrolytic electrochemical process of producing water in water takes place on the electrolyte electrodes when the current flows through the electrolyte (water is mixed with additional conductivity agents).

The external power source is connected to two electrodes (cathode and anode) immersed in an electrolyte (water mixed with other electrical conductivity enhancers). The electrode between the electrodes starts to flow through the ion-generated current. The positive ions (cations) exposed to the electrode and generated by the electrodes move to the cathode and the negative (anions) move to the anode. On the cathode, ions are reduced by attaching electrons from the cathode. During electrolysis, water is decomposed into O_2 and H_2.

The electrolyzers are hermetic to provide sufficient pressure for efficient storage of the recovered gas.

The purpose of the diaphragm in the electrolyzer is to separate the gases from the anode and the cathode (O_2 and H_2).

The reforming technology is also implemented and is efficient enough, especially from hydrocarbons. For the latter, three thermochemical methods are commonly used to decompose: catalytic degradation (reforming) with water vapor, partial oxidation, and autothermal degradation (reforming).

The first reforming method using water vapor is the most effective for it comprises a high conversion yielding hydrogen. The second partial oxidation method is less effective than the first one as it requires additional energy for heating. The aim of the autothermal reforming process is to combine the first two methods so that the low heat released during the partial oxidation is utilized to support the endothermic reactions of the reforming steam. This third method of hydrogen extraction has been successfully tested by breaking down methanol, light petroleum products, and natural gas. However, it is still an untested method for the removal of hydrogen from other waste products (fuel oil, used oil, biofuel, old tires, etc.) (Milčius, 2009).

Hydrogen for energy purposes needs to be capacitively and energy efficiently stored and safely delivered to consumers.

All countries, researching hydrogen production and use technologies, are taking real steps to achieve long-term hydrogen production from renewable energy sources or other primary energy sources that do not pollute the environment with greenhouse CO_2 gas emissions.

The European Hydrogen and Fuel Cell Technology Platform (hereafter "the Platform") provides prospects for the development of hydrogen technologies. By 2030 the platform foresees the development of research that generates important innovations to reduce the cost of hydrogen production. Intensive work is planned in the field of hydrocarbon pyrolysis. Natural gas, oil, coal, and biomass will be the main sources of energy-efficient hydrogen production.

In the long-term perspective, hydrogen production technologies will be in place by 2050. The platform sets out the main objectives to achieve such a level of hydrogen production to satisfy a large share of overall energy consumption. Hydrogen is expected to be produced mainly from renewable energy plants. In isolated energy systems, hydrogen will be stored as an energy reserve that can be used for unforeseen emergencies. By 2050 the use of hydrogen as an energy carrier is planned widely to be used in the transport sector. During this period, particular attention will be paid to the development and application of new technology safety standards, as well as to the improvement of the legislative framework, in order to accelerate the introduction of new technologies into various market sectors.

Hydrogen storage and transportation. The creation of a hydrogen storage, transportation, and distribution network is essential for the mass use of hydrogen for energy generation only in various industries and in the private sector. The Strategic Research Agenda of the Platform states that the hydrogen storage and distribution sector comprise:

- ✔ All forms of hydrogen transport (by water, by road, including by rail and pipeline);
- ✔ All forms of hydrogen storage (in high-pressure cylinders, liquid hydrogen storage systems, chemical and complex hydrides, and other systems);
- ✔ All aspects related to hydrogen storage at its extraction places, filling and distribution stations, transport, and stationary and mobile systems;
- ✔ All types of hydrogen refill systems.

Most importantly, hydrogen storage and transportation systems need to be developed in the direction of security, price reduction, and improvement of technical parameters.

As long as centralized hydrogen production centers are not developed, the largest share of the hydrogen sector of compressed and liquefied hydrogen will be transported by road, water, or rail.

There are currently four main ways of storing hydrogen: as liquid hydrogen; as compressed gas; as solid-type physical sorption on materials with a high specific surface area; and as solid-type chemical sorption storage. These methods of storage are constantly being improved (Milčius, 2009).

Liquefied hydrogen gas is usually stored in high-pressure cylinders that are additionally equipped with refrigerating equipment to maintain cryogenic temperatures, as hydrogen boils at a very low temperature ($-252.87°C$). This additional refrigerating equipment occupies up to 90% of the cylinder weight. Liquid hydrogen systems still have two major drawbacks: there is inevitable hydrogen boiling and loss of up to 1% per day; and hydrogen liquefaction consumes 30%–40% of the total energy generated by burning all compressed hydrogen.

Hydrogen storage in high-pressure cylinders also has drawbacks, especially due to low hydrogen density (at a pressure of 0.1 MPa, the hydrogen density is only 0.08,988 g/L). Conventional steel gas cylinders are tested to a pressure of 30 MPa. However, security is only up to 20 MPa. In these cylinders filled with hydrogen, it is possible to produce hydrogen up to 1% of the total weight of the cylinder. This way of storing hydrogen also has a large volume shortage, since the storage capacity of 4 kg of hydrogen must be as high as 225 L. Even if hydrogen were pressurized in carbon-reinforced cylinders with pressures above 70 MPa, hydrogen would only contain four cylinders. These hydrogen storage systems do not meet current needs, as well as safety standards. Compressing hydrogen gas as well as liquefying consumes a lot of energy.

High-pressure cylinders and cryogenic liquid hydrogen storage systems, due to the above-mentioned drawbacks, are not viable for the widespread use of hydrogen in power engineering.

Another way of storing hydrogen is to use the physico-adsorption of molecular hydrogen on materials of a large specific area. Studies have shown that the storage capacity of adsorbed hydrogen is directly proportional to the surface area of the material used (Adsorption − Lat. *ad* − to, by, in + *sarbo* − accumulation of one substance, contained in gas or liquid, on the surface of another substance, the surface of the solid body or solid micropores and capillaries). However, in practice, to maintain a higher hydrogen content a low temperature is required, but so far this form of hydrogen storage has not been achieved. Very extensive research on hydrogen storage is carried out by the so-called solid-type chemical sorption storage method. Various metal hydrides, to which catalysts are added, are used for this purpose.

As forecasted in the Strategic Research Agenda of the Platform, there should be a lot of compressed and liquefied hydrogen savings in the main urban areas by the 2030s. However, in the vast majority of public service stations, hydrogen will not be stored in large tanks but produced locally from methane using gas reforming technology.

In the same Platform, it is expected that in the period 2030–50 hydrogen technologies will widely spread and will become an important part of the energy system and social life. Hydrogen at that time should be widely used in both stationary and mobile systems. In stationary installations the combination of fuel cell systems with wind or solar power plants will permanently maintain or modulate power systems (Rifkin, 2012).

The function of the fuel cell is very similar to conventional batteries. The main difference is that the fuel cells are not discharged.

In recent years, fuel cells have been widely distributed in a variety of devices powered by electricity. The principle of functioning of fuel cells was discovered by William Grove, a scientist from Wales, in 1839. However, practical applications began in the second half of the 20th century in space technology. NASA scientists have used fuel cells to produce electricity on space ships.

The fuel element is made up of electrodes placed in a container with a liquid or solid electrolyte. Instead of electrolyte, a conductive ion membrane is often used. Oxygen (oxidizer) or air, halogens or nitric acid are supplied to the electrodes adjacent to the electrodes or directly to the electrodes. Hydrogen, methanol, ethanol, hydrazine, formic acid, and hydrocarbons (e.g., glucose or methane) are supplied as reducing agents. Electrochemical reactions occur at three-phase contact (electrode − electrolyte − reagents).

The most commonly split are H_2 and O_2 fuel cells, where hydrogen and oxygen molecules break down at the electrodes into atoms, and they ionize. Hydrogen ions differ from the positive electrode and form water (H_2O) molecules by attaching to the negative electrode. Because the fuel cells produce electricity directly, their efficiency (efficiency) amounts to 70%−80%. In addition to the above-mentioned H_2 and O_2 fuel cell, a number of other fuel cells containing acid or alkaline solutions of the electrolyte, solid electrolytes, cationic membranes, and others have been developed.

Fuel cells are quite expensive because platinum is the most commonly used catalyst for accelerating chemical reactions. Therefore, the world is looking for ways to reduce their cost. In 2011, scientists from Los Alamos National Laboratory, of California University, New Mexico, USA (the first nuclear and first hydrogen bombs were developed in this laboratory) have offered an expensive metal (platinum) replacement with free platinum catalyst. A catalyst consists of carbon (C), partially obtained from polyaniline at high temperature, and inexpensive iron (Fe) and cobalt (Co). Tests have shown that fuel cells that use $C - Fe -,Co$ catalyst generate high electricity output and are more efficient compared to fuel cells using expensive metals. It is also very important that fuel cells with $C - Fe - Co$ catalyst efficiently convert H_2 and O_2 into water without producing high levels of unwanted hydrogen peroxide (H_2O_2). Inefficient fuel conversion to H_2O, when H_2O_2 is generated, reduces the amount of electricity produced, i.e., reduces fuel cell efficiency.

However, the research is continuing to confirm the underlying theories explaining the operation of the catalyst. An additional advantage of fuel cells is that they can be used in places where electricity and water is produced (Engl. *on site*), especially for transport and military purposes. Fuel cells are a very promising source of ecologically clean energy and they have been researched for a long time (research leaders include Japan, United States, Germany, Switzerland, and France).

Worldwide, there are pilot fuel cell−generated electric cars (Daimler−Chrysler, General Motor, Toyota, Honda, Opel, etc.).

Despite all the advantages and alleged simplicity of the fuel cells, their extensive (mass) consumption is still limited, although the principle of their operation was known in the

19th century, and the current fuel cell prototypes were already used in the US Apollo space program.

For the time being, they cannot compete with conventional cogeneration plants using fossil fuels or biofuels at high fuel cell cost.

In the US Department of Energy's "Hydrogen Posture Plan" and the European hydrogen and fuel cell technology platform, it is foreseen that the combination of fuel cell systems with wind or solar cells will play an important role in stationary energy installations by 2030. During the first phases of hydrogen energy technology deployment, internal combustion engines or gas turbines may occupy their place until sufficient fuel cells are produced.

In 2050 the energy system should be decentralized. A large proportion of total energy will come from a range of renewable energy sources, other green and economical energy production systems, where fuel cells will play a key role.

Particular attention will be paid in the near future to low-power fuel cells designed for portable devices up to 50 W and portable up to 5 kW. First of all, the most developed low-temperature polymer electrolyte (PEMFC) and pure methanol (DMFC) type fuel cells should be used for commercialized and competitive market, which could use not only hydrogen as fuel but also the currently widely used liquid fuels such as liquefied propane gas or medium oxidation oil products.

Nuclear (nuclear) energy. Nuclear (nuclear) energy is an energy branch that uses nuclear energy for heat and electricity production. Nuclear energy is the atomic energy of the atom that is caused by the interaction of nucleons (protons and neutrons) (Adlys, Adlienė, 2007; Jankauskas, 2002). This energy is released in two ways: (1) by dividing chemical elements into heavy nuclei (uranium U, plutonium P, and others); and (2) by connecting chemical elements to light nuclei (deuterium D and tric T).

In the first case, in the event of a radioactive decay, during the nuclear reaction the nucleus of the atom divides into the nuclei of two lighter chemical elements (less often three or four). The kernels artificially divide when bombarded (attacked) by other particles—neutrons, high-energy protons, alpha particles, gamma quantum, and so on; spontaneous nuclear fission also takes place. Nuclear fission generates a large number of secondary neutrons (which may occur more than the nuclear fission neutrons) with high kinetic energy. Secondary neutrons further divide the other nuclei. This creates a *chain* nuclear reaction that generates large amounts of thermal energy.

The smallest amount of nuclear material in any space that is capable of a chain reaction is called *critical mass*, and the space (active zone) in which a chain reaction is possible is called *critical volume*.

This means of producing nuclear energy is used for military purposes in nuclear bombs and for peaceful purposes in nuclear reactors used in nuclear power plants.

In the second case, connecting the chemical elements to the light nuclei (hydrogen isotopes D and T) to the heavier nuclei (during *nuclear fusion*), their joining is initially hampered by the electrostatic thrust generated by the uniformly labeled electrical charges. Nuclear fusion is only possible when the kernels acquire a sufficiently high kinetic energy that overcomes the potential barrier created by the pushing forces. In the synthesis of light nuclei, much more energy is released than during the division of heavy (uranium) nuclei. This energized energy is called thermonuclear energy. Thermal fusion occurs when the substance is in a plasma state and at very high temperatures. Self-thermonuclear fusion takes place in stars where the thermonuclear energy at very high temperatures is equal to the energy emitted.

Artificial *uncontrolled* fusion for the first time on Earth was created in the United States by the explosion of a thermonuclear bomb in 1952. Devices for creating controlled thermonuclear fusion conditions began to be developed in the second half of the 20th century.

Nuclear reactors. In nuclear reactors, nuclear energy is turned into heat or neutron flux and gamma rays. This nuclear power surface is controlled by nuclear reactors. The most important part of the nuclear reactor is the active zone where nuclear fuel is placed. The active zone is surrounded by a neutron reflector reflecting the emitted neutrons. The active zone, together with the reflector, is surrounded by a protective layer along with a housing that reduces ionizing radiation.

Nuclear reactors used in nuclear power plants are divided into groups according to the following characteristics: nuclear energy sharing by neutrons (slow neutrons, intermediate neutrons, and rapid neutrons); nuclear fuel and retarder distribution in the active zone (*homogeneous*, fuel and retarder, homogeneous mixture and *heterogeneous* when the decelerator is between the nuclear fuel assemblies; design (hull, duct, and basin); nuclear reactors with heat carrier temperature greater than 700°C, called high-temperature reactors.

Nuclear reactors are also divided into:

✔ Graphite − gas with graphite as the retarder, and gas with heat carrier;
✔ Graphite − water with graphite as a retarder, and water for heat carrier;
✔ Organic reactor heat transfer is organic;
✔ Heavy water reactor retarder is heavy water;
✔ Light water reactors (pressurized water and boiling water).

The most important indicator of nuclear reactors is their thermal power.

The rate of neutron absorption and nuclear fission in the reactor is changed into an active zone by inserting or withdrawing moving rods or cassettes containing material that absorb the neutrons intensively.

The world's first nuclear reactor was released in Chicago, 1942. In Western Europe, the first low-power reactors were built in nuclear power plants in France, Great Britain, and Germany.

Worldwide, nuclear power plants are mostly built with light water reactors, in which water plays a role as both a retarder and a coolant. Nuclear fuel as the primary source of uranium dioxide UO_2, which is enriched with partial uranium isotope ^{235}U, is the most commonly used nuclear power plant. In the reactor, the primary fuel is bombarded with neutrons to become the secondary uranium isotope ^{233}U and plutonium ^{239}Pu.

In heterogeneous nuclear reactors, nuclear fuel is contained in thermal elements. Liquid nuclear fuel is used in homogeneous reactors—mixtures of uranium, plutonium, and thorium salts.

Used nuclear fuel is removed from the reactors and stored for several years in special water-filled tanks or pools. Afterward, it is recycled by separating materials from which kernels can share again. The radioactive materials obtained are used in space technology, medicine, geology, and other spheres. Discarded radioactive waste is stored in various permanent storage facilities.

However, the efficiency of the third generation of nuclear reactors for thermal energy and fuel efficiency is not high and their safety is based on the operation of complex systems.

It is expected that fourth-generation reactors with no third-generation reactor deficiencies will be put into operation between 2020 and 2030.

Following are the general operating principles of the most widely used nuclear reactors in the world.

Light water nuclear reactors. Light water nuclear reactors are of two types: pressurized water (PWR) and boiling water (BWR). Pressurized water reactors are the most common. There are around 60% of nuclear power plants in the world with this type of reactor (PWR).

The first- and second-generation nuclear reactors have been in operation worldwide for a long time. The third generation is now being built, and these are much more reliable reactors. The most important technical and safety features of these third-generation reactors are: simpler and safer equipment and longer operating time (up to 60 years). This third generation utilizes physical phenomena that enhance work reliability: natural convection, gravitational traction, compressed gas properties, and so on.

In PWR-type reactors, water is poured to prevent it from boiling. The pressurized water in these reactors acts both as a retarder of the nuclear reaction and as a heat carrier, heating the water through the secondary circuit of the steam generator. The steam that is used for turbine work is obtained. The primary water circuit (4) maintains a working pressure of 12.0–15.5 MPa water (coolant). The nuclear power plant with a pressurized water reactor has two separate water circulation systems. In the primary circuit, water circulates through the active zone (2) into the steam generator (7), in which heat is transferred to a secondary circuit where the water turns into steam.

Pressurized water heats up to 300–320°C in the primary circuit, and the water in the secondary circuit turns to steam at 260–290°C and pressure is only 4.5–7.8 MPa. The steam is transferred to a high-pressure turbine (9) and then via an intermediate steam superheater (10) to a low-pressure turbine that rotates the electric generator (12). The compressed water (PWR)–type nuclear reactors have a thermal efficiency of 32%–37%. In pressurized water reactors, uranium dioxide fuel is enriched to 4%–5% and added to the 3.5–4 m long zirconium alloy tubes.

Boiling water reactors are similar to pressurized water reactors. However, there is no steam generator in the water circuit of the BWR-type reactors. In the BWR, water circulates through the active zone. The water in this reactor is both a retarder and a heat carrier. Hot water in the reactor boils up to 300°C and produces steam. A vapor pressure of 6–7 MPa is generated inside the vessel. The fresh steam compressed from the upper part of the reactor directly enters the high-pressure turbine and then enters the low-pressure turbine, which rotates the electric generator, through an intermediate steam superheater. From the low-pressure turbine, the steam enters the condenser. From the condenser, the water from the conduit is again directed to the reactor. In this (BWR) reactor the fuel is used in the same way as in a PWR, but the fuel power density (energy per unit volume of active zone) is only half as it has a lower temperature and lower pressure. At lower pressures in the BWR-type reactor (about 7 MPa) and lower temperatures, the thermal efficiency is slightly lower than that of the PWR-type reactor.

Both types of PWR and BWR light water reactors have sealed containers (shells). PWRs and BWRs are mostly built in nuclear power plants. They account for about 80% of all reactors in the world.

Some countries use different types of nuclear reactors. For example, in Canada, nuclear power plants are equipped with CANDU (Eng.PHWR) ACR-type compressed-water reactors, and in Great Britain, mainly gas-fired reactors of the GCR and AGR type.

Pressurized heavy water reactor. The Canadian heavy water nuclear reactor CANDU (CANada Deuterium Uranium) has a core reactor element that contains a nuclear cylindrical tank called calendar filled with heavy water D_2O. Nuclear fuel is contained in the calendar fuel channels. Heavy water slows down neutrons. Heavy water is also used to cool the reactor, which under high pressure flows through the pipes into the steam generator. The simple water (H_2O) that enters the steam generator through the pump turns into steam and enters the turbine that generates an electric generator through steam channels. The already used steam is cooled in the turbine with plain water flowing through the surrounding water bodies or coolers. Heavy water reactor nuclear fuel is natural uranium or more enriched in uranium dioxide.

High-temperature reactor. The high-temperature reactor cooling system (THTR) is filled with inert helium gas. These gases are injected into the reactor and reach a temperature of 250–750°C. Hot gas transmits heat to steam generators, where the water entering the pump turns into steam. The steam enters the high-pressure turbine section and further into the low-pressure turbine section. Turbines rotate the electric generator.

The THTR reactor uses graphite-coated fuel balls with a diameter of 6 cm. Each ball contains 1 g of uranium ^{235}U and 10 g of raw fuel torium ^{232}Th particles with a diameter of 0.7 mm. One ball contains 35,000 such particles. Graphite performs neutron retardation function. The slow neutrons divide the ^{235}U uranium nuclei into energy. Toris ^{232}Th is absorbed into a new nuclear fuel ^{232}U by absorbing fast neutrons, and its nuclei are bombarded with slow neutrons to release energy.

Rapid Neutron Reactors. Depending on the atomic nuclei of the materials, the kinetic energy of the neutrons, i.e., from their speed, nuclear reactors are divided into slow neutron (previously examined) and high-speed neutron reactors. In nuclear reactions, high-speed neutrons move at a speed of 10,000 km/s and a slow neutron speed of 2.2 km/s. Rapid neutron reactors are also used to generate energy and to produce new fuel for plutonium ^{239}Pu. The active zone of the high-speed neutron reactors consists of particle and reproductive material. In this reactor, the heat carrier of the first and second contours is liquid metal (sodium), and the third circuit circulates water, passing through the steam in the steam generator.

Chapter summary

✔ The world economy has increased 3.3 times over 4 decades. Global economic growth slowed slightly: from 1971 to 1980 growth rates averaged 3.8%; between 1980 and 1990 it reached 3.0%; in the period 1990–2000 it was estimated to be 2.8%; and in 2000–11 it was already 2.6% per annum. However, the overall downward trend in global economic growth was driven by slower economic growth in OECD-based developed countries, generating GDP due to the effects of the global economic crisis in 2000–11 when it grew by an average of 1.6% per year. In 2017, global economic growth reached 3.0%, which was a sharp increase from 2.4% in 2016; and since 2011 there has been the largest recorded global growth. Labor market indicators continue to improve in many countries, hence around two-thirds of the global market in 2017 grew faster than in previous years. Global growth was expected to remain stable at 3.0% in 2018 and 2019.

✔ Primary energy consumption worldwide increased by 10% compared to 2010 and 2017. Therefore, in 2017 primary energy consumption was in regions such as Asia Pacific (5743.6), North America (2772.8), Europe (1969.5), CIS (978.0), and the Middle East (897.2) .

✔ Energy resources, especially automotive oil products and the widespread penetration of electricity in all areas of human activity, was and remains one of the most important and essential components of mankind's needs for progress and improvement of living conditions: natural gas (5.7%), nuclear (3.9%), oil (3.5%), other (includes geothermal, solar, wind, tide/wave/ocean, heat and other) consumption (1.5%) increased compared to 2010 and 2017. Oil, natural gas, and coal dominated the world primary energy balance in 2010−17; nuclear fuel consumption increased from 5.7% to 9.6% in the period 2010−17, and also increased by 0.9%−2.4% geothermal, solar, wind, tide/wave/ocean, and heat consumption.

✔ The growth of energy needs in the medium and long term may be driven by economic and population growth in developing countries. World Energy Development Scenarios for Energy Experts foresee: if current trends persist, in 2035 the primary energy world will consume 44% more than in 2011; the implementation of the new policy scenario will create preconditions for moderate energy demand growth (1.2% per year); only an efficient transformation of global energy systems can ensure a slow (on average 0.5% per year) increase in energy needs by substantially increasing the role of renewable energy sources and halting global average rise so that it does not exceed 2°C above preindustrial levels.

✔ There are three major challenges to be addressed in the energy sector on a global scale: cross-border energy policy, smart energy policy, and each state has to deal with the trilogy in a harmonious and optimal way ((how to balance energy security; solve energy availability (along with price); and harmonize environmental issues).

✔ Notwithstanding the fact that the World Energy Council in 2013 report states that fossil fuels are a very large amount of newly explored stocks, up to 2−3 times more oil, and that fossil fuels will have to rise in the future due to their heavy accessibility.

✔ The energy-intensive economy in the EU requires a lot of resources. By 2030 it is expected that primary energy consumption in the EU will grow by 1%−2% annually and reach 1950 Mtoe per year. Most energy consuming and fastest consumption occurs in the household, service sector, and transport. In the industry, energy efficiency will grow slower due to more efficient technologies. The EU's own energy supply capacity is very limited. EU oil accounts for 60% of total energy consumption. For a variety of reasons, the EU's own (local production) energy resource reserves for energy production in 2030 are: plans to reduce by almost 30% compared to 2000.

✔ It is forecast that by 2035 in both developed and developing world countries, the use of renewable energy sources will grow faster than nuclear energy, which also does not generate pollutant materials.

- In OECD countries, where energy demand has stabilized, renewable energy sources will be widely used in newly built power plants.
- The potential of renewable energy in the world is inexhaustible. Worldwide renewable energy sources in 2013 accounted for about 16% of total energy consumption. In 2012 the total installed power of power plants was 1373 GW (the power of all nuclear power plants in the world was about 300 GW). Among them: hydropower capacity (990 GW), wind power capacity (283 GW) and solar PV capacity (100 GW). Around 19.5% of the world's electricity was generated from renewable energy sources. Compared to 2012 and 2017, investment in new renewable capacity (billion USD) increased by 13%. During this period, the largest increase occurred in solar PV capacity (75%), concentrating solar thermal power capacity (49%), and wind power capacity (47%).
- The structure of renewable energy sources in the EU countries, and in particular the level of production, is uneven. In the area of renewable energy sources production, in the EU Germany is leading, followed by Spain, Italy, France, Sweden, Finland, and Austria. In 2020 in Germany, at least 40% of the electricity is expected to be produced from renewable energy sources (23.4% in Germany, 27% in Denmark, 21.1% in Spain). The EU total wind power capacity reached 117 GW and produced 257 TWh (7.8%) of electricity.
- The aim is to achieve the EU directive goals that in 2050 the postcarbon era will start in the energy sector. Currently, wind and especially solar power plants are still expensive. Installation and connection to system networks is expensive too. The maintenance of these plants, which is carried out according to strict instructions of the manufacturers, is not enough either. A great number of specially trained staff is needed for maintenance of plants with a wide range of electronic devices.
- The cost of electricity produced by nuclear, coal, and natural gas–fired power plants is rising every year. Nuclear prices rise due to growing security requirements and the increasing construction prices for new power plants. Fossil fuel power plants are becoming more expensive, in particular owing to the inclusion of CO_2 emissions costs in direct electricity production costs.
- In 2013 for the first time, the World Energy Council and the International Energy Agency confirmed that electricity generated by wind farms is cheaper than electricity produced by nuclear power plants. However, the World Energy Council recognizes that the cheapest electricity type is energy saving. Solar energy has been the most widely used in water heating technologies (with solar collectors) both in the world and in the EU.

References

Adlys, A., Adlienė, D., 2007. Branduolinė Ir Alternatyvi Energetika. Kaunas: Technologija.
Adomavičius, V., 2013. Mažosios Atsinaujinančiuju Ištekliu Energijos Sistemos. Kaunas: Technologija.
Baublys, J., 2014. Energetika krašto apsaugos sistemoje (KAS). Elektros erdvės 2 (35), 16–20.
Baublys, J., Miškinis, V., Morkvėnas, A., 2011. Lietuvos energetikos darna su gamta. Energetika 57 (2), 85–94.
BP, 2014. BP Energy Outlook 2035. Retrieved from. https://www.bp.com/energyoutlook.
BP Statistical Review of World Energy, 2018. Retrieved from. https://www.bp.com/content/dam/bp/en/corporate/pdf/energy-economics/statistical-review/bp-stats-review-2018-full-report.pdf.

Communication from the Commission to the European Parliament and the Council, 2013. Implementing the Energy Efficiency Directive — Commission Guidance, 2013. COM, p. 762. Retrieved from. http://eur-lex.europa.eu/Lex-UriServ/LexUriServ.do?uri=COM:2013:0762:FIN:LT:PDF.

Deksnys, R.P., Danilevičius, K., Miškinis, V., Staniulis, R., 2008. Energetikos Ekonomika. Kaunas: Technologija.

Directive 2009/28/EC of The European Parliament and of The Council of 23 April 2009 on the Promotion of the use of Energy From Renewable Sources and Amending and Subsequently Repealing Directives 2001/77/EC and 2003/30/EC.   Retrieved   from   https://eur-lex.europa.eu/legal-content/EN/TXT/PDF/?uri=CELEX: 02009L0028-20130701&from=LT.

Directive 2012/27/EU of the European Parliament and of the Council of 25 October 2012 on Energy Efficiency, Amending Directives 2009/125/EC and 2010/30/EU and Repealing Directives 2004/8/EC and 2006/32/EC. Retrieved from http://eur-lex.europa.eu/LexUriServ/LexUriServ.do?uri=OJ:L:2012:315:0001:0056:EN:PDF.

European Commission, 2018. EU Energy in Figures. Publications Office of the European Union. https://doi.org/10.2833/279113.

EWEA, 2003a. Wind Power Targets for Europe: 75000 MW by 2010. European Wind Energy Association — Greenpeace, Brusseles.

EWEA, 2003b. Wind Force 12. A Blueprint to Achieve 12 % of the World's Electricity from Wind Power by 2020. European Wind Energy Association — Greenpeace, Brusseles.

Forrester, J.W., 1971. World Dynamics. Pegasus Communications, Waltham, MA.

Global Wind Energy Council, 2006. Global Wind 2006 Report. GWEC, Brussels, Belgium.

Global Wind Energy Council, 2012. Global Wind Statistics 2011. GWEC, Brussels, Belgium.

Global Wind Energy Council, 2016. Global Wind Statistics 2015. GWEC, Brussels, Belgium.

Global Wind Energy Council, 2018. Global Wind Statistics 2017. GWEC, Brussels, Belgium.

IEA, 2006. CO_2 Emissions from Fuel Combustion 2006. OECD Publishing, Paris.

IEA, 2009. CO_2 Emissions from Fuel Combustion 2009. OECD Publishing, Paris.

IEA, 2012. CO_2 Emissions from Fuel Combustion 2012. OECD Publishing, Paris.

IEA, 2013a. Energy Balances of OECD Countries 2013. IEA, Paris.

IEA, 2013b. World Energy Outlook 2013. IEA, Paris.

IEA, 2014a. Energy Balances of Non-OECD Countries 2014. IEA, Paris.

IEA, 2014b. World Energy Outlook 2014. IEA Publications.

IEA, 2015. World Energy Outlook 2015. IEA Publications.

IEA, 2016. World Energy Outlook 2016. IEA Publications.

IEA, 2017. World Energy Outlook 2017. IEA Publications.

IEA, 2018. World Energy Outlook 2018. IEA Publications.

Lynch,D. JOM (2009) 61: 41. https://doi.org/10.1007/s11837-009-0166-8.

Jankauskas, V., 2002. Atominės elektrinės konkurencinėse elektros rinkose. Energetika 3, 3—11.

Kytra, S., 2006. Atsinaujinantys Energijos Šaltiniai. Kaunas: Technologija.

Meadows, D.H., Meadows, D.L., Randers, J., Behrens III, W.W., 1972. The Limits to Growth. Universe Books, New York.

Milčius, D., 2009. Vandenilis — naujos kartos energijos šaltinis. Mokslas ir technika 10, 4—6.

Miškinis, V., Baublys, J., Konstantavičiūtė, I., Lekavičius, V., 2014. Aspirations for sustainability and global energy development treds. Journal of Security and Sustainability Issues 3 (4), 17—26.

Nagavičius, M., 2014. Turime padidinti atsinaujinančios energetikos dali. Elektros erdvės 3 (36), 4—6.

REN21, 2012. Renewables 2012 Global Status Report. REN21 Secretariat, Paris.

REN21, 2013. Renewables 2013 Global Status Report. REN21 Secretariat, Paris.

REN21, 2015. Renewables 2015 Global Status Report. REN21 Secretariat, Paris.

REN21, 2017. Renewables 2017 Global Status Report. REN21 Secretariat, Paris.

Rifkin, J., 2012. Trečioji Pramonės Revoliucija. Eugrimas, Vilnius.

Statista, 2018. Installed Power Capacity in the European Union (EU-28) in 2005 and 2017. Retrieved from. https://www.statista.com/statistics/807675/installed-power-capacity-european-union-eu-28/.

Šateikis, J., 2006. Augalinės biomasės auginimo ir naudojimo kietajam kurui energetinis potencialas ir mokslinės problemos. LŽŪU, ŽŪI Instituto ir LŽŪ Universiteto moksliniai darbai 38 (3), 5—21.

Šuri, M., Huld, T.A., Dunlop, E.D., 2005. PV-GIS: a web-based solar radiation database for the calculation of PV potential in Europe. International Journal of Sustainable Energy 24 (2), 55—67.

United Nations, 2018. World Economic Situation and Prospects 2018. Retrieved from. https://www.un.org/development/desa/dpad/wp-content/uploads/sites/45/publication/WESP2018_Full_Web-1.pdf.

Vilemas, J., 2014. Prasideda atsinaujinančiosios energetikos era. Elektros erdvės 2 (35), 6—11.

World Commission on Environment and Development, 1987. Our Common Future. Oxford University Press, Oxford.

World economic Outlook database. 2018. Retrieved from https://www.imf.org/external/pubs/ft/weo/2018/01/weodata/index.aspx.

World Energy Council, 2013. World Energy Insight 2013. Official Publication of the World Energy Council to mark the 22nd World Energy Congress, Daegu, Korea.

Further reading

European Commission, 2001. Green Paper — towards a European Strategy for the Security of Energy Supply. Office for Official Publications of the European Communities, Luxembourg.

Energy consumption and greenhouse gas emissions against the background of Polish economic growth

Jacek Brożyna

Rzeszow University of Technology, The Faculty of Management, Department of Quantitative Methods, Rzeszów, Poland

Energy Transformation towards Sustainability
https://doi.org/10.1016/B978-0-12-817688-7.00002-1

Introduction

Poland has been a member of the European Union (EU) since May 2004, and thus on many levels needs to comply with the common policy and respect EU law. In the context of this work, aspects related to climate policy and sustainable development are particularly important (Tvaronavičienė and Grybaitė, 2012; Vasiliūnaitė, 2014; Vosylius et al., 2013, p. 3). The laws established by the EU described in a later part of this article are also binding for Poland in this respect.

Economic development in the modern world must include, among others, optimal management of natural resources, without harm to future generations and the natural environment. The first discussions on reducing energy consumption and taking care of the environment began in the 1960s ("First Orientation for a Common Energy Policy, Communication from the EC to the Council, 1968; Immenga, 1992; Jasiński, 1996). Oil crises lasting from 1973 to 1974 and 1979–82 (Akins, 1973; Campbell, 2005; Friedrichs, 2010; Hirsch, 2008; Hondroyiannis et al., 2002; Issawi, 1978; Kerr, 1998; Venn, 2016) made governments around the world aware of how dependent they are on mined energy sources, but also how much they exploit them. It is a process that lasted for many years, since the first step in the industrial revolution was primarily economic growth (Smil, 2010). Attention was paid to excessive consumption of goods (including energy) and growing pollution of the natural environment. Discussions in the international arena started to revolve around resources management so that it was effective and did not cause damage to the economies of countries. Such development began to be called *sustainable development*, and one of the first official documents where this term was used was the Rio Declaration of June 1992 (The Rio Declaration on Environment and Development, 1992). The basic goal of sustainable development is to strive for a stable growth of competitiveness of economies, which can only be rational when economies deal with basic social and environmental problems (i.e., environmental pollution, greenhouse gas emissions, production of energy from renewable sources) to ensure the best possible conditions for the life of societies and business development. The basic goal was and still is to create such economic plans that they would be rational for economies, and at the same time to ensure that the negative effects of business operations on the environment are restrained. Such records were included in the aforementioned Rio Declaration and Agenda 21, which were created after the Earth Summit in Rio de Janeiro in 1992.

Since the beginning of the formulation of the concept of sustainable development, European countries were actively involved in global initiatives. The states of the Old Continent were not limited only to declarations and summits of worldwide reach. European countries undertook many initiatives aimed at promoting and enabling sustainable development of the man-made economy and limited resources of the natural environment. The contemporary European Union derives from the European Coal and Steel Community. Already in the very name of the organization coal was indicated, which was an obvious symbol of economies in postwar Europe. Technological development on the one hand increased the demand for energy, but at the same time technologies that would save energy and efficiently produce it from other sources than mined fuels appeared. In most European countries, coal has been abandoned in favor of more environmentally friendly energy sources. These initiatives are reflected in the actions of individual states, but commitments at the European level are more important.

In Article 194 of the Treaty on the Functioning of the European Union of the Lisbon Treaty of 1992, provisions were included on the functioning of the energy market, including the promotion of energy efficiency and the development of new, renewable energy sources of energy. These records were reflected in the Green Book and the White Book published in the following years, and the Directives 96/92/EC and 98/30/EC. The Kyoto protocol negotiated in 1997 and ratified by 141 countries in 2005 was key to the natural environment not only in the European Union but also in the whole world. In this protocol, the signatories declared reduction of greenhouse gas emissions by 2012. The level of reduction took into account the economic specificity of individual countries. The specificity differed and was usually between 5% and 10%. There were also countries, such as Russia, for which this level was set at 0%, or Australia or Iceland, for which a possible increase of 8% and 10% was predicted. In the case of Poland, the reduction of greenhouse gases was set at 6%. The level of greenhouse gases in 1990 was determined as the baseline. However, for the former socialist countries (including Poland) 1988 was adopted as the base year. The implementation of the objectives was not as smooth as assumed, therefore in the EU all work related to sustainable development and climate policy aimed at developing tools obliging to achieve the assumed goals. Their effect was, among others adopted in 2001 in the strategy for sustainable development (A sustainable Europe for a better world: A European strategy for Sustainable Development, 2001; Strategy for sustainable development, 2001) and in 2005 in the strategy on climate change (Strategy on climate change: foundations of the strategy, 2005). The aim of these strategies was to coordinate policies at the EU level in such a way as to be consistent with the concept of sustainable development. The European Commission's proposal included also provisions on promoting the idea of sustainable development outside the borders of the European Union. The EU also developed the Energy and Climate Package in 2007 and Directive, 2009/29/EC of the European Parliament and of the Council in 2009 (Directive, 2009/29/EC of the European Parliament and of the Council of April 23, 2009 amending Directive 2003/87/EC so as to improve and extend the greenhouse gas emission allowance trading scheme of the Community (Text with EEA relevance), 2009). These directives obliged only within the EU member states to reduce greenhouse gas emissions, however, in the same way as in the Kyoto Protocol, the level of their economic development and the structure of energy systems. A 10-year Europe 2020 strategy was adopted in June 2010 (Europe 2020 Strategy, 2009). The most important goal of this strategy in the context of this work was to accelerate economic growth based on intelligent and sustainable growth. By intelligent growth, one means investing in a more innovative economy. Knowledge and innovation are to lead to build a modern society where high-tech enterprises operate and implement modern solutions that not only bring more benefits to entrepreneurs but also to the entire economy because they care more about the natural environment. On the other hand, sustainable growth is to be based largely on a rational and economical use of environmental resources, but also to limit greenhouse gas emissions in order to build a zero-emission economy, which at the same time will increase its competitiveness. These provisions were included in the Europe

2020 strategy as a 3 × 20 climate and energy package (2020 climate & energy package, 2009) where three key objectives are clearly identified:

- 20% cut in greenhouse gas emissions (from 1990 levels)
- 20% of EU energy from renewables
- 20% improvement in energy efficiency (from 2005 levels)

Because not all member states would be able to achieve the agreed goals, therefore the values given apply to the entire European Union. For Poland, these goals have been set at the level of:

- 14% cut in greenhouse gas emissions
- 15.48% energy from renewables
- 14% improvement in energy efficiency

In addition, 1990 was adopted as the base year for all countries. The assumed targets are monitored by the Commission (Eurostat) on the basis of semiannual reports submitted by member states (2009 Review of the EU Sustainable Development Strategy, 2009; National Energy Efficiency Action Plans and Annual Reports − Energy - European Commission, 2018; Progress made in cutting emissions, 2018).

Despite the fact that the 2020 targets have not yet been met, the EU has already set goals for 2030 in 2014 (2030 climate & energy framework, 2014) as:

- 40% cuts in greenhouse gas emissions (from 1990 levels)
- 27% share for renewable energy
- 27% improvement in energy efficiency

And on December 24, 2018, it increased them (New Renewables, 2018) up to:

- 40% cuts in greenhouse gas emissions (from 1990 levels)
- 32% share for renewable energy
- 32.5% improvement in energy efficiency

These objectives are in line with the long-term perspective set out in the action plan for the transition to a competitive low-carbon economy in 2050, the Energy Roadmap 2050 (2050 Energy Strategy, 2018).

The agreements on climate action, which were jointly developed by almost 200 countries of the world and are legally binding, were also adopted during the United Nations Climate Change Conference COP21 in Paris in December 2015 (Paris Agreement, 2016), and later confirmed in December 2018 at the Katowice Climate Change Conference COP24 (Katowice Climate Change Conference, 2018).

Realization of the key objectives of the climate and energy package by Poland

As a member of the EU, Poland is committed to implement the Europe 2020 strategy and three key objectives set out in the climate and energy package. This part presents some progress in this area in the last dozen or so years, and on the basis of forecasts it was estimated how to achieve the assumed goals by 2020. The ARIMA model, often used for these type of issues, was used to prepare forecasts (Albayrak, 2010; Box et al., 2015; Brożyna, Mentel and

Szetela, 2016a, 2016b; Ediger and Akar, 2007; Lee and Ko, 2011; Lotfalipour et al., 2013; Pappas et al., 2008; Siew et al., 2008; Yeboah et al., 2012). The Polish energy sector was also briefly described in order to show possible obstacles in achieving the goals.

Sources of energy in the Polish energy sector

The Polish energy industry is mainly based on fossil fuels, in particular on coal (Fig. 2.1) (Dogan and Seker, 2016; Gawlik, 2018; Scarlat et al., 2015; Stala-Szlugaj, 2016). International commitments and growing environmental awareness (da Graça Carvalho et al., 2011) are such that in many countries the transformation of the energy sector is progressing through its liberalization and the use of various types of incentives, including subsidies for investments in renewable energy sources (Bowden and Payne, 2009; Choi et al., 2018; Choi, 2018; Marques et al., 2010; Nicolini and Tavoni, 2017; Sampedro et al., 2017; Slusarczyk et al., 2013; Wang and Ye, 2017; Zhang, et al., 2015; Zhang et al., 2018). However, in Poland legislation in this area is underdeveloped and changing, and the Polish president and government even during the last world climate summit in Katowice (Katowice Climate Change Conference, 2018) tried to prove that Polish coal-based energy policy did not contradict the climate protection (Statement by the President, 2018).

As one can see in Fig. 2.1, the group of renewable energy sources also includes biofuels, biogas, and biomass, but their conversion into energy through combustion causes CO_2 emission. If such a criterion were to be applied in the future, it would turn out that the Polish power industry still has more to catch up in this area than it is currently believed. Many years of delays in the modernization of the Polish energy sector make it impossible to quickly transform it into renewable energy or at least reduce its dependence on combustible fuels, which is why the abovementioned reduction criteria are lower than $3 \times 20\%$ for the whole European

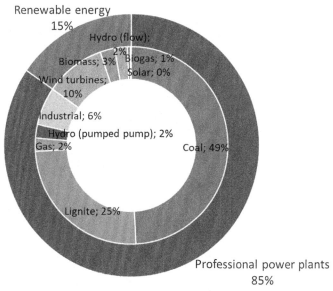

FIGURE 2.1   Energy sources in Poland in 2016. *Source: The authors' own research. Data source: Eurostat Energy balances, 2019. http://appsso.eurostat.ec.europa.eu/nui/submitViewTableAction.do.*

Union. The energy security of the country is also important as it does not allow excluding coal-fired power plants without providing energy from other sources (Ellabban et al., 2014; Fraczek et al., 2013; Li, 2005; Tvaronavičienė et al., 2015; Vosylius et al., 2013).

Greenhouse gas emissions

The most important goal from the point of view of the climate policy and the Europe 2020 strategy is the reduction of the winter gas emissions by 14% compared to 1990 levels. The greenhouse gases (GHGs) include seven gases that have a direct impact on climate change: carbon dioxide (CO_2), methane (CH_4), nitrous oxide (N_2O), chlorofluorocarbons (CFCs), hydrofluorocarbons (HFCs), perfluorocarbons (PFCs), sulfur hexafluoride (SF6), and nitrogen trifluoride (NF_3). The emission of these gases is expressed in CO_2 equivalents, and then converted into millions of tons and briefly designated as Mt. Monitoring and forecasting of greenhouse gas emissions, and in particular CO_2, is not only of interest to the European Union's supervisory authorities, but also many researchers from around the world (Azevedo et al., 2011; Chappin and Dijkema, 2009; Delarue et al., 2007; Karmellos et al., 2016; Karmellos et al., 2016; Ko et al., 2010; Krajačić et al., 2011; Pukšec et al., 2014; Tomás et al., 2010; Xu and Ang, 2013). In Fig. 2.2 greenhouse gas emissions from Poland since 1990 are presented. The same figure shows the forecast of GHG emissions by 2020 together with a 95% possible range of changes.

As Fig. 2.2 shows, greenhouse gas emissions were met by Poland in 2012 (-14.6%). Despite the short-term increase in GHGs in 2015 and 2016, the most probable forecast indicates a further reduction of these gases in the following years, thus reaching the inches assumed for 2020. In the European Union, however, there are pressures on Poland that the

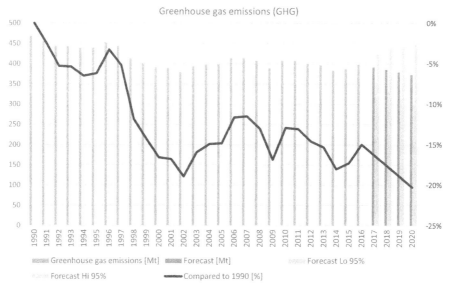

FIGURE 2.2   Greenhouse gas emissions. *Source: The author's own research. Data source: Eurostat Greenhouse gas emissions, 2019.*

reduction of greenhouse gas emissions would be greater than the assumed 14%. Greater reduction (as shown in Fig. 2.2) is possible, but due to the fact that the Polish energy sector is based to a large extent on coal (Fig. 2.1), this would require more decisive action by the Polish government, which may be difficult due to the strong mining lobby.

Renewable energy

Another goal to be achieved in 2020 is to increase Poland's share of renewable energy in gross final energy consumption to 15.48%. The data on this ratio is available in Eurostat since 2004 and is presented in Fig. 2.3.

As one can see in Fig. 2.3, it is rather impossible to achieve the goal of increasing the share of renewable energy. The most probable forecast indicates a 10.5% share of energy from renewable sources in 2020, and the range of 95% of possible deviations from this forecast indicates 7.5%–13.5%. Thus, even in the most optimistic scenario, Poland probably will not meet the 15.48% criterion by 2020. The reason is the same as in the case of greenhouse gas emissions, i.e., Polish energy based on fossil fuels, in particular on coal (Fig. 2.1). This problem concerns not only Poland but also other countries that joined the European Union in 2004 and later (Brożyna et al., 2017). To design and build new energy sources based on renewable energy, we need not only money and relevant laws, but also time. Thus, it is impossible to make up for many backlogs within renewable energy by 2020. As mentioned in Section "Sources of energy in the Polish energy sector", it is also impossible to improve this ratio by excluding, for example, a part of coal-fired power plants, because it would threaten the country's energy security.

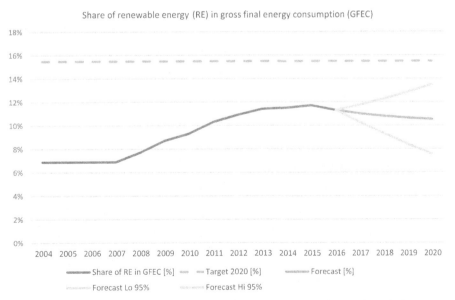

FIGURE 2.3   Share of renewable energy in gross final energy consumption. *Source: The author's own research. Data source: Eurostat, Share of Energy from Renewable Sources, 2019.*

Energy consumption and energy efficiency

In the Europe 2020 strategy and the climate and energy package, the EU committed itself to reducing energy consumption by 20% by 2020, in relation to energy consumption in 1990. This is also known as the 20% energy efficiency target (Balitskiy et al., 2016; Fresner et al., 2017; Tvaronavičienė et al., 2018). As already mentioned this value applies to the entire EU, and individual countries have different levels of reduction. Poland in this area was obliged to improve energy efficiency by 14%. The unit used in EU documents and statistics for measuring the energy consumption is million tons of oil equivalent, abbreviated as Mtoe. Energy efficiency targets were recorded in Article 3 of Directive (2012)/27/EU and concerns the reduction of two energy measures:

- primary energy consumption (PEC) to 96.4 Mtoe
- final energy consumption (FEC) up to 71.6 Mtoe

Primary energy consumption measures the total energy demand in a given country. This also includes the consumption and distribution of energy by the energy sector and the transformation of resources into energy (e.g., gas for electricity), as well as final consumption by end users. This type of energy does not include resources not used for energy production (e.g., crude oil used for the production of plastics).

The values of the PEC and energy efficiency target from 1990 to 2016 (latest available data for January 2019) were presented in Fig. 2.4.

As one can notice in Fig. 2.4, the energy efficiency target for Poland was set at a level that in theory should not be a problem in its implementation until 2020, as it had was already met in 2012. The observed increase in PEC in 2015 and 2016 did not significantly affect the

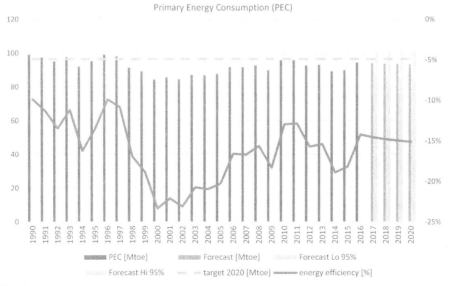

FIGURE 2.4 Primary Energy Consumption of Poland. *Source: The author's own research. Data source: Eurostat, Energy Balances, 2019.*

forecast, which indicates that in 2020 the reduction of this energy consumption will amount to 15.1%. However, there is a small risk that in the most pessimistic forecast (Hi 95%), this ratio will be slightly exceeded.

The second measure written down in the EU directive is final energy consumption, which means the total energy consumed by end users, i.e., households, industry, transport, and agriculture. Energy used for own needs by the energy sector is excluded from this energy meter. Changes in FEC and energy efficiency for Poland since 1990 are presented in Fig. 2.5.

In the case of FEC, as in the case of PEC, the most probable forecast indicates the achievement of the assumed objective since it was met already at the time of its determination and was not exceeded in subsequent years. Even in the most pessimistic forecast, 69.8 Mtoe is below the acceptable limit of 71.6 Mtoe.

The implementation of goals related to the reduction of energy consumption by the Polish economy will only be a theoretical success resulting from too low targets. In fact, as one can see in Figs. 2.4 and 2.5, energy consumption was at the level of the 1990s of the last century and was significantly higher than at the beginning of the 21st century. In the European Union, attention has been drawn to too low goals, and in the 2030 and 2050 strategies, the requirements in this area are much greater, which may pose a big problem for Poland.

An analysis of Poland's sustainable development in the context of the Europe 2020 strategy

The aim of this chapter is to analyze Poland's sustainable development in terms of obligations arising from the Europe 2020 strategy and the 3 × 20 climate and energy package. The

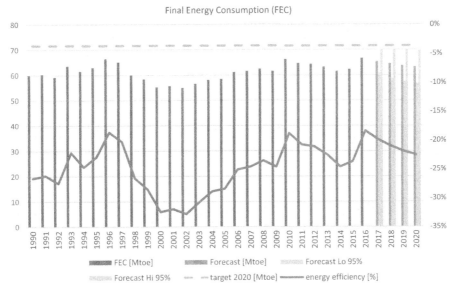

FIGURE 2.5 Final energy consumption of Poland. *Source: The author's own research. Data source: Eurostat, Energy balances, 2019).*

data on greenhouse gas emissions and energy consumption, including gross domestic product, will be presented and analyzed. These factors, which result from EU law and are monitored by it, are closely related to climate policy and sustainable development. The climate and energy package indicates the percent reduction in GHG emissions and energy consumption in relation to the entire economy. Thus, in this work, the Polish economy will be assessed as a whole, without recalculating individual values per capita or to the surface of the country.

Gross domestic product

Since its accession to the EU, Poland is still in the sixth position in terms of both population and area. In terms of gross domestic product (GDP), according to the latest Eurostat data available in 2017, it was only in the eighth position (Eurostat Gross Domestic Product, 2019) and was responsible for 3.0% of GDP of the entire EU (Campos et al., 2014; Pastor et al., 2018; Simionescu et al., 2016). It is worth mentioning that at the moment of joining the EU in 2004 in terms of GDP Poland was in the 11th position, and in the oldest Eurostat data from 1995 in the 12th position (taking into account 28 countries of the European EU, as in January 2019). A detailed list of GDP of Polish values in current prices of billion euro is presented in Fig. 2.6. In the same figure, a line is marked for the percentage that Polish GDP is in the entire EU.

The growth of Poland's position in the GDP of the European Union is related to GDP growth uninterrupted since 1995 (Fig. 2.7). Moreover, the GDP growth in Poland was always higher than the average GDP of the entire EU, even in the years of the global economic crisis (Frankel and Saravelos, 2012; Mentel et al., 2017; Simionescu, 2016).

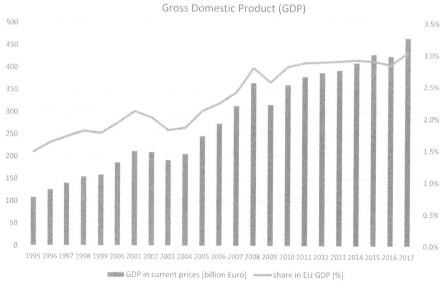

FIGURE 2.6 Gross domestic product of Poland. *Source: The author's own research. Data source: Eurostat, Gross Domestic Product, 2019.*

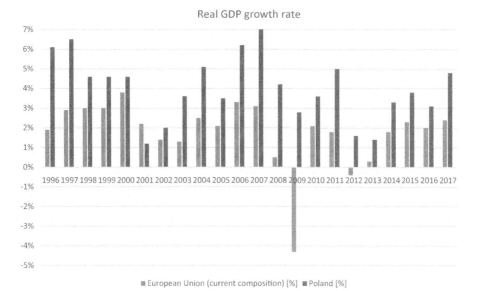

FIGURE 2.7 Real GDP growth rate of Poland and European Union. *Source: The author's own research. Data source: Eurostat, Gross Domestic Product, 2019.*

Gross domestic product is an important measure because it describes the value of all goods produced in a given country. As presented above, it allows comparing the dynamics of economic development over time and compares it with other countries. However, climate change and environmental pollution contributed to the creation of international agreements that make governments unable to care for economic growth without paying attention to the consumption of natural resources and care for the environment.

Energy intensity of the economy

There are many studies in the literature that show the relationship between economic growth and energy demand. The most common theory is that economic growth determines the energy demand, which is referred to as "conservation hypothesis" (Caraiani et al., 2015; Lee, 2006; Sadorsky, 2009; Sari et al., 2008; Shahbaz et al., 2011; Zhang and Cheng, 2009). The second of the most-emerging theory called "growth hypothesis" indicates that economic growth depends on energy production, but as research shows, energy consumption limits are imposed by government policy (Fang, 2011; Inglesi-Lotz, 2016; Kayhan et al., 2010; Kayhan et al., 2010; Menegaki, 2011; Rodríguez-Caballero and Ventosa-Santaulària, 2017). Regardless of which of these factors is a determinant, economic development needs to be sustainable to maintain a balance between the development of countries and regions, and the demand for energy and care for the natural environment (Błażejowski et al., 2016; Cîrstea et al., 2018; Kasperowicz et al., 2017; Katre and Tozzi, 2018; Peñalvo-López et al., 2017; Qoaider and Steinbrecht, 2010; Saiah and Stambouli, 2017; Zhang et al., 2017; Zhao and Guo, 2015).

Energy intensity is the ratio that measures energy consumption by the economy and its energy efficiency (Inglesi-Lotz, 2016; Markandya et al., 2006; Peñalvo-López et al., 2017; Tvaronavičienė et al., 2015; Vasiliūnaitė, 2014; Voigt et al., 2014; Wurlod and Noailly, 2018). This ratio is calculated as the ratio of gross inland energy consumption (GIEC) to GDP and is presented for Poland in Fig. 2.8.

The low value of energy intensity tells of the modern economy. Based on Fig. 2.8 one can observe that Poland made very good progress in this respect (0.2343 in 2016). However, this value is still more than twice the European average of 0.1097. It may be more and more difficult to reduce the energy intensity of the Polish economy in the coming years. The largest (positive) drop in energy intensity took place between 1995 and 2008, which was mainly due to the restructuring of the national economy by increasing the share of the services sector in national income. In the Polish industry after 1990, investments in modernization of equipment and technologies for more energy-efficient ones also began. More rational management of energy carriers and an improvement of energy processes related to energy generation and transmission were also important. After 2008 one could see a slowdown in this process as the number of enterprises requiring radical changes was getting smaller, and for those who already had relatively energy-efficient systems, the introduction of further improvements was not so economically efficient. The most probable forecasts show a very slow decline in the energy intensity of the Polish economy to 0.2274 in 2020, which will still be much higher than the average in the European Union.

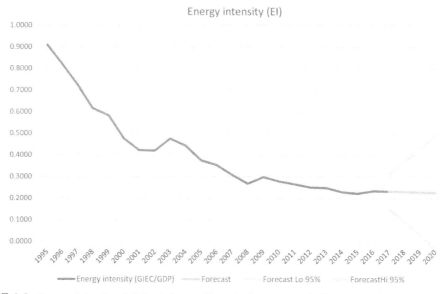

FIGURE 2.8   Energy intensity of Poland. *Source: The author's own research. Data source: Eurostat Energy balances, 2019; Eurostat Gross Domestic Product, 2019.*

Energy consumption and greenhouse gas emissions

As it was shown earlier, Poland will probably meet the requirements for reduction of energy consumption and greenhouse gas emissions by 2020, but this is due to the relatively low targets set by the EU rather than extensive efforts to this end. Fig. 2.9 shows changes during gross inland energy consumption and greenhouse gas emissions in Poland.

It can be noted that before Polish accession to the EU, there was a strong relationship between GIEC and GHG, and only after 2004 this relationship began to decline. The biggest impact this had was on the climate and energy commitments of Poland toward the EU, which contributed to an increase in the share of energy from renewable sources, as previously shown in Fig. 2.3. Unfortunately, the growth of energy from renewable sources has slowed down in recent years, and this causes greenhouse gas emissions to fall mainly when the economy becomes less energy intensive, and this is a very slow process in Poland.

An assessment of the sustainable development of the Polish economy

Sustainable development is one of the fundamental objectives of the EU and, hence, its member states. In previous chapters, the focus was put on the analysis of the sustainability aspects related to ecology and the economy through the assessment of the achievement of objectives of the Europe 2020 strategy and the climate and energy package. It was shown that the Polish energy sector was very much based on mine fuels, in particular coal (Fig. 2.1). Burning of combustible fuels causes emissions of greenhouse gases, whose reduction is also declared by Poland. The best way to reduce GHG emissions is to develop renewable energy. In Poland, however, investments in this area are too small, which means that the share of renewable energy grows too slowly to achieve the goals set (Fig. 2.3).

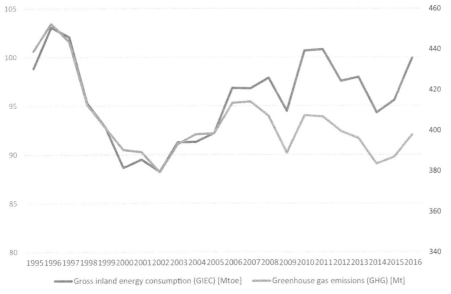

FIGURE 2.9 Gross inland energy consumption and greenhouse gas emissions in Poland. *Source: The author's own research. Data source: Eurostat Energy balances, 2019; Eurostat Greenhouse gas emissions, 2019.*

Poland's GDP is growing and has an increasing share in the GDP of the EU (Fig. 2.6). However, the Polish economy still has a lot to catch up with to be at least in the sixth position in the GDP per capita ranking, i.e., according to the position occupied in terms of the number of citizens and the area of the country. This increase, as already mentioned, must be balanced, i.e., economic development should take into account environmental protection. In industrialized countries, economic development is associated with energy. Energy intensity is the ratio that makes it possible to assess the country sustainable development in this respect. In the case of Poland, this ratio dropped very rapidly in the first years after the systemic transformation, but in recent years and forecasts (Fig. 2.8) one can see a certain limit that will be difficult to overcome. It is disturbing that this limit is more than twice as high as the EU's energy intensity.

Increased energy consumption in a growing economy, without investments in renewable energy sources, results in an increase in GHG emissions. In the case of Poland, this dependence is very well seen in Fig. 2.9, where only after accession to the EU and an increase in investments in renewable energy sources (Fig. 2.3) it can be seen that the lines depicting energy consumption and GHG emissions started to move away from each other.

Despite noticeable changes in the Polish economy and energy sector, they cannot be considered sufficient. Some of the objectives of the Europe 2020 strategy (e.g., share of renewable energy) are unlikely to be implemented, and others, such as GHG emissions, will be achieved, but mainly due to the criteria set too low and not the appropriate measures. The energy intensity ratio, already mentioned several times, is also a great example of the fact that sustainable development of Poland without endangered activities related to the energy industry and environmental protection is under threat.

Conclusions

In this study, the introduction presents the legal conditions that made the creation of the Europe 2020 strategy and the 3×20 climate and energy package whose key goals are the reduction of greenhouse gas emissions in the EU member states and an increase in the share of renewable energy and improvement in energy efficiency. Due to different levels of development of individual countries, these objectives were defined individually for each country, and for Poland are about 30% lower than the targets for the entire EU.

When analyzing the data on key objectives, the progress in this area was assessed and the forecasts for their implementation were presented by 2020. Of the three goals to be achieved, it is unlikely that the planned increase in renewable energy will reach 14%, but it will be possible to reduce energy consumption and greenhouse gas emissions. Fulfilling the last two objectives will be possible mainly because the EU at the time of defining them set Poland quite low requirements. However, an implementation of GHG emissions may be at risk if proper energy policy is not pursued. Unfortunately, this is likely because in Poland coal is three-quarters of the energy source, and the Polish government supports mining, and even during international conferences tries to convince everyone that basing the energy sector on coal does not contradict climate protection.

Climate targets were defined to reduce an excessive use of natural resources and environmental pollution. However, the goals set are not designed to slow economies but to transform them so that they continue developing, but are more energy efficient and environmentally friendly. Maintaining the balance between economic development and environmental protection is called sustainable development. In the work, an assessment of sustainable development of Poland was made by analyzing the GDP, gross inland energy consumption, and GHG emissions. The Polish economy after the system transformation of 1989 underwent very large changes that are visible in all of the three ratios. The analysis showed, however, that the energy intensity of the Polish economy increased most until 2007, and then decelerated and stabilized at a level about twice as high as the average in the EU. The forecast of energy intensity shows that in the coming years the value of this ratio will probably not change significantly.

Further sustainable development in Poland will not be possible without decisive actions of Polish governments in the field of moving away from coal-based energy and investments in renewable energy sources. The care for the energy security of the country is very important, but it cannot be at the expense of the natural environment. Building new, even modern coal-fired power plants only deepens the problem of greenhouse gas emissions and increases the long-term backlog in the modernization of the Polish energy sector, charging at the same time future generations with confronting these challenges.

Ph.D., Eng. Jacek Brożyna

In 2001 he earned a master of science degree in engineering (Rzeszow University of Technology, The Faculty of Electrical and Computer Engineering), in 2003 a bachelor's degree in marketing and management (Rzeszow University of Technology, The Faculty of Management and Marketing), and Ph.D. in 2011 (Rzeszow University of Technology, The Faculty of Mechanical Engineering and Aeronautics). He has been working at Rzeszow University of Technology, Faculty of Management at the Department of Quantitative Methods since 2002. He is the author of several dozen publications in the field of using statistical and computer methods in energy, engineering, economics, and financial issues, as well as a specialist in data analysis and forecasting.

ORCID ID: orcid.org/0000-0001-5632-7262

References

Eurostat, Energy Balances, 2019. Retrieved January 4, 2019, from. http://appsso.eurostat.ec.europa.eu/nui/submitViewTableAction.do.

2009 Review of the EU Sustainable Development Strategy, 2009. Retrieved January 2, 2019, from. https://eur-lex.europa.eu/legal-content/EN/TXT/?uri=CELEX:52009DC0400.

2020 Climate & Energy Package, 2009. Retrieved January 2, 2019, from. https://ec.europa.eu/clima/policies/strategies/2020_en.

2030 Climate & Energy Framework, 2014. Retrieved January 2, 2019, from. https://ec.europa.eu/clima/policies/strategies/2030_en.

2050 Energy Strategy, 2018. Retrieved January 2, 2019, from. https://energy/en/topics/energy-strategy-and-energy-union/2050-energy-strategy.

Akins, J.E., 1973. The oil crisis: this time the wolf is here. Foreign Affairs 51 (3), 462−490. https://doi.org/10.2307/20037995.

Albayrak, A.S., 2010. ARIMA Forecasting of Primary Energy Production and Consumption in Turkey: 1923−2006, p. 27.

Azevedo, I.M.L., Morgan, M.G., Lave, L., 2011. Residential and regional electricity consumption in the U.S. And EU: how much will higher prices reduce CO_2 emissions? The Electricity Journal 24 (1), 21−29. https://doi.org/10.1016/j.tej.2010.12.004.

Balitskiy, S., Bilan, Y., Strielkowski, W., Štreimikienė, D., 2016. Energy efficiency and natural gas consumption in the context of economic development in the European Union. Renewable and Sustainable Energy Reviews 55, 156−168. https://doi.org/10.1016/j.rser.2015.10.053.

Bowden, N., Payne, J.E., 2009. The causal relationship between U.S. energy consumption and real output: a disaggregated analysis. Journal of Policy Modeling 31 (2), 180−188.

Box, G.E.P., Jenkins, G.M., Reinsel, G.C., Ljung, G.M., 2015. Time Series Analysis: Forecasting and Control, 5 edn. Wiley, Hoboken, New Jersey.

Brożyna, J., Mentel, G., Szetela, B., 2016a. A mid-term forecast of maximum demand for electricity in Poland. Montenegrin Journal of Economics 12, 73−88. https://doi.org/10.14254/1800-5845.2016/12-1/5.

Brożyna, J., Mentel, G., Szetela, B., 2016b. Influence of double seasonality on economic forecasts on the example of energy demand. Journal of International Studies 9 (3), 9−20. https://doi.org/10.14254/2071-8330.2016/9-3/1.

Brożyna, J., Mentel, G., Szetela, B., 2017. Renewable energy and economic development in the European Union. Acta Polytechnica Hungarica 14, 11−34. https://doi.org/10.12700/APH.14.7.2017.7.2.

Błażejowski, M., Gazda, J., Kwiatkowski, J., 2016. Bayesian model averaging in the studies on economic growth in the EU regions − application of the gretl BMA package. Economics & Sociology 9 (4), 168−175. https://doi.org/10.14254/2071-789X.2016/9-4/10.

Campbell, C.J., 2005. Oil Crisis. multi-science publishing.

Campos, N.F., Coricelli, F., Moretti, L., 2014. Economic Growth and Political Integration: Estimating the Benefits from Membership in the European Union Using the Synthetic Counterfactuals Method (SSRN Scholarly Paper No. ID 2432446). Social Science Research Network, Rochester, NY. Retrieved from. https://papers.ssrn.com/abstract=2432446.

Caraiani, C., Lungu, C.I., Dascălu, C., 2015. Energy consumption and GDP causality: a three-step analysis for emerging European countries. Renewable and Sustainable Energy Reviews 44, 198−210. https://doi.org/10.1016/j.rser.2014.12.017.

Chappin, E.J.L., Dijkema, G.P.J., 2009. On the impact of CO2 emission-trading on power generation emissions. Technological Forecasting and Social Change 76 (3), 358−370. https://doi.org/10.1016/j.techfore.2008.08.004.

Choi, Y., 2018. The Asian values of Guānxì as an economic model for transition toward green growth. Sustainability 10 (7), 2150. https://doi.org/10.3390/su10072150.

Choi, G., Heo, E., Lee, C.-Y., 2018. Dynamic economic analysis of subsidies for new and renewable energy in South Korea. Sustainability 10 (6), 1832. https://doi.org/10.3390/su10061832.

Cîrstea, S.D., Moldovan-Teselios, C., Cîrstea, A., Turcu, A.C., Darab, C.P., 2018. Evaluating renewable energy sustainability by composite index. Sustainability 10 (3), 811. https://doi.org/10.3390/su10030811.

da Graça Carvalho, M., Bonifacio, M., Dechamps, P., 2011. Building a low carbon society. Energy 36 (4), 1842−1847. https://doi.org/10.1016/j.energy.2010.09.030.

Delarue, E., Lamberts, H., D'haeseleer, W., 2007. Simulating greenhouse gas (GHG) allowance cost and GHG emission reduction in Western Europe. Energy 32 (8), 1299−1309. https://doi.org/10.1016/j.energy.2006.09.020.

Directive 2009/29/EC of the European Parliament and of the Council of 23 April 2009 Amending Directive 2003/87/EC So as to Improve and Extend the Greenhouse Gas Emission Allowance Trading Scheme of the Community

(Text with EEA Relevance), Pub. L. No. 32009L0029, OJ L 140, 2009. Retrieved from. http://data.europa.eu/eli/dir/2009/29/oj/eng.

Dogan, E., Seker, F., 2016. Determinants of CO_2 emissions in the European Union: the role of renewable and non-renewable energy. Renewable Energy 94, 429−439. https://doi.org/10.1016/j.renene.2016.03.078.

Ediger, V.S., Akar, S., 2007. ARIMA forecasting of primary energy demand by fuel in Turkey. Energy Policy 35 (3), 1701−1708. https://doi.org/10.1016/j.enpol.2006.05.009.

Ellabban, O., Abu-Rub, H., Blaabjerg, F., 2014. Renewable energy resources: current status, future prospects and their enabling technology. Renewable and Sustainable Energy Reviews 39, 748−764. https://doi.org/10.1016/j.rser.2014.07.113.

Europe 2020 Strategy, 2009. Retrieved January 2, 2019, from. https://ec.europa.eu/info/business-economy-euro/economic-and-fiscal-policy-coordination/eu-economic-governance-monitoring-prevention-correction/european-semester/framework/europe-2020-strategy_en.

Eurostat, Greenhouse Gas Emissions, 2019. Retrieved January 4, 2019, from. http://appsso.eurostat.ec.europa.eu/nui/show.do?lang=en&dataset=env_air_gge.

Eurostat, Gross Domestic Product, 2019. Retrieved January 3, 2019, from. http://appsso.eurostat.ec.europa.eu/nui/show.do?dataset=nama_10_gdp&lang=en.

Eurostat, Share of Energy from Renewable Sources, 2019. Retrieved January 4, 2019, from. http://appsso.eurostat.ec.europa.eu/nui/show.do?dataset=nrg_ind_335a&lang=en.

First Orientation for a Common Energy Policy, December 18, 1968. Communication from the EC to the Council.

Fang, Y., 2011. Economic welfare impacts from renewable energy consumption: the China experience. Renewable and Sustainable Energy Reviews 15 (9), 5120−5128.

Frankel, J., Saravelos, G., 2012. Can leading indicators assess country vulnerability? Evidence from the 2008−09 global financial crisis. Journal of International Economics 87 (2), 216−231. https://doi.org/10.1016/j.jinteco.2011.12.009.

Fresner, J., Morea, F., Krenn, C., Aranda Uson, J., Tomasi, F., 2017. Energy efficiency in small and medium enterprises: lessons learned from 280 energy audits across Europe. Journal of Cleaner Production 142, 1650−1660. https://doi.org/10.1016/j.jclepro.2016.11.126.

Friedrichs, J., 2010. Global energy crunch: how different parts of the world would react to a peak oil scenario. Energy Policy 38 (8), 4562−4569. https://doi.org/10.1016/j.enpol.2010.04.011.

Fraczek, P., Kaliski, M., Siemek, J., 2013. The modernization of the energy sector in Poland vs. Poland's energy security. Archives of Mining Sciences 58 (2), 301−316. https://doi.org/10.2478/amsc-2013-0021.

Gawlik, L., 2018. The Polish power industry in energy transformation process. Mineral Economics 31 (1−2), 229−237. https://doi.org/10.1007/s13563-017-0128-5.

Hirsch, R.L., 2008. Mitigation of maximum world oil production: shortage scenarios. Energy Policy 36 (2), 881−889. https://doi.org/10.1016/j.enpol.2007.11.009.

Hondroyiannis, G., Lolos, S., Papapetrou, E., 2002. Energy consumption and economic growth: assessing the evidence from Greece. Energy Economics 24 (4), 319−336. https://doi.org/10.1016/S0140-9883(02)00006-3.

Immenga, U., 1992. The developement of European energy policy: from ECSC treaty to the internal market. In: Mestmäcker, E.J. (Ed.), Natural Gas in the Internal Market. London; Boston: Baden-Baden: Kluwer Law International.

Inglesi-Lotz, R., 2016. The impact of renewable energy consumption to economic growth: a panel data application. Energy Economics 53, 58−63. https://doi.org/10.1016/j.eneco.2015.01.003.

Issawi, C., 1978. The 1973 oil crisis and after. Journal of Post Keynesian Economics 1 (2), 3−26. https://doi.org/10.1080/01603477.1978.11489099.

Jasiński, P., 1996. Polityka Energetyczna Wspólnot Europejskich - Tło Historyczne (P. Jasiński, T. Skoczny). Warszawa: CE UW.

Karmellos, M., Kopidou, D., Diakoulaki, D., 2016. A decomposition analysis of the driving factors of CO_2 (Carbon dioxide) emissions from the power sector in the European Union countries. Energy 94, 680−692. https://doi.org/10.1016/j.energy.2015.10.145.

Kasperowicz, R., Pinczyński, M., Khabdullin, A., 2017. Modeling the power of renewable energy sources in the context of classical electricity system transformation. Journal of International Studies 10, 264−272. https://doi.org/10.14254/2071-8330.2017/10-3/19.

Katowice Climate Change Conference, 2018. Retrieved January 2, 2019, from. https://unfccc.int/katowice.

Katre, A., Tozzi, A., 2018. Assessing the sustainability of decentralized renewable energy systems: a comprehensive framework with analytical methods. Sustainability 10 (4), 1058. https://doi.org/10.3390/su10041058.

Kayhan, S., Adiguzel, U., Bayat, T., Lebe, F., 2010. Causality relationship between real GDP and electricity consumption in Romania (2001−2010). Journal for Economic Forecasting (4), 169−183.

Kerr, R.A., 1998. The next oil crisis looms large–and perhaps close. Science 281 (5380), 1128−1131. https://doi.org/10.1126/science.281.5380.1128.

Ko, F.-K., Huang, C.-B., Tseng, P.-Y., Lin, C.-H., Zheng, B.-Y., Chiu, H.-M., 2010. Long-term CO_2 emissions reduction target and scenarios of power sector in Taiwan. Energy Policy 38 (1), 288−300. https://doi.org/10.1016/j.enpol.2009.09.018.

Krajačić, G., Duić, N., Zmijarević, Z., Mathiesen, B.V., Vučinić, A.A., da Graça Carvalho, M., 2011. Planning for a 100% independent energy system based on smart energy storage for integration of renewables and CO_2 emissions reduction. Applied Thermal Engineering 31 (13), 2073−2083. https://doi.org/10.1016/j.applthermaleng.2011.03.014.

Lee, C.-C., 2006. The causality relationship between energy consumption and GDP in G-11 countries revisited. Energy Policy 34 (9), 1086−1093. https://doi.org/10.1016/j.enpol.2005.04.023.

Lee, C.-M., Ko, C.-N., 2011. Short-term load forecasting using lifting scheme and ARIMA models. Expert Systems with Applications 38 (5), 5902−5911. https://doi.org/10.1016/j.eswa.2010.11.033.

Li, X., 2005. Diversification and localization of energy systems for sustainable development and energy security. Energy Policy 33 (17), 2237−2243. https://doi.org/10.1016/j.enpol.2004.05.002.

Lotfalipour, M.R., Falahi, M.A., Bastam, M., 2013. Prediction of CO_2 emissions in Iran using grey and ARIMA models. International Journal of Energy Economics and Policy 3 (3), 229−237.

Markandya, A., Pedroso-Galinato, S., Streimikiene, D., 2006. Energy intensity in transition economies: is there convergence towards the EU average? Energy Economics 28 (1), 121−145. https://doi.org/10.1016/j.eneco.2005.10.005.

Marques, A.C., Fuinhas, J.A., Pires Manso, J.R., 2010. Motivations driving renewable energy in European countries: a panel data approach. Energy Policy 38 (11), 6877−6885. https://doi.org/10.1016/j.enpol.2010.07.003.

Menegaki, A.N., 2011. Growth and renewable energy in Europe: a random effect model with evidence for neutrality hypothesis. Energy Economics 33 (2), 257−263. https://doi.org/10.1016/j.eneco.2010.10.004.

Mentel, G., Brożyna, J., Szetela, B., 2017. Evaluation of the effectiveness of investment fund deposits in Poland in a time of crisis. Journal of International Studies 10 (2), 46−60. https://doi.org/10.14254/2071-8330.2017/10-2/3.

National Energy Efficiency Action Plans and Annual Reports − Energy − European Commission, 2018. Retrieved January 2, 2019, from. https://energy/en/topics/energy-efficiency/energy-efficiency-directive/national-energy-efficiency-action-plans.

New Renewables, Energy Efficiency and Governance Legislation Comes into Force on 24 December 2018, 2018. Retrieved January 2, 2019, from. https://ec.europa.eu/info/news/new-renewables-energy-efficiency-and-governance-legislation-comes-force-24-december-2018-2018-dec-21_en.

Nicolini, M., Tavoni, M., 2017. Are renewable energy subsidies effective? Evidence from Europe. Renewable and Sustainable Energy Reviews 74, 412−423. https://doi.org/10.1016/j.rser.2016.12.032.

Pappas, S.S., Ekonomou, L., Karamousantas, D.C., Chatzarakis, G.E., Katsikas, S.K., Liatsis, P., 2008. Electricity demand loads modeling using AutoRegressive Moving Average (ARMA) models. Energy 33 (9), 1353−1360. https://doi.org/10.1016/j.energy.2008.05.008.

Paris Agreement, November 23, 2016. Retrieved January 2, 2019, from. https://ec.europa.eu/clima/policies/international/negotiations/paris_en.

Pastor, J.M., Peraita, C., Serrano, L., Soler, Á., 2018. Higher education institutions, economic growth and GDP per capita in European Union countries. European Planning Studies 1−22, 0(0). https://doi.org/10.1080/09654313.2018.1480707.

Peñalvo-López, E., Cárcel-Carrasco, F.J., Devece, C., Morcillo, A.I., 2017. A methodology for analysing sustainability in energy scenarios. Sustainability 9 (9), 1590. https://doi.org/10.3390/su9091590.

Progress Made in Cutting Emissions, 2018. Retrieved January 2, 2019, from. https://ec.europa.eu/clima/policies/strategies/progress_en.

Pukšec, T., Mathiesen, B.V., Novosel, T., Duić, N., 2014. Assessing the impact of energy saving measures on the future energy demand and related GHG (greenhouse gas) emission reduction of Croatia. Energy 76, 198−209. https://doi.org/10.1016/j.energy.2014.06.045.

Qoaider, L., Steinbrecht, D., 2010. Photovoltaic systems: a cost competitive option to supply energy to off-grid agricultural communities in arid regions. Applied Energy 87 (2), 427–435. https://doi.org/10.1016/j.apenergy.2009.06.012.

Rodríguez-Caballero, C.V., Ventosa-Santaulària, D., 2017. Energy-growth long-term relationship under structural breaks. Evidence from Canada, 17 Latin American economies and the USA. Energy Economics 61, 121–134. https://doi.org/10.1016/j.eneco.2016.10.026.

Sadorsky, P., 2009. Renewable energy consumption and income in emerging economies. Energy Policy 37 (10), 4021–4028.

Saiah, S.B.D., Stambouli, A.B., 2017. Prospective analysis for a long-term optimal energy mix planning in Algeria: towards high electricity generation security in 2062. Renewable and Sustainable Energy Reviews 73, 26–43. https://doi.org/10.1016/j.rser.2017.01.023.

Sampedro, J., Arto, I., González-Eguino, M., 2017. Implications of switching fossil fuel subsidies to solar: a case study for the European Union. Sustainability 10 (1), 50. https://doi.org/10.3390/su10010050.

Sari, R., Ewing, B.T., Soytas, U., 2008. The relationship between disaggregate energy consumption and industrial production in the United States: an ARDL approach. Energy Economics 30 (5), 2302–2313.

Scarlat, N., Dallemand, J.-F., Monforti-Ferrario, F., Banja, M., Motola, V., 2015. Renewable energy policy framework and bioenergy contribution in the European Union — an overview from national renewable energy action plans and progress reports. Renewable and Sustainable Energy Reviews 51, 969–985. https://doi.org/10.1016/j.rser.2015.06.062.

Shahbaz, M., Tang, C.F., Shahbaz Shabbir, M., 2011. Electricity consumption and economic growth nexus in Portugal using cointegration and causality approaches. Energy Policy 39 (6), 3529–3536.

Siew, L.Y., Chin, L.Y., Wee, P.M.J., 2008. ARIMA and integrated ARFIMA models for forecasting air pollution index in Shah Alam. Selangor 12 (1), 7.

Simionescu, M., 2016. The Relation between Economic Growth and Foreign Direct Investment during the Economic Crisis in the European Union (SSRN Scholarly Paper No. ID 2803140). Social Science Research Network, Rochester, NY. Retrieved from. https://papers.ssrn.com/abstract=2803140.

Simionescu, M., Dobeš, K., Brezina, I., Gaal, A., 2016. GDP rate in the European Union: simulations based on panel data models. Journal of International Studies 9 (3), 191–202. https://doi.org/10.14254/2071-8330.2016/9-3/15.

Slusarczyk, B., Brzezinski, S., Kot, S., 2013. Electricity market liberalization in Poland and Romania. Metalurgia International 18 (11), 31.

Smil, V., 2010. Energy Transitions: History, Requirements, Prospects. Praeger, Santa Barbara, Calif.

Stala-Szlugaj, K., 2016. Trends in the consumption of hard coal in Polish households compared to EU households. Gospodarka Surowcami Mineralnymi 32 (3), 5–22. https://doi.org/10.1515/gospo-2016-0024.

Statement by the President at the Official Opening of the Climate Summit and the COP24 Leaders' Summit, 2018. Retrieved January 2, 2019, from. http://www.president.pl/en/news/art,915,statement-by-the-president-at-the-official-opening-of-the-climate-summit-and-the-cop24-leaders-summit.html.

Strategy for Sustainable Development, 2001. Retrieved January 2, 2019, from. https://eur-lex.europa.eu/legal-content/PL/TXT/?uri=LEGISSUM:l28117.

Strategy on Climate Change: Foundations of the Strategy, 2005. Retrieved January 2, 2019, from. https://eur-lex.europa.eu/legal-content/EN/TXT/?uri=LEGISSUM:l28157.

A Sustainable Europe for a Better World: A European Strategy for Sustainable Development, 2001. Retrieved January 2, 2019, from. https://eur-lex.europa.eu/legal-content/EN/TXT/?uri=CELEX:52001DC0264.

The Rio Declaration on Environment and Development, 1992, p. 19.

Tomás, R.A.F., Ramôa Ribeiro, F., Santos, V.M.S., Gomes, J.F.P., Bordado, J.C.M., 2010. Assessment of the impact of the European CO_2 emissions trading scheme on the Portuguese chemical industry. Energy Policy 38 (1), 626–632. https://doi.org/10.1016/j.enpol.2009.06.066.

Tvaronavičienė, M., Grybaitė, V., 2012. Sustainable development and performance of institutions: approaches towards measurement. Journal of Security and Sustainability Issues 1 (3), 167–175. https://doi.org/10.9770/jssi/2012.1.3(2).

Tvaronavičienė, M., Mačiulis, A., Lankauskienė, T., Raudeliūnienė, J., Dzemyda, I., 2015. Energy security and sustainable competitiveness of industry development. Economic Research-Ekonomska Istraživanja 28 (1), 502–515. https://doi.org/10.1080/1331677X.2015.1082435.

Tvaronavičienė, M., Prakapienė, D., Garškaitė-Milvydienė, K., Prakapas, R., Nawrot, Ł., 2018. Energy efficiency in the long-run in the selected European countries. Economics & Sociology 11 (nr 1), 245—254. https://doi.org/10.14254/2071-789X.2018/11-1/16.

Vasiliūnaitė, R., 2014. Sustainable development: methodological approaches toward issues. Journal of Security and Sustainability Issues 3 (3), 69—75. https://doi.org/10.9770/jssi.2014.3.3(6).

Venn, F., 2016. The Oil Crisis. Routledge. https://doi.org/10.4324/9781315840819.

Voigt, S., De Cian, E., Schymura, M., Verdolini, E., 2014. Energy intensity developments in 40 major economies: structural change or technology improvement? Energy Economics 41, 47—62. https://doi.org/10.1016/j.eneco.2013.10.015.

Vosylius, E., Rakutis, V., Tvaronavičienė, M., 2013. Economic growth, sustainable development and energy security interrelations. Journal of Security and Sustainability Issues 2 (3), 5—14. https://doi.org/10.9770/jssi.2013.2.3(1).

Wang, Z.-X., Ye, D.-J., 2017. Forecasting Chinese carbon emissions from fossil energy consumption using non-linear grey multivariable models. Journal of Cleaner Production 142, 600—612. https://doi.org/10.1016/j.jclepro.2016.08.067.

Wurlod, J.-D., Noailly, J., 2018. The impact of green innovation on energy intensity: an empirical analysis for 14 industrial sectors in OECD countries. Energy Economics 71, 47—61. https://doi.org/10.1016/j.eneco.2017.12.012.

Xu, X.Y., Ang, B.W., 2013. Index decomposition analysis applied to CO_2 emission studies. Ecological Economics 93, 313—329. https://doi.org/10.1016/j.ecolecon.2013.06.007.

Yeboah, S.A., Ohene, M., Wereko, T.B., 2012. Forecasting aggregate and disaggregate energy consumption using arima models: a literature survey. Journal of Statistical and Econometric Methods 1 (2), 71—79.

Zhang, X.-P., Cheng, X.-M., 2009. Energy consumption, carbon emissions, and economic growth in China. Ecological Economics 68 (10), 2706—2712.

Zhang, H., Zheng, Y., Zhou, D., Zhu, P., 2015. Which subsidy mode improves the financial performance of renewable energy firms? A panel data analysis of wind and solar energy companies between 2009 and 2014. Sustainability 7 (12), 16548—16560. https://doi.org/10.3390/su71215831.

Zhang, Y., Zhao, X., Zuo, Y., Ren, L., Wang, L., 2017. The development of the renewable energy power industry under feed-in tariff and renewable portfolio standard: a case study of China's photovoltaic power industry. Sustainability 9 (4), 532. https://doi.org/10.3390/su9040532.

Zhang, L., Xue, B., Liu, X., 2018. Carbon emission reduction with regard to retailer's fairness concern and subsidies. Sustainability 10 (4), 1209. https://doi.org/10.3390/su10041209.

Zhao, H., Guo, S., 2015. External benefit evaluation of renewable energy power in China for sustainability. Sustainability 7 (5), 4783—4805. https://doi.org/10.3390/su7054783.

3

National growth and regional (under)development in Brazil: the case of Pará in the Brazilian Amazon

Martina Iorio, Salvatore Monni

Università degli Studi Roma Tre, Rome, Italy

Introduction

In postcolonial Brazilian history, the Amazon has always had a central economic role. The intensive exploitation of its water resources dates back to the sectorial development policies of national interest started in the 1970s with the *grandes projetos* and it is still in progress today (Farias, 2016; Sousa & Santos, 2006; Maglhaes Filho, 1987). However, before becoming an energy frontier, Pará had been a nerve center for rubber production, a hub for the trade of the Brazilian chestnut (*castanha do Pará*), and the first Latin American market for tropical fruits (Prado Júnior, 2004). Although national energy policies have strongly affected the area for almost 50 years, the results in terms of regional and local development do not seem satisfactory (FAPESPA, 2016; UNDP FJP & IPEA, 2013). The aim of this chapter is therefore to

Energy Transformation towards Sustainability
https://doi.org/10.1016/B978-0-12-817688-7.00003-3

examine the causes of the backwardness of such a rich area in resources and to highlight how this area, which represents a crucial element for the development of the country, paradoxically fails to develop itself. With the objective to give a closer view of the regional and local Amazonian reality, in the second paragraph we propose a detailed description of the federal state of Pará, one of the main states of the Brazilian Amazon for both energy production and economic growth, which is taken as the main reference for the observation of the issues described. In the third paragraph, we will try to examine as thoroughly as possible the issue of the use of water for energy production and the role of the Amazon as a supplier of energy for the country. In the fourth paragraph the possible effects of this specific use of water resources will be commented on. Eventually, this overview of the crucial issues of today's debate on the economic and social role of the Amazon will, therefore, support paragraph five in which the main effects on regional and local development of the economic engagement of Pará will be identified. Some suggestions and challenges modeled on a careful analysis of the *paraense* (meaning of the state of Pará) context are reported here since they could be useful in the analysis of similar contexts within the Amazon area (Pinto, 2011, 2017; Rocha, 2008).

Pará as the Amazonian power station

The Brazilian Amazon (*Amazônia Legal*) is a region that corresponds to 64% of the Brazilian territory. Thanks to its immense natural resources, and its mining potential, it is considered a strategic area for the economic development of the country (Moretto et al., 2012; Coelho et al., 2011). *Amazônia Legal* includes nine Brazilian states, including, Acre, Amapá, Amazonas, Pará, Rondônia, Roraima, Tocantins in the Northern Region, Mato Grosso in the Centerwest Region, and part of the state of Maranhão in the Northeast Region (IBGE, 2017). Among the Brazilian states, Pará is the richest in natural resources and it is the most involved in mining and energy activities, which are strategic for national economic development (ANEEL, 2019; FAPESPA, 2018). With a territorial extension of 1,247,954,320 km^2 it is the second largest federal unit within the Amazon; it had an estimated population of 8,175,113 million inhabitants in 2015 and a density of 6.07 hab./km^2 (IBGE, 2017). Beyond the high concentration of investment in mining and hydroelectric energy, it retains the greatest potential for future exploitation (SIPOT, 2016), that's why it seems to be perfectly suitable to be representative of the evolution of the economic growth and development processes of the *Amazônia Legal* as a whole (Figs. 3.1–3.4)

Since the late 1970s, after discovering important mineral veins (such as *Sierra Carajás* in 1967) in the federal state of Pará, Brazilian Amazon has been considered the new frontier of capital expansion in Latin America, since it was an attractive target for international capital (da Silva, 2016). According to the national strategy, the attraction of large public and private capital, including international ones, would have accelerated the economic organization of the Brazilian Amazon in a (only theoretically) holistic perspective (Farias, 2016). In fact, the economic development of the region would have taken place thanks to the industrialization of the territory driven by the flourishing mining activity. Thus, it should have generated some specific side effects such as the creation of infrastructure (e.g., roads) and generation of electricity (hydropower) (Rocha, 2016). These are the so-called side investments, encouraged by tax incentives conceived for all economic activities in the territories affected by extractive activity and

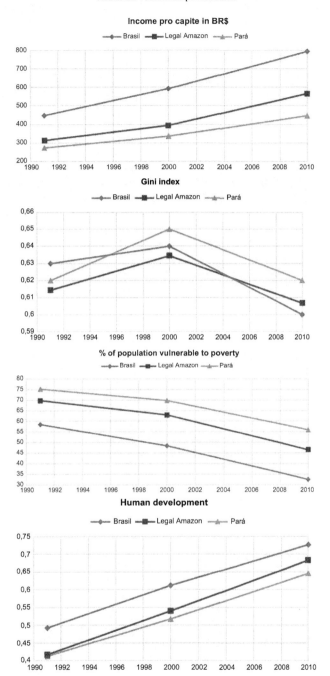

FIGURES 3.1–3.4 Figures on socioeconomic performance: from Brazil to Pará. *Source: Personal elaboration from UNDP, FJP & IPEA, 2013*

not only for companies in the metallurgical sector (da Silva, 2016). As in a virtuous cycle, the increase in the population due to the immigration of the labor force makes it necessary to create infrastructure, by putting pressure first on the construction of motorways and railway lines for the transport of goods. Then, it becomes necessary to install plants to produce energy with the dual aim of supplying electricity to metallurgical companies and to new urban centers. It is clear, then, that tax incentives, e.g., to infrastructure and energy production, was not an attempt to achieve specific development objectives but rather a part of a wider economic development project of the region or even of the nation (Rocha, 2016).

Considering the above-described phenomena, it is possible to say that since the 1970s the construction of the hydroelectric power plant of Tucuruí in Pará has totally changed the territory of the state as well as its socioeconomic structure (Rocha, 2008). Therefore, although it suffered all socioeconomic and environmental negative effects of resource exploitation, it is undeniable that the big projects era has given impetus to the growth of Pará so much so that its main city, Belém, is nowadays considered the capital of the entire Brazilian Amazon (Rocha, 2008). However, researchers point out that the attempt of the national policy to create an Amazonian development pole in the state of Pará has failed precisely because of *Great Carajás* kind of program, which was aimed at the exploitation of the *Sierra Carajás*, within Pará. Actually, large investors, such as Vale S.A. in the mineral/hydroelectric sector, were the great driver of economic development in the Southeast region of Pará (da Silva, 2016). However, the rapid economic development, together with the sudden increase in migration flows (e.g., labor force) and the rapid stimulation of urbanization, caused the rise of new economic elites represented by foreign investors, an obstacle to real integrated development of the territory. The installation of the pole primarily attracted a large number of workers from other states or other areas of the state, and this drove either the rapid expansion of existing urban centers or the creation ex novo of built-up areas (Rocha, 2008). The needs of the growing urban structures put pressure on the supply of infrastructure, the construction of which attracted in turn a labor force, thus triggering a spiral of growth (also accompanied by the exodus from the countryside). Therefore, although it was a pivotal tool for the economic growth, the fast urbanization had some social and environmental side effects, adverse to human development, such as, inadequate transport and basic sanitation infrastructures (at least in the first phases of the project), health problems, disease, and prostitution (Figueiredo and Saraiva, 2018; Hazeu, 2015), as well as the increase of consumption, waste, and contamination of natural resources (de Souza et al., 2016). Moreover, the population of Pará also endured a disconnection with the economic elites that were external to the Amazonian context (i.e., foreign capitals) and they were not interested in the creation of added value or in generation of employment (da Silva, 2016). Actually, they just used the *sierra* as their own "garden" (Pinto, 2017). Thus, in Pará growth based on the export of precious metals has never been really associated with a robust development strategy alternative to pure mining, especially due to the weak political intervention of mitigation played by the institutions (Rocha and Neves, 2018), that is, beyond the nation, the state, and municipalities. As a result, the centralized approach of the nation, aimed at the homogenization of regional development policies, generated the widening of inequalities between different areas of the same region of intervention, in the state of Pará as well as in the Amazon as a whole (Farias, 2016). The critiques to the process of definition of region of intervention—which combines areas with totally different socioeconomic potential just following the criterion of the best geographical strategy in order to facilitate the flux of goods, the accumulation of capital, and the domination of lands and peoples (interests of economic elites based on private and international capital)—should be totally endorsed. As a matter of fact, this idea of regionalizing only exacerbates geographically the dynamics of specialization of labor, increasing inequalities through the homogenization of public policies of regional development (Farias, 2016). Eventually, it is noteworthy to investigate these causes of the *impasse* of the development process of Pará and of Amazonia in general, starting from the assumption that the weak development of the state represents in itself an obstacle to its future in a sort of path-dependence pattern (Farias, 2016; Piketty, 2000).

The role of water resources in Brazil

The Amazon Basin, measuring 6,110,000 km^2, was the only navigable area within the country at the time of the occupation; it rises from the Andes and flows into the Atlantic Ocean, and this gave it a central logistical role for both trade and exploration (Prado Júnior, 2004). The port city of Belém, the current capital of the state of Pará, located at the point in where the Amazon river flows into the Atlantic Ocean, has always been a strategic hub collecting tropical products destined to exportation, such as the natural rubber (Prado Júnior, 2004). In particular, the cycle of rubber was outstanding for its contribution to the development of the Brazilian Amazon, despite not being very long-lasting (it just reproduced the blind attitude that has marked the short-term parable of the golden age) (Prado Júnior, 2004). It mainly involved the state of Pará, and it was responsible for the rise of the village of Belém that turned itself into a flourishing capital. Almost the same occurred with Manaus, the capital city of Amazonas (Galvani, 1948). In the first half of the 20th century, once the rubber era had passed and Brazil started pursuing domestic industrialization based on the use of national manganese and coal, the country had the opportunity to strategically locate plants halfway between the mines and the harbors (Galvani, 1948). In fact, since the colonization era, the settlements and the installation of the prevailing economic activities throughout the country have taken place starting from the port cities, given that the lines of expansion mainly depended on climate conditions and commercial hubs (Prado Júnior, 2004). While the population process of the Northeast region followed the evolution of livestock farming, and the occupation of the Centresouth was linked to the presence of mineral reserves (especially *Minas Gerais* and *Ouro Preto*), the occupation of the North, starting from Belém, took the form of a slow infiltration into the Amazonian valley, which offered natural products, fish, and wood for export (Prado Júnior, 2004).

Moreover, without forgetting its mining potential (e.g., diamonds and other metals), the Brazilian Amazon has an undeniable endowment of water resources. It is not a case that about two-thirds of Brazil's hydroelectric potential is located in the northern region, which hosts four sub-basins—Xingu, Tapajos, Madeira, and Negro—two of which flow within Pará. Technological innovation did not delay in making its contribution, allowing the exploitation of the enormous hydrokinetic potential of the area through the installation of hydroelectric plants (Monni et al., 2018). In general, the exploitation of hydroelectricity in Brazil started in the first half of the 20th century; in 1941 the country with the largest water basin in the world had a total of about 907 hydroelectric power plants of different sizes, including the *Serra de Cubatão* power plant, which at the time was the seventh largest power plant in the world (Galvani, 1948). However, in the 1970s, a more intensive phase of exploitation began, starting from the Amazon, which still remains an area of great potential at the national level, thus having the role to energetically support mining activities. The dual vocation (mining and energy) of the Amazon was fully supported in the period of the great economic miracle, coinciding with the military dictatorship (1964–84). In fact, it was the dictatorship that implemented the *Grande Carajás* project in the state of Pará, with the dual aim of boosting the national economy through the exploitation of the great potential of resources and also of showing the magnificence and power of the dictatorship itself (Pinto, 2011; Amparo & Porto, 1987). The main role of the Amazon was to provide the natural resources and also the geographical space (with an originally very low level of urbanization) for the installation of economic poles mainly related to the mining sector. The creation of these poles took place when, infrastructure (highways and urban facilities) and energy production began to develop around the areas of extraction and processing of iron ore, as part of the so-called side investments (Rocha, 2016). In fact, when the Tucuruí power plant was built in 1984, its role was mainly to supply electricity to the heavy metal extraction and refining industries, and this was feasible due to the creation of large conglomerates of state-owned enterprises such as the Elettrobras (Centrais Elétricas Brasileiras S.A.). It was born officially on June 11, 1962, as the largest company in the electric sector both in Brazil and Latin America, whose major shareholder is the Brazilian federal government (Sousa & Santos, 2006). On June 20, 1973, Eletronorte (Centrais Elétricas do Norte do Brasil S.A.) was created as a joint-stock company of Eletrobras, which had the aim to generate and supply electricity to the nine states of the Legal Amazon and to provide energy to buyers from other regions of the country (Sousa & Santos, 2006). According to the data published by the Agência Nacional de Energia Elétrica (ANEEL) it is interesting to note that currently the Legal Amazon contributes to more than 24% of the country's electricity production that consists of around 60% made by hydroelectricity (ANEEL, 2019). In this context, Pará had a total installed capacity of roughly 19,275,209 kw in 2018, that is, it was responsible for roughly half of the generation of the entire region and it was also the second largest electricity producer in Brazil, with 11.83% of the total electricity produced in the country. The interesting fact, however, is that 96.95% of the electricity produced in Pará is produced by only five large power plants, all powered by hydroelectric energy (ANEEL, 2019) (Fig. 3.5).

Territorial Units	Electricity		Hydroelectricity*		
	Installed potential (kw)	% on national potential	Installed potential (kw)	% on territorial unit potential	number of LHP**
Acre	115.727,80	0,07	0,00	0	0
Amapá	1.037.816,20	0,64	941.950,00	90,76	4
Amazonas	2.190.897,64	1,34	274.710,00	12,54	2
Maranhão	3.899.122,43	2,39	1.087.000,00	27,88	1
Mato Grosso	3.086.137,98	1,89	1.180.870,00	38,26	10
Pará	19.277.808,76	11,83	18.690.333,00	96,95	5
Rondônia	8.311.378,60	5,1	7.608.250,00	91,54	1
Roraima	285.568,38	0,18	0,00	0	0
Tocantins	1.937.803,00	1,19	1.644.450,00	84,86	3
North Region	33.157.000,38	20,34	29.159.693,00	87,94	18
Legal Amazon	40.142.260,79	24,63	31.427.563,00	78,29	29
Brasil	162.974.079,84	100	98.286.811,00	60,31	217

*one is taking into account just large hydro-plants.
**plants placed at the border are considered once.

FIGURE 3.5   Figures on electricity and hydroelectricity installed potential (per territorial unit) in 2019. *Source: Personal elaboration from ANEEL, 2019.*

Eventually, the political and economic national strategy deliberated that the northern aluminum pole would strategically depend on the installation of a hydroelectric power plant (i.e., Tucuruí). In this context, the state of Pará, which stood and still stands today between two river basins, the Amazonas and the Tocantins, was in a strategic position. In fact, despite the wicked invasion and unsustainable exploitation of other natural resources, the abundance of water and iron ore was not too affected by anthropogenic pressures. However, the same cannot be said of the quality of and the access to water (UNDP, FJP&IPEA, 2013).

The unavoidable impact of a dam

Hydropower, which peaked in popularity in the 1970s (the period in which the largest power stations and the largest number of projects in the world were built), collapsed in popularity in the 1990s, when press and literature surveys began to highlight the shortcomings of a technology that clearly had an impact in terms of both the environment (Barros et al., 2011; Manyari and de Carvalho Jr., 2007; Fearnside, 1995) and society (Figueiredo and Saraiva, 2018; Diamond and Poirier, 2010; Fearnside, 1999; Scudder, 1981), especially due to two elements: the dam and the reservoir (Iorio, 2015). Therefore, not only the integrated development process starting with the implementation of major projects but also some of the side investments generated were found, in the light of ex post impact analysis, to be detrimental to local development objectives.

In fact, although it was expected that hydroelectric plants would have less impact than thermoelectric ones, it is not possible to deny the negative evidence that emerged ex post from the analysis of large plants in Brazil such as that of Tucuruí (Pinto, 2011; Rocha, 2008; Fearnside, 1999), Belo Monte (Figueiredo and Saraiva, 2018; Jaichand and Sampaio, 2013; Fleury and Almeida, 2013; Fearnside, 2006), Samuel (Fearnside, 2005), or Madeira Complex (Werner, 2011). Usually the environmental and social costs, whether demonstrated and evaluated, mainly affect the population living in the surrounded area, while the energy produced is fed into the *Sistema Interligado Nacional* (SIN), which is an electricity transmission system. It consists of a set of facilities and equipment that enable the supply of electricity in the regions of the country that are electrically interconnected and its main users are the large urban centers (especially in the extreme south of the country) and the industrial poles (ANEEL, 2019). Thus, the distribution of costs and benefits related to energy production is definitely uneven and it is also emphasized by the high costs for the extension of the transmission grid to isolated small and medium-sized communities, which therefore remain isolated (Aresti & Micangeli, 2016; Iorio, 2015).

As highlighted in the third paragraph, the objective of hydroelectric power generation in Pará was originally to provide energy to the Amazonian extraction poles for both national economic utility and local development. Nowadays, despite the impacts it generates, the hydroelectric is the predominant source among renewables both at the state and at the country level. However, both the state and the country enjoy an extreme economic advantage in the generation of this type of energy due to the political and economic choice of disregarding the real environmental and social costs (that remain hidden) (Buarque, 1987). Moreover, the imbalance of costs and benefits (i.e., the extraction costs and the benefits of using the exploited resource), also resulted in an internal dynamic north-south within the country (Furtado, 2000). This is nowadays particularly evident when considering the reckless use of Amazon water for electrically supply of the state of São Paulo (ANEEL, 2017), while the middle Tocantins region in the state of Pará became a symbol of the process of destructuring and restructuring of the territory due to the cumbersome presence of Tucuruí Large Hydroelectric Plant (LHP) (Rocha, 2016). This is, as a matter of practice, what usually occurs in the Brazilian Amazon in the event of the installation of hydroelectric projects. Precisely because of the impact the Tucuruí LHP has generated since its installation, the area originally called *Médio Tocantins* started to be officially named Tucuruí Lake Region of Integration (named for the reservoir of the plant) (Rocha, 2016). The decomposition of the territory occurred mainly as a result of social and environmental impacts such as hydrological deviations and consequent damage to animal and plant biodiversity and changes in the economic system. These changes, both during the military and the democratic phases, have always been compulsory and full of contradictions. From a socioeconomic point of view, the alteration of the watercourse for the construction of the artificial lake by means of a dam, generating forced displacement of both indigenous people and riverside dwellings, created an alteration in the socioeconomic life of the communities concerned. In the case of the population living along the Tocantins river, they moved from chestnuts and diamond extraction to agriculture, farming, and energy production (Rocha, 2016). Changes of this kind altered the social economic equilibrium of affected communities, sometimes slowing down the development process due to the difficulties to forcefully adapt to new subsistence activities. Moreover, the investments attracted the labor force, which populated the new urban centers, creating demand for the construction of additional infrastructures, as if they were an extension of the production line of the main investments (Rocha, 2016). This second phase is precisely the reconstruction of a socioeconomic order (which follows the deconstruction of the previous order), which generates impacting urban changes. In the case of the Tucuruí hydroelectric power station, the flooding of the artificial lake in the first phase of construction in the 1980s affected not only the riverside communities but also part of the infrastructure necessary for trade of the Brazilian chestnuts coming from the extraction centers (there existed the "chestnut polygon" consisting of Marabá, Tucuruí, and Belém) such as the railroad or the Tocantins river line. On the one hand, to compensate for the flowed rail line, the motorway network (asphalted road) began to be developed; on the other hand, new urban centers were built, some of which were created to accommodate the displaced people, others were built from scratch. Among the new cities stand out both the temporary cities for the workforce and the company towns. The latter were cities provided with urban facilities already, made available by the main investor company. As a matter of fact, the big project was the only one responsible for the deconstruction and consequent reorganization of the territory and this type of urbanization process was pioneering in that area (Rocha, 2016). In the case of Belo Monte, another power station that started operating in 2016 in the state of Pará, the rapid installation of urban centers at the service of the big reference project revealed socioeconomic impacts, including negative impacts such as prostitution, with regard to the new urban centers (Figueiredo & Saraiva, 2018; Hazeu, 2015). This also raises important issues such as the lack of bottom-up participation by indigenous peoples and riverside dwellings whose protected areas are affected by territorial modification interventions due to the modification of the normal watercourse due to the dam (da Silva, 2016).

According to more recent literature, it is therefore expected that, once eliminated the elements considered to be the main cause of impact (i.e., the dam and the reservoir) or simply reduced in size (scale reduction), a positive evaluation of the small hydroelectric plants can be reached (Bagher et al., 2015). Off-grid solutions, although more expensive and less efficient than a normal connection to the *Sistema Interligado Nacional* (SIN), may become even more competitive from an economic perspective. This situation occurs whether the connection to the national grid of particularly remote areas is complicated by the absence of infrastructure or the real social/environmental costs of major plants are actually taken into account (Aresti and Micangeli, 2016). In this case, the proliferation of small hydroelectric plants, beyond being compatible with the objectives of energy security (diversification) and sustainability (no environmental/social impact), could possibly achieve a third development objective, namely widespread energy access by means of rural electrification (Sánchez et al., 2015; Schiffer & Swan, 2018).

Amazonian aftermaths of national projects

In countries, or simply in areas of late industrialization such as Brazil, it is common to encourage industrialization through the implementation of big projects. In fact, these projects are often an excuse used to penetrate regional and local context from an economic point of view (Becker, 2005).

In this cases, investments in the richest areas of the country in terms of resources are mostly focused on the mining industry, often followed by investment in infrastructure, crucial for the development process and therefore extremely welcome. However, the investment activity is mainly in the hands of the extractive industry or of the agribusiness as well (da Silva, 2016).

In the Brazilian case, the Legal Amazon is a space for the concentration of major projects, both of mining and of power generation, that are an incentive for economic growth but always involve huge and often irreversible changes in terms of both territory and population (Caravaggio et al., 2017).

The natural favorable condition of the Amazon region generated its long-lasting economic boom that has never stopped since it started in the 16th century. However, in the case of Pará it has had multiple phases due to the peak in the use of different resources (e.g., the *rubber era*), up to the present "era of water." Nowadays the specialization of Pará in energy production at the national level is indisputable. Despite its valuable endowment, the state of Pará was able to reach neither the economic standard of the nation nor a desirable human development level (UNDP, FJP&IPEA, 2013). As a matter of fact, according to the last three censuses, its income per capita is more than 50% lower than the income per capita of the Union (Fig. 3.1). Moreover, it has a medium *Municipal Human Development Index* (MHDI) while the Union performed a high MHDI in 2010 (Fig. 3.4) (UNDP, FJP&IPEA, 2013). Despite being well below the national level, Pará's performance over the last 20 years considered in terms of average income and vulnerability to poverty shows an undeniable improvement in economic wealth (Fig. 3.3). However, it maintained stable high levels of inequality (Fig. 3.2) that slowed down the reduction in the percentage of the population vulnerable to poverty, which was less than proportional than the increase in per capita income (Iorio et al., 2018).

This demonstrates that the economic growth at the country level, initially boosted by the plenty of resources (i.e., the exploitation of minerals first and then water resources) from the so-called periphery of the nation, has not been able to lead an inclusive process of development with respect to its remote areas as well, while it exacerbated the center-periphery dynamics within the nation (Furtado, 2000).

The economic growth by means of extractive poles also represented the main impediment to the development of Pará, exacerbating the dynamics of specialization within the country, and widening the inter- and intraregional inequalities especially through the formation of new economic elites externally and not culturally or physically linked to the territory (Farias, 2016; da Silva, 2016; Rocha, 2016). In fact, in the modern Pará, both private and also public capital, which mainly finances major projects, are usually more interested in focusing on the expected revenues from the energy sector than providing basic welfare services for locals (i.e., clean water and sanitation and affordable energy), thus undermining the process of universalizing both access to water and access to energy, drivers of development, especially in the case of foreign capital (de Souza et al., 2016). Moreover, the poor engagement of institutions in cultivating human capital also within rural areas through basic education and specific technical training couldn't start a virtuous circle. For example, lack of both entrepreneurial ability and cooperative management due to poor institutions represents a hindrance to the electrification, preventing the emancipation of the area from the rest of the country thanks to the energetic, economic, and finally cultural independence (Rocha & Neves, 2018; Shiffer & Swan, 2018; Sanchez et al., 2015).

In conclusion, Brazilian energy strategy, endangering environmental conservation and threatening local culture and economic activities, transformed the issue of the resource exploitation for energy purposes of both Pará and the whole Amazon region from a technical scientific issue into a sociopolitical matter (Iorio, 2019).

Conclusions

From the analysis carried out in these pages, we tried to outline some strategic guidelines for the elaboration of state or municipal interventions in support of local development.

First of all, it is necessary to recognize that the causes of the *impasse* of the Amazonian development may be in part traced back to the incorrect definition of "region of intervention." The definition of a geographic unit as the recipient of a public policy must be characterized by a counter-hegemonic process that can no longer allow the homogenization of development policies. At the same time, the regionalizing of areas of intervention (which may coincide with the administrative boundaries of federal states or even go beyond them, sometimes depending on physical and environmental and nonadministrative criteria) must eliminate inequalities while safeguarding differences and specificities (Farias, 2016). Once this is done, it will finally be possible to act with an integrated and participatory approach, and no longer from a sectorial or top-down approach (da Silva, 2016).

Secondly, it cannot be denied that, in the absence of combined interventions at federal, national, and municipal levels, the rapid and substantial industrialization and consequent uncontrolled urbanization only induce economic growth and not local development. As a matter of fact, it becomes a vehicle for problems such as uncontrolled consumption and pollution of water, which is recognized worldwide as a primary good and a human right (de Souza et al., 2016).

Thirdly, it is necessary to observe that the use of water for the electrification process cannot favor a real process of development of the area when it does not lead to a widespread access to electricity. In fact, although the Brazilian Amazon hosts one of the largest water reserves in the world, guaranteeing an enormous hydroelectric potential to its country, the lack of rural electrification continues to represent a serious problem, even after the installation of large hydroelectric projects such as Tucuruí or Belo Monte (Tundisi at al., 2014). In this perspective, federal programs such as *PROINFA* (2002) and *Luz para Todos* (2003) play the role, respectively, to encourage the installation of small power plants for the remote production of electricity from renewable energy in isolated areas and to promote access to capillary energy for the population (Sánchez et al., 2015; MME, 2002; MME, 2003). Thus, rural electrification programs must be supported with the threefold objective of diversifying sources (energy security), reducing environmental impact, and, most of all, guaranteeing access to energy (economic scarcity) (Cavalcanti, 2017).

The main criticism made against the federal government is that it was (deliberately?) short-sighted at the time of both the analysis of the territory and the identification of opportunities and threats, strengths, and weaknesses. From a regional/local point of view, it can be criticized for not having been able to exploit the economic opportunity of such an abundance in terms of both water and precious metals, and for having transformed the strength of a region into its weakness (da Silva, 2016). More attention has been paid to short-term earning opportunities than to the long-term development strategy, which has overshadowed the needs of the inhabitants of the occupied and exploited territories (indigenous people, *quilombolas* and riverside communities) and this has always been too attentive to the needs of private and foreign capital. Even worse, it has intervened only with policies of welfare support, often assisted by the economic support of the federal states, without preventing forced and irreversible changes.

The federal state (in our case, Pará) has been criticized for not being able to mediate between national and local interests. That is, it has not been able to structure development strategies that would have enabled the state to benefit from the trickle-down originated by the large investments of national interest.

Finally, the criticism of the individual municipalities is that they have lost sight of the interests of the weakest, giving in to the seduction of the new economic elites who, not being in any way linked to the territory, have exploited it only for their own economic interests.

In conclusion, although the undoubted usefulness for the Union, in economic terms, of the economic growth occurred in the Amazon, and given the results achieved at a regional and local level in this area, the construction of extraction poles and infrastructures for energy purposes has revealed itself as just a new form of occupation (and exploitation) of the Amazon (Becker, 2005). Therefore, it should be a moral obligation of researchers, scientists, and professionals to try to make its occupation more and more arduous, advising policy makers with the aim to induce awareness on the requirement for environmental constraints and also with the purpose of avoiding the deliberate overlooking of social movements and indigenous people.

References

Amparo, P.P., Porto, E., 1987. Breve descrição e apreciação de alguns programas na Amazônia. In: Costa, J.M.M., Castro, E.M.R. (Eds.), Os grandes projetos da Amazônia: Impactos e Perspectivas — Cadernos NAEA, vol. 9. Falangola, Belém, pp. 39—57.

ANEEL - Agência Nacional de Energia Elétrica, 2019. Plano de Dados Abertos 2018-2019. Available at: http://www.aneel.gov.br/dados.

Aresti, M., Micangeli, A., 2016 — November. Minigrid technical Solution and Business Model. Poster Session Presented at the ATC Integration of Renewable Energy Solutions in the Mediterranean Electricity Markets, Milan, Italy.

Bagher, A.M., Vahid, M., Mohsen, M., Parvin, D., 2015. Hydroelectric energy advantages and disadvantages. American Journal of Energy Science 2 (2), 17—20.

Barros, N., Cole, J.J., Tranvik, L.J., Prairie, Y.T., Bastviken, D., Huszar, V.L.M., del Giorgio, P., Roland, F., 2011. Carbon emission from hydroelectric reservoirs linked to reservoir age and latitude. Nature Geoscience 4 (9), 593.

Becker, B.K., 2005. Geopolitica da Amazonia. Estudos Avançados 19 (53).

Buarque, C., 1987. Notas para uma metodologia de avaliação aos grandes projetos da Amazônia. In: Costa, J.M.M. (Ed.). In: Castro, E.M.R. (Ed.), Os grandes projetos da Amazônia: Impactos e Perspectivas - Cadernos NAEA, vol. 9. Falangola, Belém, pp. 104—127.

Caravaggio, N., Costantini, V., Iorio, M., Monni, S., Paglialunga, E., 2017. The challenge of hydropower as a sustainable development alternative. Benefits and controversial effects in the case of the Brazilian Amazon. In: Fadda, S., Tridico, P. (Eds.), Inequality and Uneven Development in the Post-Crisis World. Routledge, London, pp. 213—242.

Cavalcanti, E., 2017 - May. Água e Recursos Hídricos: de Direitos Fundamentais a Commodities. Poster session presented at the Preparatório da Engenharia e da Agronomia para o 8° Fórum Mundial da Água, Manaus, AM.

Coelho, M., Miranda, E., Wanderley, L.J., Garcia, T.C., 2011. Questão energética na Amazônia: disputa em torno de um novo padrão de desenvolvimento econômico e social. Novos Cadernos NAEA 13 (2).

da Silva, J.M.P., 2016. Dinâmica Territorial da Mineração na Messoregião Sudeste do Estado do Pará — Região Norte Do Brasil. In: Rocha, G.M., Teisserenc, P., Vasconcellos Sobrinho, M. (Eds.), Aprendizagem territorial: dinâmicas territoriais, participação social e ação local na Amazônia. NUMA/UFPA, Belém, pp. 63—76.

de Souza, C.F., Rocha, G.M., Vasconcellos Sobrinho, M., 2016. Água e desenvolvimento humano, vol. 2. Revista do Instituto Histórico e Geográfico do Pará, pp. 69—75. https://doi.org/10.17553/2359-0831/ihgp.v2n2p69-75, 02.

Diamond, S., Poirier, C., 2010. Brazil's native peoples and the Belo Monte Dam: a case study. NACLA Report on the Americas 43 (5), 25—29.

FAPESPA - Fundação Amazônia de Amparo a Estudos e Pesquisas, Diretoria de Estudos e Pesquisas Ambientais, 2018. Pará em Números. Belém. Available at: http://www.fapespa.pa.gov.br/produto/relatorios/172?&mes=&ano=2018.

FAPESPA - Fundação Amazônia de Amparo a Estudos e Pesquisas, Diretoria de Estudos e Pesquisas Ambientais, 2016. Barômetro da Sustentabilidade da Amazônia. Belém. Available at: http://www.fapespa.pa.gov.br/upload/Arquivo/anexo/1126.pdf?id=1557227969.

Farias, A.L.A., 2016. Politica Estadual de Integração Regional do Pará: Limites, Contradições e Possibilidades de Desenvolvimento Territorial na Amazônia. In: Rocha, G.M., Teisserenc, P., Vasconcellos Sobrinho, M. (Eds.), Aprendizagem territorial: dinâmicas territoriais, participação social e ação local na Amazônia. NUMA/UFPA, Belém, pp. 121—136.

Fearnside, P.M., 2006. Dams in the Amazon: Belo Monte and Brazil's hydroelectric development of the Xingu river basin. Environmental Management 38 (1), 16.

Fearnside, P.M., 2005. Brazil's Samuel Dam: lessons for hydroelectric development policy and the environment in Amazonia. Environmental Management 35 (1), 1—19.

Fearnside, P.M., 1999. Social impacts of Brazil's Tucuruí dam. Environmental Management 24 (4), 483—495.

Fearnside, P.M., 1995. Hydroelectric dams in the Brazilian Amazon as sources of 'greenhouse' gases. Environmental Conservation 22 (1), 7—19.

Figueiredo, A.C.P., Saraiva, L.J.C., 2018. A prostituição em grandes projetos na Amazônia: o impacto do grande capital nos fluxos de mão de obra na UHE de Belo Monte. Nova Revista Amazônica 6 (4), 69—77.

Fleury, L.C., Almeida, J.P.D., 2013. A construção da Usina Hidrelétrica de Belo Monte: conflito ambiental e o dilema do desenvolvimento. Ambiente & Sociedade 16 (4), 141—158.

Furtado, C., 2000. Teoria e politica do desenvolvimento econômico. Paz e Terra, São Paulo.

Galvani, L., 1948. Brasile Moderno. Terra incantata. Cavallotti, Milano.

Hazeu, M., 2015. O não-lugar do outro: sistemas migratórios e transformações sociais em Barcarena. Doctoral dissertation. NAEA/UFPA, Belém, PA.

Iorio, M., 2019. Brazilian development: dependency, endowments and the energy issue. Doctoral Dissertation. In: Three Essays on the Development of Brazilian Amazon. Roma Tre University, Rome, Italy.

Iorio, M., Monni, S., Brollo, B., 2018. The Brazilian Amazon: a resource curse or renewed colonialism? Entrepreneurship and Sustainability Issues 5 (3), 438–451.

Iorio, M., 2015. Management of Water Resources in the Amazon Region – Who Might Benefit and Who Might Loose from Dams Expansion. Environmental and Social Damages and Economic Compensation for the Existence of Dams. A Cost Benefit Analysis on Tucuruí Hydroelectric Plant (UHE/TUC). Department of Economics, Roma Tre University, Rome, Italy. Master thesis.

IBGE - Instituto Brasileiro de Geografia e Estatística, 2017. IBGE cidades. Available at: https://cidades.ibge.gov.br/.

Jaichand, V., Sampaio, A.A., 2013. Dam and be damned: the adverse impacts of Belo Monte on indigenous peoples in Brazil. Human Rights Quarterly 35, 408–447.

Magalhães Filho, F., 1987. Grandes progetos ou grande projeto? In: Costa, J.M.M. (Ed.). In: Castro, E.M.R. (Ed.), Os grandes projetos da Amazônia: Impactos e Perspectivas - Cadernos NAEA, vol. 9 Falangola editora, Belém, pp. 17–26.

Manyari, W.V., de Carvalho Jr., O.A., 2007. Environmental considerations in energy planning for the Amazon region: downstream effects of dams. Energy Policy 35 (12), 6526–6534.

MME - Ministério de Minas e Energia, 2003. Programa Luz para Todos. Available at: http://www.mme.gov.br/web/guest/6-programa-luz-para-todos.?inheritRedirect=true.

MME - Ministério de Minas e Energia, 2002. Programa de Incentivo às Fontes Alternativas de Energia Elétrica – PROINFA. Available at: http://www.mme.gov.br/web/guest/acesso-a-informacao/acoes-e-programas/programas/proinfa.

Moretto, E.M., Gomes, C.S., Roquetti, D.R., Jordão, C.D.O., 2012. Histórico, tendências e perspectivas no planejamento espacial de usinas hidrelétricas brasileiras: a antiga e atual fronteira Amazônica. Ambiente & Sociedade 15 (3), 141–164.

Monni, S., Iorio, M., Realini, A., 2018. Water as freedom in the Brazilian Amazon. Entrepreneurship and Sustainability Issues 5 (4), 812–826.

Piketty, T., 2000. Theories of persistent inequality and intergenerational mobility. Handbooks in Economics 16, 429–476.

Pinto, L.F., 2017 - June. Internacionalizar para não internacionalizar a Amazônia. Seminar presented at Projeto Descolonizar #Amazonia, Belém, PA. Available at: https://descolonizar.tumblr.com/amazonia.

Pinto, L.F., 2011. Tucuruí: a barragem da ditadura. Jornal Pessoal, Belém.

Prado Júnior, C., 2004. História Econômica Do Brasil. Brasiliense, São Paulo, p. 45.

Rocha, G.M., Neves, M.B., 2018. Hydroelectric projects and territorial governance in regions of the State of Pará, Brazilian Amazon. Entrepreneurship and Sustainability Issues 5 (4), 712–723.

Rocha, G.M., 2016. Hidrelétrica, Dinâmica Populacional e Mudança Espacial na Região de Integração Lago Tucuruí (1970–2010). In: Rocha, G.M., Teisserenc, P., Vasconcellos Sobrinho, M. (Eds.), Aprendizagem territorial: dinâmicas territoriais, participação social e ação local na Amazônia. NUMA/UFPA, Belém, pp. 77–102.

Rocha, G.M., 2008. Todos convergem para o lago. Hidrelétrica de Tucuruí: municípios e territórios na Amazônia. Numa-UFPA, Belém.

Sánchez, A.S., Torres, E.A., Kalid, R.D.A., 2015. Renewable energy generation for the rural electrification of isolated communities in the Amazon Region. Renewable and Sustainable Energy Reviews 49, 278–290.

Schiffer, A., Swan, A., 2018. Water security: a summary of key findings exploring islands in Brazil. Security and Sustainability Issues 7 (4), 855–860. https://doi.org/10.9770/jssi.2018.7.4(20.

Scudder, T., 1981. What it means to be dammed: the anthropology of large-scale development projects in the tropics and subtropics. Engineering and Science 44 (4), 9–15.

SIPOT – Sistema de Informações do Potencial Hidreletrico Brasileiro, 2016. Potential Hidrelétrico brasileiro – usinas acima de 50 MW. Available at: http://eletrobras.com/pt./AreasdeAtuacao/geracao/sipot/Mapa%20Sipot%202016_novo27.pdf#search=sipot.

Souza, R.C.R., Santos, E.C.S., 2006. Estado e desenvolvimento regional: a falta de compromisso com o setor elétrico na Amazônia. In: Scherer, E., Oliveira, J.A., orgs (Eds.), Amazônia: políticas públicas e diversidade cultural. Garamond, Rio de Janeiro, pp. 61–85.

Tundisi, J.G., Goldemberg, J., Matsumura-Tundisi, T., Saraiva, A.C., 2014. How many more dams in the Amazon? Energy Policy 74, 703–708.

UNDP - United Nation Development Programme, FJP - Fondação João Pinheiro, IPEA - Instituto de Pesquisa Econômica Aplicada, 2013. Atlas do desenvolvimento Humano no Brasil. Available at: http://www.atlasbrasil.org.br/2013/pt./.

Werner, D., 2011. Desenvolvimento regional e grandes projetos hidrelétricos (1990-2010): o caso do Complexo Madeira (Doctoral thesis, Universidade Estadual de Campinas, Campinas, São Paulo, Brazil).

4

Energy consumption and income distribution. International policies for mitigating social exclusion

Dainius Genys

Energy Security Research Center at Vytautas Magnus University, Kaunas, Lithuania

Energy security to sociologists is attractive for many reasons, one of which is its constantly evolving definition. Not long ago the world was concerned with energy availability and affordability (Yergin, 1988; Ang et al., 2015), later with infrastructure's reliability and resistance to threats (Kruyt et al., 2009; Winzer, 2012) then efficiency and ecology, and finally, with its relation to social justice (Sovacool et al., 2014; Sovacool et al., 2014). Notable growth of social science research focusing on energy studies recently has fostered the emergence of a new concept—energy justice (Hall, 2013; Jenkins et al., 2016), which explores justice principles to energy policy, energy production, consumption, and impact on society. Even though it requires comprehensive studies and empirical research, it serves as an indication of a fundamental shift in energy studies—moving from technical, economic, and political questions to its interrelations with social arrangements.

This chapter discusses energy security as a social problem and raises such questions as: What is the relation between energy security and distribution of "life chances" among people that are in different places in the social hierarchy? What are the differences of energy security impact among social groups in regard to consumption, access, integration, participation, mobility, influence, and recognition? In this chapter energy security is related not only to the availability of services, material deprivation but also with the quality of life (poverty policy), so attention is drawn to the following analytical dimensions as belonging, inclusion, participation, recognition of the legitimacy and by covering both attitudinal as well as behavioral levels. The chapter is dedicated to the analysis of energy security implications on public behavior by reviewing various policies for mitigating social exclusion and presenting a Lithuanian case study as empirical evidence. By presenting the unique case of experimental research we contribute an integrative view of energy security's implications on behavioral patterns in contemporary Central and East European democracies.

The chapter consists of five parts. The first part discusses the specific relations among energy security and social cohesion. The second focuses on peculiarities of the region with particular focus on post-Soviet legacy. The third presents a theoretical framework that is used in empirical analysis and explains operationalization of the theoretical model and concrete indicators that are used in the analysis. The fourth part explains the results and emphasizes main discoveries. The fifth part summarizes discussion. Lastly, the chapter ends with concluding remarks.

Energy security as a social problem

The definition of energy security is variable and depends on different countries' different interests, but it is possible to grasp a more consistent approach that focuses not only on an energy system's ability to provide energy for the consumers at acceptable prices but also its ability to withstand technical, environmental, economic, political, and social threats (Zhang et al., 2016; Sovacool et al., 2016). Energy security, being closely related to economy, with no doubt has a huge impact on society, but there is still lack of evidence on its contribution to social justice. Considering the consequences, there is a growing concern in academic literature to expand the definition by involving societal aspects (both perception and participation) (Stirling, 2008; Knox-Hayes et al., 2013; Demski et al., 2014). For example, it is important to take into account not only the efficiency (strategic or economic) of a concrete energy project itself but also its impact on social exclusion. Previous research has shown that in the quest for strategic long-term goals sometimes it is inevitable to raise the price of energy, but from a sustainable development point of view this might lead to the fragmentation of society and even to the growth of anxiety (Genys, Krikstolaitis 2015). To compensate for this negative balance it is important to gain as much public support as possible, hoping that public consciences will lead endurance of financial burden for tax payers as a consequence of investments in long-term goals.

Sociologists (Taket et al., 2009; Jehoel-Gijsbers, Vrooman, 2007; Jenson, 1998; Bernard, 1999) analyzing social cohesion accurately notice that material deprivation, which is vusually defined as inability to satisfy essential goods (such as decent living conditions—heat, cold/hot water, housing, etc.—as well as decent level of quality of life), impoverishes people's lives. However, material deprivation shouldn't be considered as the only indicator of misery (Bourdieu, 1999, p. 4). By focusing solely on objective processes and analyzing economic and political consequences of energy security do we not forget to analyze less visible but not less important aspects of sociocultural dimensions (such as attitudes, norms, values, and power relations)?

Therefore the sociological interpretation of energy security should be linked to two aspects: (1) its capacity to balance possible opposition between its aim and public attitude to it; (2) the actual effect of the pursuit of energy security policy and its impact on public behavior, i.e., state interest versus public concern, development of security scenarios versus public support for concrete projects, and efficiency and balance between investments versus social justice.

The concept of social exclusion helps to understand the life experiences stemming from multiple forms of deprivation and inequality experienced by people in different places of the social hierarchy. It also reveals the reduced abilities of participation in society, consumption, mobility, integration, and influence of particular individuals or social groups (Taket et al., 2009, p. 3). Energy security consist of two most important aspects—price and security—and in the post-Soviet society it serves as a check objectively dividing society between those who are willing to feel securer (despite the costs) and those who are willing to save expenses (by risking its security). By paraphrasing the classical insight of M. Weber, it might be stated that energy security in Lithuania (as well as pretty much in other Baltic countries) means the distribution of "life chances" among different social groups living in different social conditions.

There are plenty of various conceptualization (Taket et al., 2009; Jehoel-Gijsbers, Vrooman, 2007; Duhaime et al., 2004; Martin, 2004) and operationalization (Burchardt et al., 2009; Chan et al., 2006; Rajulton et al., 2007) differences of social exclusion. The most important aspects of social exclusion usually are distinguished as the following: participation, consumption, mobility, access to services, integration, influence, and recognition. Four main dimensions are distinguished in the analysis of social exclusion: consumption (ability to buy goods and services), production (participation in activities that are considered economically and socially valuable), political participation (participation in decision-making at local and national level), and social interaction (relationships with family, friends, and the community). Deprivation of any of these dimensions can lead to social exclusion (Burchardt et al., 2009, p. 31).

Social exclusion is similarly conceptualized by another group of researchers, who say that social exclusion consists of multiple dynamic processes driven by unequal power relations between the four (economic, political, social, and cultural) main dimensions that have different impacts on individual or group, community, national, or global scale (Popay et al., 2008, p. 2).

Such interpretation is similar to the theoretical framework of social cohesion offered by Jenson (1998) and Bernard (1999) and their suggested six analytical dimensions. Social exclusion as a result of the lack of resources or its high prices and inability of some people to acquire them, or because of that it affects dignity and position in social hierarchy or creates obstacles for some people to participate and maintain normal social relations, has an impact not only on quality of life but also affects public perception of social justice and social cohesion (Levitas et al., 2007, p. 9).

The analysis of energy security from a social exclusion point of view is supplemented by the sociocultural aspect. It includes such dimensions as insufficient social integration (energy security impact on participation in formal/informal social networks, including leisure activities, and social support as well as social isolation) and insufficient cultural integration (noncompliance with norms and values of active citizenship, i.e., indifferences for interest representation, alienation from energy security policy, misinterpretation of social justices, abuse of the state privileges for poor, etc.).

Tracing the relations between energy security and social exclusion, it can be said that the pursuit of energy security is associated with social justice, which could be operationalized by the following questions: How public interest is recognized, defined, and represented in energy security policy? How (and if at all) the interests of smaller social groups (environmentalists, pensioners, poor, etc.) are recognized, defined, and represented? Whether energy security policy acknowledges the interests of poor, deprived and disenfranchised individuals or addresses solely active and powerful (from consumption point of view) individuals? Finally, how existing energy security policy treats and fosters to feel vulnerable groups?

Peculiarities of Post-Soviet society

Usually, energy is one of the most important pillars of modern economy, and society is interested that it would be efficient and work smoothly; these were the main concerns in the development of energy sectors for many years in many countries. However, energy security is not a simple "reliable supply of energy resources at affordable prices" (Yergin, 1988, p. 111) problem in Lithuania. It does face some extra challenges due to its historical and geopolitical aspects (Augutis et al., 2013, 2014, 2015). Lithuania inherited an energy sector that was neither efficient nor developed to respond to an independent country's needs (for more than 2 decades it was totally dependent on Russian energy). Lithuania had to fundamentally restructure its energy sector; therefore energy security became a huge and expensive challenge that was laid on the shoulders of a relatively poor society (Genys, Krikštolaitis, 2015).

Like all modern industrial societies, the Soviet state inevitably had parallel ways to develop various closely related sectors of society (industrialization, urbanization, bureaucratization, public education, and health care systems, etc.). The energy subsystem had to be one of the basic areas of state development. Typologically, they were the necessary conditions of all modern industrial societies' development, so we can talk about Western-type society's modernization, which took place during the Soviet period (Castells, 2007, pp. 18–70). The construction of communist society was accompanied by industrialization, homogenization, and urbanization, therefore energy infrastructure became an important part of Soviet modernization. Soviet modernization was framed by totalitarian politics and ideology, and most of modernization's negative aspects (environmental, bureaucratic, economic, energy, etc.) as well as the aspects of risk for society were more expressed than in Western democratic industrial societies, where it was possible to discuss the risks produced by modernization (Leonavičius and Genys 2012, Leonavičius and Genys 2014a).

After the transition from planned economy to liberal market economy, the post-Soviet societies faced the challenge to switch their understanding about the energy sector and adapt to the current situation. A few decades ago, energy was cheap, provided by the state and the former Soviet Union, was understood as a guarantee of safe and cheap energy supply. Today, energy has become quite an expensive challenge; the Lithuanians have to pay the price dictated by Gazprom, and Russia has become the biggest threat.[1] After its declaration of independence, Lithuanian society was confronted not only by the change of political and economic systems but by the new balance of geopolitical powers in the region, and also by the change of symbolic meaning of energy sector—due to its inevitable structural modernization, it became an expansible challenge, an anchor of the young country's economy, which finally lay down on the shoulders of taxpayers (Leonavičius et al., 2015).

Even just a few years ago Lithuania (along with other Baltic states) used to be deservedly referred to as an "energy island". Drastic but positive change happened in 2015, when the LNG terminal "Independence" was launched at the end of the year and successful completion of electricity links with Poland and Sweden have helped to dispose of the "island" status. Figuratively speaking, persistent efforts, careful planning, and smart politics have helped Lithuania to emerge as a new bright energy security star among many Eastern European countries. However, this change was accompanied with lots of worries and anxiety from a public perception point of view.

The research showed that for years the most important aspect of energy security for Lithuanians was price (Leonavicius et al., 2015, 2018). Such perception is quite understandable in a post-Soviet country where residents of the former Soviet empire were used to cheap energy prices. However, sometimes public will might be inadequate or contradict the strategic interests of the country (i.e., when government wants to invest and society—strive for cheap energy). Even though the majority of Lithuanian society agrees that energy independency from other countries is an important (important or very important; 71.8% agreed) aspect of energy security, 68.7% mentioned that "the state should be concern[ed] with and do more about cheap energy instead of energy security," and only 30.8% agreed that "the state should be concern[ed] with energy independence despite the requirement for bigger investments" (Augutis et al., 2015, p. 23). Even though it is almost impossible to ensure the supply of cheap energy without achieving independence of the energy sector from a monopolistic system, the society does not intend to support this goal at the expense of personal wealth. The implementation of energy policy is based on the rationality of society and its trust in the public interest.

Economic differentiation is quite vivid in Lithuania (Lisauskaitė, 2010; Zabarauskaitė, Blažienė, 2012), therefore energy prices have different effects on different social groups. The welfare of a large part of Lithuanian society depends on centralized supply of energy resources (gas, electricity, district heating), poor quality of energy infrastructure, inability to make individual decisions, and especially prices (Leonavičius and Genys, 2014b). It is obvious that the part of society with lower income is particularly vulnerable not only because of increasing energy prices, disruption in supply, or other risks of the energy system but also because of the growing financial burden that occurs due to the quest for energy security. Therefore, the cost for energy security (the same as labor market, low income, unemployment, health care, etc.) might become the reason for increase of social exclusion.

---

[1]The situation has changed with the beginning of 2015 when LNG terminal was put into operational use and Lithuania finally obtained actual alternative (from monopoly supply from Russia) for gas import.

The energy system is important for the economy of every country because its efficiency or inefficiency, respectively, has a positive or negative impact on country's economy and creates (or not) conditions for sustainable development of society (Blum, Legey, 2012). Comparative research[2] shows that those societies that are cohesive and mobilized usually overcome challenges faster and more successfully. And in contrast, fragmented and unorganized societies face some extra challenges. If a vivid inequality is present in society and the burden of prices is experienced unequally, opportunity appears to manipulate the public attitude toward particular projects or even foster fragmentation in society. That is why the implementation of any reform or specific energy infrastructure projects should be based not only on economic benefit but also on the potential impact on social cohesion.

Theoretical framework of social cohesion and its operationalization

To define and to measure social cohesion is not an easy task as there can be many definitions as well as many measurements. The main challenge, either in defining or measuring the concept, lays in its multilevel and multidimensional nature. There are plenty of various research models of social cohesion, but despite the differences occurring in conceptualizing (Berger, 1998; Kearns, Forrest, 2000; Berger Schmitt, 2002; Duhaime et al., 2004) and operationalizing (Jenson, 1998; Bernard, 1999; Chan et al., 2006; Rajulton et al., 2007), the concept of cohesion serves as a kind of frame that allows to understand what in some societies, even in times of crisis (or challenges), leads to cohesion growth and what leads toward fragmentation.

This chapter adapts the model of social cohesion proposed by P. Bernard. The integrated conceptual scheme of social cohesion provided by Bernard (1999) is based on three activity spheres: economic, political, and sociocultural, and on a twofold nature of relations: the formal, attitudinal (how people perceive) and substantial/behavioral (how people act). These two theoretical facets lead to the conceptualization of the following dimensions: affiliation/isolation, insertion/exclusion, participation/passivity, acceptance/rejection, legitimacy/illegitimacy, and equality/inequality. Such theoretical framework helps to form empirical dichotomies of both activity spheres (Table 4.1) and such analysis helps to identify the relation between attitudinal and behavioral aspects of society toward energy security. This model is successfully used in comparative social cohesion analysis of various countries (Dickes et al., 2008, 2009).

Bernard (1999) was concerned with analyzing social cohesion and searching for the most appropriate approach; here are some examples of the guidelines of indicators provided by the author:

Economic sphere. The items of formal/attitudinal dimensions suppose to help to identify the attitude of society toward existing insertion/exclusion mechanisms. Meanwhile the items of substantial/behavioral dimensions suppose to reveal the existing equality/inequality balance of society in reality.

---

[2]The Global Competitiveness Report 2014–15 (http://reports.weforum.org/global-competitiveness-report-2014-2015/), http://www.eurofound.europa.eu/sites/default/files/ef_publication/field_ef_document/ef1472en.pdf.

TABLE 4.1   Bernard's integrated conceptual scheme of social cohesion.

Sphere	Nature of relations	
	Formal/attitudinal	Substantial/behavioral
Economic	**Insertion/exclusion**	**Equality/inequality**
The analogy of empirical items for energy security research	The formal/attitudinal items of economic dimension cover various questions with the aim to reveal the societal attitude toward the evaluation of the burden of energy security as well as its social justice and evaluation of public opinion of particular projects.	The substantial/behavioral items of economic dimension cover various questions with the aim to reveal the real economic burden experienced by the society, its impact to the distances (economic and social) between different groups of society, and the approval of concrete projects.
Political	**Legitimacy/illegitimacy**	**Participation/passivity**
The analogy of empirical items for energy security research	The formal/attitudinal items of political dimension cover various questions with the aim to reveal societal trust in various organizations and institutions as well as private companies (including foreign) related with energy security, assigned responsibility, and atittudes toward safety of concrete energy projects.	The substantial/behavioral items of political dimension cover various questions with the aim to reveal factually society's civic activity and involvement as well as their knowledge about various aspects of energy security.
Sociocultural	**Acceptance/rejection**	**Affiliation/isolation**
The analogy of empirical items for energy security research	The formal/attitudinal items of sociocultural dimension cover various questions with the aim to reveal public perception of social justices of energy security politics and personal readiness to contribute to public interest in energy security as well as perception of energy security (whether it is based on self-interest or societal interest).	The substantial/behavioral items of sociocultural dimension cover various questions with the aim to analyze whether the existing effect of energy system on society maintains the possibility to remain autonomous and ability to individually defend oneself from energy threats.

Political sphere. The items of formal/attitudinal dimensions suppose to help to identify society's trust in various governmental institutions and organizations, its legitimacy and efficiency in representing public interest. Meanwhile the items of substantial/behavioral suppose to reveal factual participation and activity of society in democratic governance. *Sociocultural sphere.* The items of formal/attitudinal dimensions suppose to help to identify the attitude of society toward openness and respect for diversity. Meanwhile the items of substantial/behavioral suppose to reveal the dominated values and their diversity through the belonging of the society to various organizations.

The conceptual framework borrowed from Bernard needs modification because in this case it is used to test the impact of energy security on social cohesion (but not social cohesion itself). To assess the level of social cohesion usually both subjective and objective metrics are used, which cover formal/attitudinal as well as substantial/behavioral nature of relations needed for a conceptual framework. The following Table 4.1 describes conceptual dichotomies between formal/attitudinal and substantial/behavioral nature of relation and sets the guidelines for analogical empirical items of each dimension for energy security research.

A representative survey was conducted by the public opinion research company "Vilmorus" in June 2014. Number of respondents: N = 1009; interviewed 18 years old and older residents of Lithuania. The method of survey: questioning respondents at home using premade questionnaires. Method of selection: multistage, probabilistic sampling. Selection of respondents was prepared so that each resident of Lithuania should have an equal chance of being questioned. The results reflect the opinion of the entire population of Lithuania and distribution by age, sex, place of residence, education, and purchasing power. Error of survey results: 3% (probability — no less than 97%).

The five-point Likert scale was used for the data analysis and interpretation. Respondent disapproval of a particular issue was marked "1," indecisiveness/not knowing was "3," and approval was "5." All questions are formulated in the way that the increased average of the responses (e.g., when responses' average is approaching "5") means a higher importance of the particular aspect from the point of respondents' opinion and, conversely, lower average, lower importance (e.g., when responses' average is approaching "1").

The content of social cohesion: what works in Lithuania?

The statistical analysis showed that from the public attitude point of view the impact of energy security on social cohesion in Lithuania has almost neutral effect—the average of indicators from formal/attitudinal dimension is 3.05 ("1" — very negative; "3" — neutral; "5" — very positive). Meanwhile the actual impact of energy security on social cohesion has negative effect—the average of indicators from substantial/behavioral dimension is 2.65. Finally, the aggregated average of the indicators from both nature of relations show—the overall impact of energy security on social cohesion in Lithuania has negative effect—2.85 (Fig. 4.1).

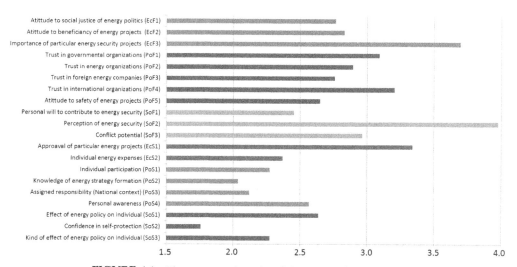

FIGURE 4.1   The aggregated results of the impact of every indicator.

This picture illustrates the diverse impact of energy security on social cohesion in Lithuania. The results show that the impact of empirical items has different impact not only between different dimensions but also within it. Having this picture we could draw an assumption that energy security policy lacks consistency.

With step-by-step detailing of the results of every dimension we'll try to identify the most positive and the most negative aspects of energy security from a social cohesion point of view. The analysis of the economic dimension of formal/attitudinal nature of relations shows that respondents positively evaluate the importance of particular energy projects (3.66) but have doubts about their benefits (2.83) as well as the implementation of energy security according to social justice (2.77). The interesting fact is that it seems the society agrees on the direction of energy security and legitimates the necessity of concrete energy projects, but the majority of society disagrees that "the problem of energy security in Lithuania is addressed taking into consideration the interests of all social groups"; this reveals the existing fragmentation among society, and it seems energy security does not contribute to its decrease. It comes with surprise that the analysis of economic dimension of substantial/behavioral nature of relations demonstrates better results. The high score (3.34) of the "approval of particular energy projects" once again illustrates public support. It seems that the amount of individual expenses on energy products also has positive impact (contrary to what might have been expected) on social cohesion (thus the comparable high level of standard deviation 0.84 speaks about the diversity of impact of this indicator).

The analysis of political dimension of formal/attitudinal nature of relations shows quite diverse impact of these indicators. Trust in international organizations attracted more public support (3.21) than trust in governmental organizations (3.09). Interestingly enough "Attitude to safety of energy projects/PoF5" became the worst evaluated indicator (2.65), which speaks about the need of further legitimation in public view. The analysis of political dimension of substantial/behavioral nature of relations demonstrates even worse results. If quite passive individual participation in civic activities (2.27) comes with no surprise, thus the negative evaluation of knowledge on the formation of energy strategy (2.04) as well as personal awareness about energy problems in general and advantages as well as disadvantages of particular projects (2.57) becomes a surprise. It indicates the existing quite a big gap in public knowledge on energy affairs. Another indicator, "Assigned responsibility (National context)/PoS3," is also eloquent; its low score (2.12) reveals that society is very critical of the implementation of energy security policy as well as it is critical of the subjects in charge of the process.

Finally, the analysis of sociocultural dimension of formal/attitudinal nature of relations shows that bigger part of respondents understands and agrees with the importance of energy security (3.99), however, a much smaller part is willing to contribute to this aim personally (2.45). The last indicator in this dimension illustrates that Lithuanians are not a very conflicting society and the "Conflict potential/SoF3" remained indifferent (2.97). The analysis of sociocultural dimension of substantial/behavioral nature of relations demonstrates much worst results. It seems that the biggest damage to social cohesion derives from society's total dependence on existing energy infrastructure, i.e., only a small part of respondents believes in self-protection from increasing energy prices (1.76). A little bit better (but still negative) is the evaluation to kind of effect of energy policy on individual (2.28) as well as "Effect of energy policy on individual/SoS1" (2.64).

Discussion

Based on the theoretical framework of social cohesion we created 20 operational indicators that cover all dimensions of Bernard's integral theoretical scheme. It allowed to calculate different scores of energy security impact on social cohesion. Statistical analysis proved the usefulness of the research yet stressed the need for additional clarification of conceptual as well as empirical nature of methodology due to specificity of Lithuanian context.

We should admit that adaptation of the model of social cohesion in post-Soviet context influenced the purity of the conceptual model. This aspect needs improvement not only from an empirical (by picking most actual variables) but also from a conceptual (by recognizing the features of authentic model of social cohesion in Lithuania) point of view in future research. For example, we have included individual expenses for energy, however, the indicators of economic dimensions require "hard" data such as income, employment rate. The political dimension fully covers attitudinal nature (by analyzing the legitimation of concrete aspects of energy security), however, the substantial nature requires to better grasp not only general civic participation activity and depth but also peculiarities of the energy sector. Public participation in energy security is quite controversial because it requires a specific set of knowledge that not everybody has. Even though there are spaces where people can engage directly (i.e., personal support for particular projects, involvement in public debate, representation of personal will during a referendum, etc.), usually the interest of society is represented through various delegates (politicians, NGOs, entrepreneurs, experts or media, etc.). Finally, the sociocultural dimension would benefit if it would be possible to support with some objective data (confidence in self-protection is more attitudinal than behavioral indicator).

The research indicated different tendencies of energy security impact on social cohesion at the attitudinal (aggregated result − 3.05) and behavioral (aggregated result − 2.65) nature of relations, which indicates the inconsistency of Lithuanian energy security policy estimated from a social cohesion development point of view. Based on the results we might conclude that main goals of energy security policy have been legitimized at the attitudinal level and a large part of Lithuanian population understands the importance of energy security and the need to protect from possible risks. However, the behavioral level indicated that society is not satisfied with the implementation of the energy security policy, and its effect on society does not have a positive impact from a social cohesion development point of view. The overall impact of energy security on social cohesion in Lithuania has negative effect of −2.85 and it tends to increase the distance among society rather than bridge it.

The aspects that glue Lithuanian society from an attitudinal nature of relations are "The importance of particular energy security projects," "Trust in international organizations," and "Perception of energy security," which suggests that the main goals of energy security policy mobilize society. Interestingly enough, the aspects that have the most positive impact from a behavioral nature of relations are related with economical dimensions: "approval of particular energy projects" and surprisingly "Individual energy expenses," which might be explained by the fact that the poorer part of society receives compensation from the states and therefore amortizes the attitude and behavior of the poor.

Speaking about the aspects that divide society, we see the mismatch when on the one hand we see the wishful need to democratize the issue of energy security but on the other hand the total ignorance of such a right among society (society is very passive in participation, involvement, and interest in issues of energy security). The large part of society thinks that energy security is more oriented toward the interests of the rich rather than all social groups (both representative from business and government are evaluated badly). Another important aspect is related to the fact that a larger part of society does not believe in the economic benefit of particular energy projects (i.e., Visaginas Nuclear Power Plant, shale gas development in Lithuania). Society does not trust the impact on energy security of the Lithuanian Parliament, municipality and national as well as private companies. Even worse results derive from mistrust of international energy companies (with the exception of Scandinavian companies). The worst results come from the fact that a bigger part of society feels "chained" to existing energy infrastructure, which limits their ability for an individual maneuver.

As demonstrated by the research, energy security, at least in Lithuania, goes beyond economic optimization, political independence, and infrastructure efficiency questions. As a social issue it penetrates classic understanding and conceptualization. Among the relatively poor and on energy infrastructure-depending society, energy security implications could be traced to behavioral patterns of society and its particular groups. Due to the constantly changing geopolitics, global prices, and technological evolution, this interrelation is also changing, however, energy security as a social problem remains.

Further research needs to be carried out to distinguish the particular, most common model of social cohesion in Lithuania. It is possible to grasp some aspects of liberal, social democrat, and conservative models, but at the same time some aspects contradict each other. Having in mind society's skepticism toward businessmen it is difficult to talk about the adaptation of a liberal model. Witnessing very passive society and its poor involvement in energy issues it is not possible to expect the adoption of a social democratic model and rely on civil society and its solidaristic values contributing to social cohesion. The most notable feature is strong support by the government (but also criticism of its performance); it seems Lithuanian social cohesion is mostly rooted in institutional distribution of social justice.

Conclusions

The effectiveness of energy security policy in mitigating social exclusion should be linked to its ability to respond to the public interest and to reduce socioeconomic distances between members of society as well as to involve different social groups in policy making. The existence of polarization of the public interest may result in the opposite effect for different groups in society. As we saw in the Lithuanian case, even though a large part of society understands the meaning of energy security and the need to protect against possible risks, the research results show that this goal does not promote but actually reduces social cohesion.

Having such results it is easy to notice different impacts of energy security on social cohesion from attitudinal and behavioral nature of relations. It seems that the main goals of energy security policy have been legitimized at the attitudinal level (which could be understandable due to the specific post-Soviet Lithuanian context). A large part of the Lithuanian population understands the importance of energy security and the need to protect from possible risks. However, the aggregated statistical results indicate the insufficiency of such support in order for it to have an overall positive effect. It seems that the pursuit of energy security is still a heavy burden for society at the behavioral level. Therefore we might presume that various social groups in society experience this burden differently.

The research showed that the aggregated average of the indicators from both nature of relations show that the overall impact of energy security on social cohesion in Lithuania has a negative effect of −2.85. It seems it tends to increase the distance among society rather than bridge it. Thus from the attitudinal point of view the impact of energy security has almost a neutral effect, 3.05 (mean of formal/attitudinal nature), and with a little improvement it might lead to the positive effect. However the actual effect of energy security on society needs to be improved more consistently. The actual impact of energy security has negative effect −2.65 (mean of substantial/behavioral nature).

References

Ang, B.W., Choong, W.L., Ng, T.S., 2015. Energy security: definitions, dimensions and indexes. Renewable and Sustainable Energy Reviews 42, 1077–1093.

Augutis, J., Krikštolaitis, R., Genys, D., Česnakas, G., 2013. Lietuvos Energetinis Saugumas. Metinė Apžvalga. 2011–2012 (Lithuania's Energy Security. Annual Review) (Kaunas: VDU).

Augutis, J., Krikštolaitis, R., Genys, D., Pečiulytė, S., Česnakas, G., Martišauskas, L., 2014. Lietuvos Energetinis Saugumas. Metinė Apžvalga. 2012–2013 (Lithuanian Energy Security. Annual Review. 2012–2013) (Kaunas: VMU).

Augutis, J., Krištolaitis, R., Leonavičius, V., Pečiulytė, S., Genys, D., Česnakas, G., Martišauskas, L., Juozaitis, J., 2015. Lietuvos energetinis Saugumas. Metinė Apžvalga. 2013–2014 (Lithuania's Energy Security. Annual Review) (Kaunas: VDU).

Berger, P., 1998. The Limits of Social Cohesion: Conflict and Mediation in Pluralist Societies : A Report of the Bertelsmann Foundation to the Club of Rome. Westview Press, Buolder, Colo.

Berger-Schmitt, R., 2002. Considering social cohesion in quality of life assessments: concept and measurement. Social Indicators Research 58, 403–428.

Bernard, P., 1999. Social Cohesion: A Critique. Canadian Policy Research Networks Discussion Paper No. F09. Canadian Policy Research Networks, Ottawa.

Blum, H., Legey, L.F.L., 2012. The challenging economics of energy security: ensuring energy benefits in support to sustainable development. Energy Economics 34.

Bourdieu, P., 1999. The Weight of the World: Social Suffering in the Contemporary Society. Stanford University Press, Stanford trans., P. Parkhurst Ferguson et al.

Burchardt, T., Le Grand, J., Piachaud, D., 2009. Degrees of exclusion: developing a dynamic, multidimensional measure. In: Hills, J., Le Grand, J., Piachaud, D. (Eds.), Understanding Social Exclusion. Oxford University Press, Oxford, pp. 30–43.

Castells, M., 2007. Informacijos Amžius: Ekonomika, Visuomenė Ir Kultūra. Tūkstantmečio Pabaiga (End of Millennium, the Information Age: Economy, Society and Culture). Poligrafija ir informatika, Kaunas. T. III.

Chan, J., To, H., Chan, E., 2006. Reconsidering social cohesion: developing a definition and analytical framework for empirical research. Social Indicators Research 75 (2), 273–302.

Demski, C., Poortinga, W., Pidgeon, N., 2014. Exploring public perceptions of energy security risks in the UK. Energy Policy 66, 369–378.

Dickes, P., Valentova, M., Borsenberger, M., 2008. Social cohesion: measurement based on the EVS micro data. Italian Journal of Applied Statistics 20 (2), 77–91.

Dickes, P., Valentova, M., Borsenberger, M., 2009. Construct validation and application of a common measure of social cohesion in 33 european countries. Social Indicators Research 98, 451–473.

Duhaime, G., Searles, E., Usher, P.J., Myers, H., 2004. Social cohesion and living conditions in the Canadian artic: from theory to measurement. Social Indicators Research 66, 295–317.

Genys, D., Krikštolaitis, R., 2015. Interrelations between energy security economics and social cohesion: analysis of a Lithuanian case. The Journal of Baltic Law and Politics 8 (2), 46–70. https://doi.org/10.1515/bjlp-2015-0018.

Hall, S., 2013. Energy justice and ethical consumption. Local Environment 18 (4).

Jehoel-Gijsbers, G., Vrooman, C., 2007. Explaining Social Exclusion: A Theoretical Model Tested in the Netherlands. the Netherlands institute for social research, The Hague.

Jenkins, K., McCauleya, D., Heffronb, R., Stephanc, H., Rehner, R., 2016. Energy justice: a conceptual review. Energy Research & Social Science 11, 174–182.

Jenson, J., 1998. Mapping Social Cohesion: The State of Canadian Research. Canadian Policy Research Networks Inc, Ottawa.

Kearns, A., Forrest, R., 2000. Social cohesion and multilevel urban governance. Urban Studies 37 (5–6), 995–1017.

Knox-Hayes, J., Brown, A.M., Sovacool, K.B., Wang, Y., 2013. Understanding attitudes toward energy security: results of a cross-national survey. Global Environmental Change 23 (3), 609–622.

Kruyt, B., van Vuuren, P.D., de Vries, H.J.M., Groenenberg, H., 2009. Indicators for energy security. Energy Policy 37 (6), 2166–2181.

Lisauskaitė, V, 2010. "Lietuvos gyventojų pajamų ir vartojimo diferenciacija" [Differentiation of income and consumption of the Lithuanian population]. Verslas: teorija ir praktika 11, 266–278.

Leonavičius, V., Genys, D., 2012. Rizikos visuomenė: centralizuoto šildymo sistemos atvejis Lietuvoje (Risk society: the case of district heating in Lithuania). Filosofija Sociologija 23 (4), 237–245.

Leonavičius, V., Genys, D., 2014a. Daugiabučių namų renovacija: socialinis ir ekonominis aspektai (Renovation of multi-apartment Houses: social and economic aspects). Filosofija Sociologija 25 (2), 98–108.

Leonavičius, Genys, 2014b. Sociologinis energetinio saugumo tyrimas (Sociological energy security research). In: Augutis, J., Krikštolaitis, R., Genys, D., Pečiulytė, S., Česnakas, G., Martišauskas, L. (Eds.), Lietuvos Energetinis Saugumas. Metinė Apžvalga. 2012–2013. Kaunas: VMU, pp. 15–21 [Lithuanian Energy Security. Annual Review. 2012–2013].

Leonavičius, V., Genys, D., Krikštolaitis, R., 2015. Public perception of energy security in Lithuania. Journal of Security and Sustainability Issues 4, 311–322.

Lisauskaitė, V., 2010. "Lietuvos gyventojų pajamų ir vartojimo diferenciacija" [Differentiation of income and consumption of the Lithuanian population]. Verslas: teorija ir praktika 11, 266–278.

Leonavičius, V., Genys, D., Krikštolaitis, R., 2018. Public perception of energy security in Lithuania: between material interest and energy independence. The Journal of Baltic Studies 49 (2), 157–175. https://doi.org/10.1080/01629778.2018.1446033.

Levitas, R., Pantazis, C., Fahmy, E., Gordon, D., Lloyd, E., Patsios, D., 2007. The Multi-dimensional Analysis of Social Exclusion. Bristol Institute for Public Affairs University of Bristol. https://dera.ioe.ac.uk/6853/1/multidimensional.pdf.

Martin, S., 2004. Reconceptualising social exclusion: a critical response to the Neo-liberal welfare reform agenda and the underclass thesis. Australian Journal of Social Issues 39, 79–94.

Popay, J., Escorel, S., Hernandez, M., Johnston, H., Mathieson, J., Rispel, L., 2008. Understanding and Tackling Social Exclusion. Final Report to the WHO Commission on Social Determinants of Health from the Social Exclusion Knowledge Network. Access on line: http://www.who.int/social_determinants/knowledge_networks/final_reports/sekn_final%20report_042008.pdf.

Rajulton, F., Ravanera, Z.R., Beaujot, R., 2007. Measuring social cohesion: an experiment using the Canadian national survey of giving, volunteering and participating. Social Indicators Research 80, 661-492.

Sovacool, B.K., Dworkin, M., 2014. Global Energy Justice: Problems, Principles, and Practices. Cambridge University Press, Cambridge.

Sovacool, B.,K., Sidortsov, B.,V., Jones, B.R., 2014. Energy Security, Equality and Justice. Routledge.

Sovacool, B.K., Bazilian, M., Toman, M., 2016. Paradigms and poverty in global energy policy: research needs for achieving universal energy access. Environmental Research Letters 11, 1–6.

Stirling, A., 2008. Opening up and "Closing Down": power, participation, and pluralism in the social appraisal of technology. Science, Technology & Human Values 33 (2), 262–294.

Taket, A., Crisp, B.R., Nevill, A., Lamaro, G., Graham, M., Barter-Godfrey, S. (Eds.), 2009. Theorising Social Exclusion, 1st Edition. Routledge, p. 238.

Winzer, C., 2012. Conceptualizing energy security. Energy Policy 46, 36–48.

Yergin, D., 1988. Energy security in the 1990s. Foreign Affairs 67 (1), 111. Fall.

Zabarauskaitė, R., Blažienė, I., 2012. Gyventojų pajamų nelygybė ekonominių ciklų kontekste [Income inequality in the context of economic cycles]. Verslas: teorija ir praktika 13, 107–115.

Zhang, L., Yu, J., Sovacool, B., Ren, J., 2016. Measuring energy security performance within China: toward an inter-provincial prospective. Energy 125, 825–836.

What does energy security mean?

Tadas Jakstas

Energy Security Expert at NATO Energy Security Centre of Excellence, NATO Civilian
Expert on Energy Resilience and Crisis Management

Definitions of energy security

Energy security is a complex term with its implications in a wide range of spheres: political, economic, environmental, social, technical, etc. According to L. Chester (2010, p. 891), it is difficult to describe "energy security" as it is rather a concept resembling an abstract idea, than a policy or a term. This concept can differ by institutional, national, personal, etc. perspectives, meaning that it depends on the subject that uses it (Sovacool and Brown, 2010, p. 80).

Already for several decades researchers have tried to conceptualize energy security in a universal and applicable way. Even though the concept of energy security appeared in academic literature already in the 1960s (Lubell, 1961), energy security as a study subject emerged in the background of the 1970s' oil crises. At that time, as Linda Miller jokingly notes, "even casual newspaper readers have become aware that there are links between energy, security and foreign policy" (Miller, 1977, p. 111). One of the most prominent early works on energy security is Willrich's (1976) article on international energy issues and options, where the author presents a thorough analysis of the international energy sector and what implications energy issues could have to countries on a national level, the world economy, and global environment. Willrich points out that energy issues are somehow different for each interested party, may it be oil importing or producing country, and he makes a distinction between "security of supply" and "security of demand" (Willrich, 1976, p. 746). From this point during the end of the 20th century energy security studies were mostly conducted in a political economy framework focused on supply (Deese, 1979; Yergin, 1988). Researchers and energy policy makers were mostly concerned about the diversification of supply, uninterrupted flow of supply, and affordability of energy. As a result, the concept of security of supply became and still is one of the main definitions of energy security.

Nowadays energy security studies has shifted from a classic approach and become an interdisciplinary field. Climate change, globalization, and uncertain future of fossil fuels have added new dimensions, such as sustainability, energy efficiency, mitigation of greenhouse gas emissions, accessibility of energy services (energy poverty), etc. Thus, the concept of energy security became interconnected with other environmental, social, political, and security issues. One of the aims at capturing multidimensional essence of energy security is international research about differing perceptions of energy security (Sovacool et al., 2012). Instead of trying to conceptualize a universal term, it presents 16 distinct dimensions of energy security, such as affordable energy services, equitable access to the energy, energy efficiency, etc. The authors broaden the term and include such dimensions as energy education and ensuring transparency in energy projects. Based on these dimensions, they arrange a questionnaire and conduct a survey on people's perception of the energy security. The authors argue that this perception differs based on education, age, gender, culture, etc. To go further, Sovacool adds the concept of "cultures" within the energy sector. He argues that different perceptions could be explained by a culture to which a person belongs (Sovacool, 2016). In addition, Sovacool (2016) note that at least five different cultures, such as national, economic, political, professional, and epistemic could exist. The author shows the multidimensional essence of energy security, but also adds to the discussion how this matter is differently perceived by different subjects based on their own preferences.

Another effort to conceptualize the term is a study by Cherp and Jewell (2014). They focus on the concept of security per se by emphasizing that when talking about energy security we need to ask "what to protect?" They describe energy security in a broad term as "low vulnerability of vital energy systems." "Vulnerability" rises from exposure to the risks, be it natural or from other social actors, and resilience. The authors trace vulnerabilities within different energy systems, including energy infrastructure, energy services, renewable energy sources, etc. This term helps to define energy security in a universal way, but as authors themselves note, it is still very abstract and is dependent on the actor using it (Cherp and Jewell, 2014, p. 420). Therefore, energy security is an ambiguous concept.

Perspectives of energy security

The concept of energy security differs not only across the academic literature but also among subjects/actors that apply and use it, depending whether it is used by energy importing or exporting countries, international organizations, military environment, etc. The broader meaning of energy security is largely shaped by measures used to preserve or achieve energy security. As it was discussed before, energy security is a complex concept with multilayered dimensions that interconnects different subject areas. In the following part of this chapter, I will analyze different perspectives of energy security in order to capture how various national or international actors understand energy security.

Energy importers

For a country that is a net importer in the energy sector (imports more than exports), energy security first of all means security of energy supply. A states has to meet its citizens' needs, especially by ensuring uninterrupted functioning of critical sectors like transport, heating, and electricity, which affect everyday life (Kisel et al., 2016, p. 3). Mason Willrich describes it as an "assurance of adequate energy supplies to maintain the national economy at a normal level" (Willrich, 1976, p. 747). Thus, importing countries are mostly concerned with policies and actions to ensure stable supply of energy by diversifying energy supplies, developing indigenous supplies (e.g., renewables), improving energy efficiency, and developing new technologies, etc. (Willrich, 1976; Kisel et al., 2016).

For example, the International Energy Agency (IEA), an international organization that has 29 member countries, largely consisting of major energy importers, defines energy security "as the uninterrupted availability of energy sources at an affordable price" (IEA, 2014, p. 13). As the IEA was created on the Organisation for Economic Cooperation and Development (OECD) framework, its definition of energy security corresponds to that of the European Union's (EU). Nevertheless, the agency also conducts a variety of research on environmental issues.

Europe is a major energy importing region. Energy security measures were traditionally in the hands of individual member states and national initiatives still remain. Nevertheless, as an intergovernmental organization, the EU has become increasingly involved in energy security, especially since the 2000s (Biesenbender, 2015). While Europe generally shares American concerns over rising energy imports, the EU does not necessarily seek to maximize energy self-sufficiency and rather stresses supply source diversification (European Commission, 2001). Although there is no clear definition of energy security in EU policy papers, one of the most characteristic features about the European energy security concept is the explicit linkage between energy security of supply, competition policy, and environmental agenda (European Commission, 2001). This is based on the conviction that an effective market is the lowest cost method of addressing energy policy challenges. According to EU Energy Security Strategy 2014, "The key to improved energy security lies first in a more collective approach through a functioning internal market and greater cooperation at regional and European levels, in particular for coordinating network developments and opening up markets, and second, in a more coherent external action" (European Commission, 2014a,b).

The EU Energy Security Strategy discusses both short-term and long-term energy security measures. The short-term measures mainly focus on resilience to energy supply disruptions, and the long-term means take into account the reduction of energy dependency on external energy supplies. An ability to overcome supply disruptions could be achieved by coordinating risk assessments (e.g., creation of reserves), protecting strategic infrastructures (especially with a focus on cyber security), and building an integrated internal market (construction of key interconnectors between member states) as well as increasing cooperation with new suppliers and building new routes for energy transit (e.g., development of the Southern Gas Corridor) (European Commission, 2015).

The United States is the largest energy importer and the largest energy producer and consumer in the world (EIA, 2018). It is a growing consensus that the shale revolution is transforming the United States into a net exporter of oil and gas. In addition, U.S. Energy Information Administration projects that the United States will become a net energy exporter in 2022 (EIA, 2018). In the US context, energy security has traditionally been aimed at so-called energy independence, especially in oil. However, as Sovacool (2014a, pp. 2–3) points out, neither the Food and Energy Security Act of 2007 nor the Energy Independence and Security Act of 2007 offer any definition of energy security. Seemingly, a paper from the Executive Office of the President of the United States (Energy Charter Secretariat, 2015, p. 11) is the only official document in recent years that refers to the term, by stating "energy security is used to mean different things in different contexts, and broadly covers energy supply availability, reliability, affordability, and geopolitical considerations." Thus, the paper generally follows the UN and IEA definitions. It is a growing consensus that the shale revolution is transforming the United States into a net exporter of oil and gas.

Being almost totally dependent on imported energy, Japan, for example, has always been concerned about energy supply security. The country will remain an important player in the global energy market but is nervous about import dependence (Butler, 2018). The oil embargoes of the 1970s opened Japanese eyes to the risks of relying on oil from the Middle East. The nuclear accident at Fukushima in March 2011, followed within weeks by the closure of the entire fleet of Japanese nuclear reactors, led to a rapid growth in imports of liquefied natural gas and a surge in Asian gas prices (Butler, 2018). Tokyo's major energy policy paper, the "Strategic Energy Plan 2018," does not offer a definition of energy security. Nevertheless, the document explicitly discusses Japan's increasingly severe insecurity of energy supply, especially after the Fukushima nuclear accident in 2011, and offers comprehensive measures that include domestic and overseas supply expansion of fossil fuels, accelerated energy efficiency, renewables expansion, restructuring nuclear policy, market liberalization, and emergency responses (Ministry of Economy, Trade and Industry, 2018).

Energy exporters

Energy security for energy exporting countries, given their nature as a supplier, could be defined by security of demand. According to Mason Willrich, security of demand to energy exporter means "guaranteed access to foreign markets" (Willrich, 1976, p. 752). This draws connotations from the concept of security of supply. It is important for an energy exporter to keep its energy revenue stable and for achieving that it needs to cooperate with more than one importer of the energy, which in a sense could as well be called diversification of demand. One more aspect could be added to security of demand, naming supply-demand balance. It is in energy exporters' interests that demand for energy products would remain stable, which is related with the Western world's invested interests and development of nonfossil energy sources.

The Organization of the Petroleum Exporting Countries (OPEC) is an intergovernmental organization consisting of 15 oil exporting states. Even though some of the huge suppliers, for example, Russia, are not members of this organization, in 2017 OPEC members together produced 42.4% of global oil, making it the largest energy exporter (BP, 2018).

OPEC energy policies are outlined in the OPEC Long-term Strategy. The main objectives laid out here are mainly related to ensuring long-term petroleum revenues by stabilization of fair prices of energy, and overall securing world oil demand for producers. As it is noted in the strategy, prevailing attitudes in regard to future oil demand requirements are one of OPEC's energy security issues. Another important objective presented in the strategy is security of supply. But here its meaning is different than discussed above. For OPEC, security of supply means ensuring an efficient, economic, and regular supply of petroleum to consuming nations (OPEC, 2010).

But it must be noted that even though OPEC perceives energy security mainly from an energy exporter perspective, it shares some of the energy security concerns with the rest of the world. As OPEC notes, there are environmental issues such as climate change that affect the global arena and need multilateral cooperation between all of the energy sector actors (OPEC, 2017).

Energy transit countries

Some of the global energy transportation routes, such as the Hormuz and Malakka straits, pass through territorial waters of key choke points. Global oil, LNG supplies are exclusively focused on sea transport. Energy flows within the CIS countries and from Russia to Europe involve land transit countries like Ukraine and Belarus. Energy security for transit countries like Ukraine and Belarus could bear substantial similarity to energy supply security for importing countries. In its "Energy Strategy 2030," the Ukrainian government cites enhancing energy security as one of the goals and objectives of energy strategy (Energy Charter Secretariat, 2015, p. 16). This Strategy defines energy security as "the attainment of a technically reliable, stable, competitive and environmentally sound supply of energy resources for the economy and social sphere of the country" (Energy Charter Secretariat, 2015, p. 16). This is a very similar definition as that of importing countries, and has no elements relating to energy transit in the country. Ukraine's policy measures for energy security are therefore similar to the ones of importing countries.

Geopolitics of energy dependencies

Since the industrial revolution, the geopolitics of energy, including who supplies it and securing access to those supplies, has been a driving factor in global security and prosperity (Pascual, 2008). The energy geopolitics is shaped by both the size and location of own and other natural resources, how available they are, who controls them, their cost, alternative transportation routes, how regional and global markets balance, market mechanisms and regulations, political decisions, and prices in general. Furthermore, due to intertwined national and international energy issues, there are different stakeholders, including state and nonstate, which affect political outcomes. The geopolitical role of a country is influenced by the scale and scope of the dependence it represents for other actors (businesses, countries) (Pascual, 2008). Resources affect national policy making by acting upon domestic actors, which in turn affect the domestic political system through associations, state structure, and ideology and, hence, business-to-business and business-to-government relations, must be included in the analysis (Austvik and Lembo, 2017, p. 663–666). Energy and geopolitics have been closely linked in both old and new formulations. Countries have made and make national strategies and geostrategies to meet their energy needs, reach markets, and secure national positions and interests. The securitization of energy policy has contributed to the development of national security policies.

In recent decades, climate change concerns and growing debates on decarbonization have added to the politicization of the energy sector and created worldwide pressures and policies for improved energy efficiency, more renewable energy, and less dependence on fossil sources. The climate debate has contributed to the complexity of the energy industry, not least because fossil energy represents as much as 87% of world energy usage and is the main source of global CO_2 emissions (Cusick, 2013).

Military and energy security

Energy is a fundamental enabler of military capability. The ability of the armed forces to project and sustain depends on the assured delivery of this energy (US DOD, 2016). For this reason, energy security considerations are a growing component of military planning and operations. Multiple forces are driving this trend, including developments in the world oil market, the growing concerns about global climate change, the challenges of delivering fuel to forward base units, and the vulnerability of military installations to disruptions in electric power. Military often define energy security as operational energy, which is energy required for training, moving, and sustaining military forces and weapons platforms for military operations (US DOD, 2011). Therefore, armed forces consider operational energy to be the energy used in military operations, in direct support of military operations, and in training that supports unit readiness for military operations, to include the energy used at nonenduring locations (contingency bases).

The disruption of energy supply could affect the security of societies of NATO allies and partners, and have an impact on NATO's military operations. While these issues are primarily the responsibility of national governments, NATO allies continue to consult on energy security and further develop NATO's capacity to contribute to energy security, concentrating on areas where it can add value. Even though NATO does not have any common energy security policy, the alliance's energy agenda could be divided into three broad areas (NATO, 2008):

- NATO seeks to enhance its strategic awareness of energy developments with security implications by sharing intelligence on energy development, providing political consultations among allies and partners, and exchanging expertise with outside experts (NATO, 2008).
- NATO aims to develop its competence in supporting the protection of critical energy infrastructure, which is mainly about sharing best practices among experts, organizing training courses, and inserting energy-related scenarios into exercises (NATO, 2008).
- NATO works toward significantly improving the energy efficiency of the military by sharing of national best practices, demonstrating energy-efficient equipment, and developing military energy efficiency standards (NATO, 2008).

Energy security challenges, risks, and vulnerabilities

Energy security is known for its unpredictable and constantly changing nature and subsequent broad range of definitions depending on different perspectives and views. Instead of attempting to grasp its essence in search for a common definition, we offer a critical reflection that problematizes energy security itself.

After a short historical and methodological analysis of different perspectives of energy security, we unpack energy security challenges, risks, and vulnerabilities. These include an analysis of armed conflicts and protection of critical energy infrastructure (CEI), energy crises, natural disasters, energy poverty, and political and regulatory risks. Each of these are discussed to show how energy security works, how it is shaped, and what role it plays within a larger security environment.

Armed conflicts and protection of critical energy infrastructure

Critical infrastructure and resources could be regarded as a question of national security, especially during times of conflict. The well-being and security of our societies have historically been dependent on different resources for their survival, such as iron ore and coal. History has proved that deprived of critical resources, countries have lost wars and societies collapsed. For example, the evolution of warfare and development of musketry, cannons, and battleships consumed ever-increasing volumes of gunpowder as Europe's quarrelsome monarchies battled for supremacy over the continent.

The use of critical resources requires infrastructure that allows them to be produced and distributed. For instance, coal must be mined, transported, and burned in power stations to generate electricity or heat. Onshore oil extraction requires drilling rigs, pipelines, tankers, and refineries before crude oil is transformed. All these factors contribute to the security of supply of critical resources. The evolution of the targeting of CEI, from the realization of its importance in World War I (WWI), to the extensive aerial bombardment during World War II (WWII), to ever increasingly precise targeting as technology, understanding, and the pervasiveness of CEI in society increased (Jakson et al., 2016). During WWI and the emergence of oil geopolitics, the great powers first understood the importance of oil in military operations. During WWII, the Allies prioritized the CEI above others.

During the major conflicts of the Cold War period, attacks against CEI were considered one of the priorities. The targeting of oil infrastructure in the Iran—Iran War turned out to be a prevalent tactic of both the belligerents, aimed at cutting the enemy's source of revenues and also the external support from third parties (Jakson et al., 2016). The significance of CEI targeting in a conventional conflict between major energy producers also sheds light on potential involvement of third parties in these conflicts.

Apart from the Iran—Iraq war, the wars in Korea and Vietnam show the importance of CEI targeting in third-party interventions to counter aggression and pressure sides into negotiations. CEI targeting was later replicated in coalition military operations against CEI in the former Yugoslavia, which were designed to pressure the enemy to cease hostilities and engage in negotiations (Jakson et al., 2016). As energy is essential to modern economies, destroying the CEI and cutting the population off from power supplies can bring any country to a standstill.

During the armed conflicts, attacks against CEI have played an important role. CEI has been vulnerable in any conventional conflict, therefore substantial efforts were made to protect infrastructure. Efforts must also be made to understand the implications of targeting multiple CEIs, and the broader role energy plays in triggering and steering conflicts.

Energy crises

Historically energy crises have been caused by energy shortages, conflicts, and market manipulations. In addition, some actions like tax hikes, nationalization of energy companies, and regulation of the energy sector shift supply and demand of energy away from its economic equilibrium. A crisis could also emerge due to industrial actions like protests and government embargoes. The cause may be overconsumption, aging infrastructure, choke point disruption or bottlenecks at oil refineries, and port facilities that restrict fuel supply. An emergency may emerge during very cold winters due to increased consumption of energy.

Infrastructure failures may cause minor interruptions to energy supplies. A crisis could possibly emerge after infrastructure damage from severe weather. Attacks by terrorists or militia on important infrastructure as discussed previously are pose serious challenges for energy consumers. Political events, for example, regime change, may create shortages by disrupting oil and gas production. Fuel shortage can also be due to the excess and wasteful use of the fuels.

An energy crisis is a significant bottleneck in the supply of energy resources to an economy. For example, the 1970s energy crisis was a period when the major industrial countries of the world faced substantial petroleum shortages. The 1970s oil crisis knocked the wind out of the global economy and helped trigger a stock market crash, soaring inflation, and high unemployment (Macalister, 2011). The two worst crises of this period were the 1973 oil crisis and the 1979 energy crisis.

During the 1973 Arab−Israeli War, Arab members of OPEC imposed an embargo against the United States in retaliation for the U.S. decision to resupply the Israeli military and to gain leverage in the postwar peace negotiations (Office of the Historian, 2018). Arab OPEC members also extended the embargo to other countries that supported Israel, including the Netherlands, Portugal, and South Africa. The embargo both banned petroleum exports to the targeted nations and introduced cuts in oil production (Office of the Historian, 2018). Several years of negotiations between oil-producing nations and oil companies had already destabilized a decades-old pricing system, which exacerbated the embargo's effects.

The 1979 oil crisis occurred due to decreased oil output in the wake of the Iranian Revolution (Verleger, 1979). In 1980, following the outbreak of the Iran−Iraq War, oil production in Iran nearly stopped, and Iraq's oil production was severely cut as well. Economic recessions were triggered in the United States and other countries. Oil prices did not subside to precrisis levels until the mid-1980s.

The 1990 oil price shock occurred in response to the Iraqi invasion of Kuwait on August 2, 1990. Iraq's invasion and the ensuing embargo totally changed conditions in the oil market. The initial concern of consuming-country government and oil company officials was the loss of crude oil supplies (Verleger, 1990). Lasting 9 months, the price spike was less extreme and of shorter duration than the previous oil crises of 1973−74 and 1979−80, but the spike still contributed to the recession of the early 1990s.

The consequences of oil shock in the 2000s were quite different in comparison to the crises in the 1970s and 1990s. Despite the importance of external pressure on the price of oil, such as hurricanes in the Gulf of Mexico in September 2005, turmoil in Nigeria in 2006—08, and ongoing strife in Iraq, global oil production in the 2000s has been remarkably stable. The primary cause for a hike in energy prices was disequilibrium in market fundamentals. Demand growth, especially in developing countries, in particular China[1] and India, together with lower supply growth and market speculation activities created conditions where the price of oil spiked to all-time highest levels, reaching $145.61 per barrel in July 2008 (Yergin, 2006). According to Yergin (2006, p. 72) "The result was the tightest oil market in three decades (except for the first couple of months after Saddam's invasion of Kuwait in 1990)."

Natural hazards

Natural hazards, such as earthquakes, floods, typhoons, hurricanes, tornados, and landslides, ice storms, volcanic eruptions, and even wildfires could cause many different types of damage to energy systems. The recovery time may be prolonged compared to failures not generated by natural events, because multiple systems may be affected simultaneously. Power outages are a common occurrence during and after major earthquakes. Earthquakes cause widespread structural damage to power generation, transmission, and distribution subsystems. Electricity is typically interrupted during or immediately after the shake. The epicentral area is affected, but adjacent areas may also suffer outages. The duration of the outages in any area is a function of the level of damage to the substations and power lines that serve the area. Following the earthquake, power is restored progressively. For example, following a major earthquake, a 15-meter tsunami disabled the power supply and cooling of three Fukushima Daiichi reactors, causing a nuclear accident on March 11, 2011. The nuclear accident at the Fukushima Daiichi nuclear power station affected both short- and long-term energy security in Japan, resulting in crisis-driven, ad hoc energy policy and, because of the decision to shut all nuclear reactors, increased the country's demand for fossil fuels, primarily natural gas (Hayashi and Hughes, 2013). However, the effects of the accident on energy security were not restricted to Japan; for example, the worldwide availability and affordability of liquefied natural gas were affected by Japan's increased demand; while the accident itself resulted in the loss of public acceptability of nuclear power and led countries, such as Germany and Italy, to immediately shut down some of the nuclear reactors or abandon plans to build new ones (Hayashi and Hughes, 2013).

Floods are commonly associated with power outages. The longest blackouts were caused during floods associated with hurricanes (Cornell, 2014). Because water is a very good conductor of electricity, some electrical equipment items can suffer catastrophic failures in the presence of even minute quantities of moisture and dirt. In addition, the flooding also causes damage to oil and gas operations, including ruptured flow lines and storage tanks (Cornell, 2014). For instance, in summer 2015 hurricanes Katrina and Rita brought destruction and misery to hundreds of thousands of people living in New Orleans and elsewhere

---

[1]China's demand in 2004 rose by an extraordinary 16% compared to 2003, driven partly by electricity bottlenecks that led to a surge in oil use for improvised electric generation. U.S. consumption also grew strongly in 2004, as did that of other countries.

along the Gulf Coast of the United States. No less of a shock came from the storms' impact on the U.S. energy economy. More than 100 oil and gas production platforms were destroyed by Katrina and Rita. The country's only deep-water oil import facility, the Louisiana Off-shore Oil Port ('the LOOP'), ceased operation for several days after Katrina hit, and again after Rita. As Yergin (2006, p.74) pointed out "hurricanes Katrina and Rita shut down 27 percent of U.S. oil production as well as 21 percent of U.S. refining capacity. As late as January 2006, U.S. facilities that before the hurricanes had produced 400,000 barrels of oil a day were still out of operation." In addition, Katrina had significant effects on natural gas production, oil and gas transport, refineries, and electricity in one of the most energy intensive areas in the world (Chevalier, 2006). Therefore, the highly interconnected energy system was hit.

Energy poverty

Energy security is often viewed from the national and regional levels. However, energy security could also be seen from an individual household perspective, looking at socioeconomic dimensions of energy scarcity. From this angle, energy insecurity could be defined as a lack of sufficient, affordable energy of the type and quantity necessary for healthy life. According to Pachauri (2011, p. 191), "Security from household perspective requires that supplies of energy be regular, reliable, and of standard quality, that is, uninterrupted and unadulterated."

Energy poverty affects nearly half the world's population, though its effects are disproportionately felt in South Asia and sub-Saharan Africa. In addition, nearly 2.7 billion people do not have access to clean cooking facilities, relying instead on biomass, coal, or kerosene as their primary cooking fuel. Energy poverty could be seen as a threat multiplier. It enables terrorism and violence, such as poverty, environmental degradation, political instability, and social tension. Energy poverty can also breed power theft and "electricity cartels" that illegally divert electricity from the grid to areas that power companies do not service (Robert Strauss Center, 2016).

Political and regulatory risks

Economic risks mainly cover erratic fluctuations in the price of energy products on markets. Price variations can be due to actual or anticipated imbalances between supply and demand, but they can also result from speculative movements and market power abuse. On the one hand, the rise in fuel prices creates monetary and trade imbalances between energy producing and consuming countries. On the other hand, decreasing prices of energy sources tend to diminish capacity-enhancing investment in energy producing countries, creating new bottlenecks to oil and gas supply. In addition, economic risks may include regulatory risks. Government-regulated policies and market interventions could diminish the level of future investments in production as well as distort prices. According to Checchi et al. (2009, p. 3), "Geopolitical risks concern potential government decisions to suspend deliveries because of deliberate policies, war, civil strife and terrorism. Energy industries in most supplier countries are subject to extensive government interference, and do not necessarily function in a competitive market framework. This adds to the fears that energy will increasingly be used as a political weapon." For example, the Russian Federation has used energy (in the form of gas supplies) as a weapon multiple times. Russia's use of energy coercion began in early 1990s when it interrupted oil supplies to the Baltic states in an effort to crush the region's independence movement. According to Collins (2017, p. 2), "Russian energy companies — presumably with Kremlin's blessing — have gone on to make multiple attempts over the past 25 years to use energy supplies to gain leverage over Russia's neighbors and advance Moscow's strategic priorities." In at least 15 instances Russia used price and energy supplies — often amid political tensions — to pressure consumers in Central and Eastern Europe (Collins, 2017). In addition, security of supply is threatened by political instability of exporting regions where civil wars, local conflicts, and terrorism have often been the cause of temporary damage to energy facilities and infrastructures. Although the concept of geopolitical risk generally refers to the oil and gas sector, cross-border trade of renewables-generated electricity could also raise similar concerns.

Renewables and energy security

Renewable energy systems could reduce the risk of energy supply disruptions and the reliance on imported fuels (Olz et al., 2007). Renewable energy sources could provide alternative choices for generating electricity, producing heat, and manufacturing transport fuels. In addition, significant greenhouse gas reductions and various other cobenefits can be obtained. As Miguel Arias Cañete (2015), EU commissioner for Climate Action and Energy, pointed out, "For years we have talked about energy security in terms of secure external suppliers and secure routes for fossil fuels — this is important but only half of the story. The secret to true energy security lies closer to home: clean, locally-produced renewable energy." Despite some positive effects, unbalanced and uncontrolled renewable energy expansion could also create energy insecurity in the form of power loop flows when a country's grid infrastructure is inadequate to handle new renewable production so the power is diverted through neighboring countries' grids, thus contributing to congestion and instability of power grids in neighboring countries (EU, 2013). For example, the extensive development of renewable energy in northern Germany could lead to energy security vulnerabilities for Austria, Poland, and the Czech Republic as constant unwanted supply of electricity flows into the grids of these countries. As a result, loop flows could knock out transmission systems and cause blackouts (Eckert, 2014).

Climate change and energy security

Climate change—caused precipitation, rise in sea levels, and the frequency and severity of extreme events is likely to have a significant effect on energy security. A warmer climate could decrease the efficiency of power production for many existing fossil fuel and nuclear power plants because these plants use water for cooling. In addition, higher air and water temperatures could reduce the efficiency with which these plants convert fuel into electricity (US EPA, 2018). Moreover, energy and water systems are connected. Energy is needed to pump, transport, and treat drinking water and wastewater. Cooling water is needed to run many of today's power plants. Hydroelectricity (electricity produced by running water) is itself an important source of power. Changes in precipitation, increased risk of drought, reduced snowpack, and changes in the timing of snowmelt in spring could affect the security of energy supplies. For example, hydroelectric power plants are sensitive to the volume and timing of stream flows (US CCSP, 2008). In some regions, especially during times of increased rainfall, dam operators may have to allow some water to bypass the electric turbines to prevent downstream flooding (IPCC, 2014).

Energy infrastructure could be disrupted by sea level rise, and storm surge could increase the risk of energy supply disruptions. For example, fuel ports and the generation and transmission lines that bring electricity to major urban coastal centers could be at risk. Changes in the frequency and severity of storms may also damage energy infrastructure, resulting in energy shortages that harm the economy and disrupt peoples' daily lives. Moreover, offshore oil drilling platforms are vulnerable to extreme weather events. As discussed previously, hurricanes Katrina and Rita damaged more than 100 platforms and 558 pipelines in 2005 (US GCRP, 2009).

Flooding and intense storms can damage power lines and electricity distribution equipment. These events may also delay repair and maintenance work. Electricity outages can have serious impacts on other energy systems as well. For example, oil and gas pipeline disruptions following extreme weather events are often caused by power outages rather than physical damage to the infrastructure (US CCSP, 2008).

Railways and marine transportation that move large amounts of oil and coal in the United States are also vulnerable to climate change. More intense rainfall and storms can threaten railways by washing out railway beds. Changes in precipitation could affect marine transportation by reducing the navigability of rivers.

Conclusion

There is a lack of a single definition of energy security due to the fact that energy security is a complex concept with implications in a wide range of spheres: political, economic, environmental, social, and technical. Although there is a vast literature and much discussion about what constitutes "energy security," the view of energy security depends upon the perspective from which one is approaching the concept, whether it is energy importer, energy exporter, business, transit countries, or military. At the most basic level, energy security means having access to the requisite volumes of energy at affordable prices. There is also an implicit assumption that access to the required energy should be impervious to disruptions—that alternative supplies should be readily available at affordable prices and sufficient with respect to both available volume and time required for distribution.

Instead of attempting to grasp the essence in search for a common definition, we offer a critical reflection on global energy security challenges, risks, and vulnerabilities. When energy security is viewed from this perspective as involving armed conflicts and protection of critical energy infrastructure, natural hazards, energy crises, political and regulatory risks, energy poverty, climate change, and renewables, a number of stark energy security challenges become apparent.

References

Austvik, O., Lembo, C., 2017. EU-Russia gas trade and the shortcomings of international law. Journal of World Trade 51 (4), 645–674.

Biesenbender, S., 2015. The EU's energy policy agenda: directions and developments. In: Tosun, J., Biesenbender, J., Shultze, K. (Eds.), Energy Policy Making in the EU: Building the Agnda. Springer, UK.

BP, 2018. BP Statistics Review of World Energy. BP [online], 65, pp, pp. 1–53. Available from: https://www.bp.com/content/dam/bp/business-sites/en/global/corporate/pdfs/energy-economics/statistical-review/bp-stats-review-2018-full-report.pdf [seen 2018-11-34].

Butler, N., 2018. Japan Is Nervous About its Energy Security. Financial Times [online], 9 July. Retrieved from: https://www.ft.com/content/66c37158-801a-11e8-bc55-50daf11b720d [seen 2018-10-01].

Cañete, M., 2015. True energy security lies in renewable energy. Speech at the conference. In: EU Leading on Renewable Energy Policy. News [online]. Retrieved from: https://ec.europa.eu/energy/en/news/true-energy-security-lies-renewable-energy-–-arias-cañete [seen 2018-12-12].

Checchi, A., Behrens, A., Egenhofer, C., 2009. Long-Term Energy Security Risks for Europe: A Sector-specific Approach [online], Janurary. CEPS Working Document, pp. 1–52. Retrieved from: http://aei.pitt.edu/10759/1/1785.pdf [seen 2018-11-25].

Cherp, A., Jewell, J., 2014. The concept of energy security: beyond the four as. Energy Policy 75, 415–421.

Chester, L., 2010. Conceptualising energy security and making explicit its polysemic nature. Energy Policy 38, 887–895.

Chevalier, J., 2006. Security of energy supply for the European Union. European Review of Energy Markets 1 (3), 1–20.

Collins, G., 2017. Russia's Use of the "Energy Weapon" in Europe. Baker Institute for Public Policy Issue Brief. 18 July, pp. 1–8.

Cornell, K., 2014. Environmental exposure: flood risk in the oil & gas industry. Insurance Journal [online], 7 April. Retrieved from: https://www.insurancejournal.com/magazines/mag-features/2014/04/07/325072.htm.

Cusick, D., 2013. Fossil Fuel Use Continues to Rise. Climate wire [online], 25 October. Retrieved from: https://www.scientificamerican.com/article/fossil-fuel-use-continues-to-rise/ [seen 2018-11-23].

Deese, D., 1979. Energy: economics, politics, and security. International Security 4 (3), 140–153.

Doman, L., Kahan, A., 2018. United States Remains the World's Top Producer of Petroleum and Natural Gas Hydrocarbons. U.S. Energy Information Administration [online], 21 May. Retrieved from: https://www.eia.gov/todayinenergy/detail.php?id=36292 [seen 2018-10-12].

Eckert, V., 2014. Austria Burdened by Germany's Shift to Green Power. Reuters [online], 12 February. Retrieved from: https://www.reuters.com/article/us-austria-energy-germany/austria-burdened-by-germanys-shift-to-green-power-idUSBREA1B0ZL20140212 [seen 2018-12-02].

EIA, 2018. The United States Is Now the Largest Global Crude Oil Producer. U.S. Energy Information Agency [online], 12 September. Retrieved from: https://www.eia.gov/todayinenergy/detail.php?id=37053 [seen 2018-10-12].

Energy Charter Secretariat, 2015. International Energy Security: Common Concept for Energy Producing, Consuming and Transit Countries [online]. Retrieved from: https://energycharter.org/fileadmin/DocumentsMedia/Thematic/International_Energy_Security_2015_en.pdf [seen 2018-10-24].

European Commission, 2001. Green Paper: Towards a European Strategy of the Security of Supply. Office for Official Publications of the European Communities [online]. Retrieved from: https://iet.jrc.ec.europa.eu/remea/sites/remea/files/green_paper_energy_supply_en.pdf [seen 2018-10-05].

European Commission, 2014. Q&A on Gas Stress Tests. Retrieved from: http://europa.eu/rapid/press-release_MEMO-14-593_en.htm [seen 2018-12-04].

European Commission, 2014b. Energy Security Strategy. https://eur-lex.europa.eu/legal-content/EN/ALL/?uri=CELEX:52014DC0330&qid=1407855611566 [seen 2018-10-05].

European Commission, 2015. Energy Union Strategy. Retrieved from: https://eur-lex.europa.eu/legal-content/EN/ALL/?uri=CELEX:52014DC0330&qid=1407855611566 [seen 2018-10-05].

European Union, 2013. Loop Flows – Final Advice. THEMA Report 2013-36, pp. 1–58.

Hayashi, M., Hughes, L., 2013. The Fukushima nuclear accident and its effect on global eenrgy security. Energy Policy 55, 102–111.

IEA, 2014. Energy Supply Security: Emergency Response of IEA Countries. https://www.iea.org/publications/freepublications/publication/ENERGYSUPPLYSECURITY2014.pdf [seen 2018-10-05].

IPCC, 2014. Climate Change 2014: Impacts, Adaptation, and Vulnerability. Contribution of Working Group II to the Fifth Assessment Report of the Intergovernmental Panel [online]. Retrieved from: https://www.ipcc.ch/site/assets/uploads/2018/02/WGIIAR5-PartA_FINAL.pdf.

Jakson, H., et al., 2016. Energy in conventional warfare. In: Energy in Conflict Series, pp. 7–42 [seen 2018-12-15].

Kisel, E., Hamburg, A., Härm, M., Leppiman, A., Ots, M., 2016. Concept for energy security matrix. Energy Policy 95, 1–9.

Lubell, H., 1961. Security of Supply and Energy Policy in Western Europe, World Politics, vol. 13(3). Cambridge University Press, pp. 400–422.

Macalister, T., 2011. Background: What Caused the 1970s Oil Price Shock? The Guardian [online], 3 March. Retrieved from: https://www.theguardian.com/environment/2011/mar/03/1970s-oil-price-shock [seen 2018-12-12].

Miller, L., 1977. Energy security and foreign policy: a review essay. International Security 1 (2), 111–123.

Ministry of Economy, 2018. Trade and Industry of Japan.

NATO, 2008. NATO's Role in Energy Security. NATO [online]. Retrieved from: https://www.nato.int/cps/en/natohq/topics_49208.htm# [seen 2018-11-14].

Office of the Historian, 2018. Oil Embargo, 1973–1974. U.S. Department of State [online]. Retrieved from: https://history.state.gov/milestones/1969-1976/oil-embargo [seen 2018-11-24].

Olz, S., Sims, R., Kirchner, N., 2007. Contribution of Renewables to Energy Security. IEA Information Paper, pp. 1–72.

OPEC, 2010. OPEC Long-Term Strategy. https://www.opec.org/opec_web/static_files_project/media/downloads/publications/OPECLTS.pdf [seen 2018-10-14].

OPEC, 2017. Annual Report 2017. https://www.opec.org/opec_web/static_files_project/media/downloads/publications/AR%202017.pdf [seen 2018-10-14].

Pachauri, S., 2011. The energy poverty dimension of energy security. In: Sovacool, B. (Ed.), The Routledge Handbook of Energy Security. Routledge, UK.

Pascual, C., 2008. The Geopolitics of Energy: From Security to Survival. Brooking Report [online], 8 January. Retrieved from: https://www.brookings.edu/research/the-geopolitics-of-energy-from-security-to-survival/ [seen 2018-11-20].

Robert Strauss Center, 2016. Energy Poverty. The University of Texas at Austin [online]. Retrieved from: https://www.strausscenter.org/energy-and-security/energy-poverty.html [seen 2018-11-25].

Sovacool, B., 2014a. Differing cultures of energy security: an international comparison of public perceptions. Renewable and Sustainable Energy Reviews 55, 811–822.

Sovacool, B., Brown, M., 2010. Competing dimensions of energy security: an international perspective. Annual Review of Environment and Resources 35 (10), 77–108.

Sovacool, B., Valentine, S., Bambawale, M., Brown, M., Cardoso, T., Nurbek, S., Suleimenova, G., Li, J., Xu, Y., Jain, A., Alhajji, A.F., Zubiri, A., 2012. Exploring propositions about perceptions of energy security: an international survey. Environmental Science & Policy 16, 44–64.

Sovacool, B., 2014b. Introduction: defining, measuring and exploring energy security. In: Sovacool, B. (Ed.), The Routledge Handbook of Energy Security. Routledge, UK.

Sovacool, B., 2016. Differing cultures of energy security: an international comparison of public perceptions. Renewable and Sustainable Energy Reviews 55, 811–822.

United States Climate Change Science Program, 2008. Effects of Climate Change on Energy Production and Use in the United States. A Report by the U.S. Climate Change Science Program and the Subcommittee on Global Change Research [online]. Retrieved from: https://downloads.globalchange.gov/sap/sap4-5/sap4-5-final-all.pdf [seen 2019-01-05].

United States Environmental Protection Agency, 2018. Climate Impacts on Energy. Climate Change Impacts [online]. Retrieved from: https://19january2017snapshot.epa.gov/climate-impacts/climate-impacts-energy_.html [seen 2019-01-05].

US DOD, 2011. Energy for the Warfighter: Operational Energy Strategy [online]. Retrieved from: http://www.secnav.navy.mil/eie/ASN%20EIE%20Policy/DODOperationalEnergyStrategy.pdf [seen 2018-11-12].

US DOD, 2016. 2016 Operational Energy Strategy [online]. Retrieved from: https://www.acq.osd.mil/eie/Downloads/OE/2016%20DoD%20Operational%20Energy%20Strategy%20WEBc.pdf [seen 2018-11-12].

US GCRP, 2009. Global Climate Change Impacts in the United States. United States Global Change Research Program. Cambridge University Press, US [seen 2019-01-05].

Verleger, P., 1979. The U.S. Petroleum crisis of 1979. Brookings Papers on Economic Activity 2, 464–476.

Verleger, P., 1990. Understanding the 1990 oil crisis. Energy Journal 11 (4), 15–33.

Willrich, M., 1976. International energy issues and options. Annual Review of Energy 1, 743–772.

Yergin, D., 1988. Energy security in the 1990s. Foreign Affairs 67 (1), 110–132.

Yergin, D., 2006. Ensuring Energy Security. Foreign Afairs, pp. 69–82, 1 March.

Further reading

OECD, 2011. Energy security. In: Better Policies for Development: Recommendations for Policy Coherence. OECD Publishing, Paris.

6

Weather risk and energy consumption in Poland

Grzegorz Mentel[1], Joanna Nakonieczny[2]

[1]Department of Quantitative Methods, Rzeszow University of Technology; [2]Department of Finance, Banking and Accountancy, Rzeszow University of Technology

OUTLINE

Recent research on risk diversification is characterized by a strict dependence of risk on the natural environment. At present, the relationship between the humans and the environment is the most important subject of consideration, and any changes taking place in the environment are the subject of an increased interest in it. Indeed, the concept of risk and decision-making in conditions of uncertainty about future events is still important, but in the mainstream, the relationship of risk with the environment in which humans have live and the treating the risk from the point of view of the global perspective becomes more important. Contemporary risk management is therefore characterized by increased social awareness and multilayered issues, while at the same time, the research is interdisciplinary. Despite this awareness, it has not yet been possible to develop a single risk research concept or a universal definition of risk. We can safely say that it is just the opposite. Risk research has to a large extent become an area of formulating contradictory research paradigms and contradictory goals and assessments. This does not mean, however, that in a few years this will not lead

Energy Transformation towards Sustainability
https://doi.org/10.1016/B978-0-12-817688-7.00006-9

to positive solutions in this area, both for the benefit of humans and the environment in which they live.

Essence and characteristics of weather risk

Starting from the assumption that today's economic activity is largely determined by the situation of the natural environment in which humans live, it is worth referring explicitly to the report *The Human Choice and Climate Change* (HCCC) (Rip and Kemp, 1998). This document analyzes global changes from different perspectives, and its main premise stems from the statement that continuous human intervention in nature has contributed to the occurrence of threats. Thanks to the global reach and irreversible nature of changes in the threat, they require the development of specific defense strategies and, most importantly, early warning mechanisms. However, the HCCC is also a kind of starting point or compendium for the development of effective 'weather' risk-management systems. Moreover, and this seems to be the most important thing, the research in this report has been carried out taking into account the achievements of social sciences, which in fact testifies to the interdisciplinary nature of contemporary research, which has already been mentioned. An attempt has therefore been made to fill some of the gaps typical of natural sciences, mainly due to a lack of ability to predict the actions to be formally taken when it is possible to predict the dynamics and effects of climate change. Thanks to such an approach, the so-called synergy effect was achieved.

The HCCC is a four-volume document. Its first two parts, *The Social Framework* (Vol. 1) and *Resources and Technology* (Vol. 2), address general issues such as indicating the interaction between science, ubiquitous politics, and society, or the relationship between climate change and activity, human behavior. The essence of the document is in *The Tools for Policy Analysis* (Vol. 3) and *What Have We Learned?* (Vol. 4). Vol. 3 analyzes traditional risk management instruments and assesses them in terms of their effectiveness in solving climate change problems. Vol. 4, in turn, is not only a synthesis of research but also asks to what extent confronting climate change is a challenge for the social sciences.

It is worth noting that the risk research carried out already in the 21st century is of a different nature than that carried out previously. They concern both its individual, institutional, and social level. Moreover, there is a clear shift toward interdisciplinarity in this research. As far as their context in the relationship between environmental change and climate change is concerned, there is also a political aspect to it. In this case, international organizations such as the Intergovernmental Panel on Climate Change, the World Climate Research Program, the International Council of Scientific Unions, and the World Meteorological Organization are working out agreements and directions of research. However, these organizations focus more on the analysis of data by unifying and standardizing measurements and less on the direct fight against threats. This issue is dealt with by the Weather Risk Management Association (WRMA), an organization of experts in this field established to raise public awareness of weather risk management (Majewska, 2013).

Apart from the aforementioned issues of new directions of research on risk and trends that we can currently observe in this area, it is worth focusing on the problem posed in this monograph, namely focusing exclusively on the weather risk rather than the broadly understood human-environmental relationship. It is therefore important to briefly characterize this type of risk, which is becoming increasingly important over the years. It should be remembered that the weather risk relates to losses both in production volume and in production capacity, which occur as a result of certain atmospheric phenomena, such as precipitation, winds, or temperature, which may affect economic activity, which cannot be underestimated.

The common feature of these phenomena is their variability, which makes them even more similar to classical risks and increases the possibility of quantitative methods interfering in an effective risk management process. Such an approach allows their course to be described using, for example, a normal distribution (Dischel, 1999).

In addition to the integrating element, there are also differences that have the basis in the so-called fat tails, responsible for the description of unusual, extreme atmospheric phenomena. Thus, if these changes are quite regular, they can be distinguished as a separate risk factor, as in the case of very strong winds, hurricanes.

Phenomena such as temperature or precipitation, i.e., within the limits of the three sigma rule, can be attributed a continuous character. They have a direct impact on the volume of production, and thus on the use of production capacity. This in turn strongly determines the financial result.

As far as the typically extreme and violent phenomena in the play of these "fat tails" is concerned, they affect the destruction of fixed assets owned by the company and the destruction of assets owned by the population. Typical examples are floods or hurricanes.

The impact of atmospheric phenomena on the energy sector is not insignificant. For example, during the analysis of temperature it becomes apparent that the temperature results into the demand for energy carriers. Too light winters cause a decrease in demand for energy resources. Fluctuations in demand also make energy prices fluctuate and undergo trends (Hull, 2010).

Throughout this analysis, the problem of the consequences of weather risk, which is the variability of prices, is also important. As a result of a decrease in production volume or a more destructive destruction of assets, prices change (Fig. 6.1). This is a result of a decrease in supply. When analyzing the nature of weather risk, it is worth noting that these factors may entail substitution effects. In the longer term, drought may relatively influence the increase in demand for renewable energy sources, and thus create the level of prices of these carriers.

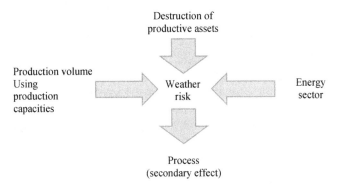

FIGURE 6.1   Weather risk dependencies. Source: *Own elaboration.*

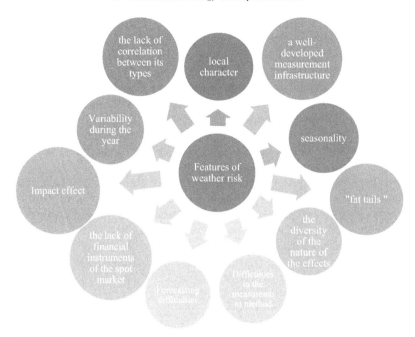

FIGURE 6.2   Characteristic features of the weather risk. Source: *Own elaboration.*

In the case of weather risk characteristics (Fig. 6.2), one may be tempted to separate them as a result of the analysis of the nature of atmospheric phenomena.

For example, features such as local character or seasonality can be mentioned. The first one is clear in its interpretation, since the phenomena in question relate to a specific geographical territory and are therefore characteristic mainly for specific corners of the globe. This applies equally to, for example, the numerous tornadoes or hurricanes on the east coast of the United States, or other phenomena that occur in a given area. The second feature considered is related more to the possible cyclicality in the repetition of their occurrence. When analyzing some of them, over the years, the seasonal character is clearly visible, mainly referring to the seasons. This does not mean, however, that within given seasons weather phenomena are predictable but only that there are certain features specific to a given period. For example, frost occurs in winter, but neither its strength nor the exact period of its occurrence is known (Czekaj, 2016).

Another significant feature is the so-called "fat tails" They are important because they refer to the possibility of extreme phenomena that do not fall within the classic limits. In addition, which is equally important, the probability of their occurrence is very low, but the possible effects, losses may be very high.

In addition, some of the changes that occur are of a continuous nature, such as temperature, while others are of a discontinuous nature (Fig. 6.3). Here we include the values characteristic of the tail areas. Generally speaking, it can be said that the feature of weather risk in this case is the so-called diversity of the nature of the effects of weather phenomena (Banks, 2002).

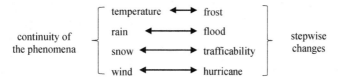

FIGURE 6.3   Continuity and discontinuity of weather phenomena. Source: *Own elaboration.*

A very important issue is the difficulty in measuring in this respect and converting into monetary values. The solution to the problem is to "transfer" atmospheric phenomena to the level of financial markets, and thus, possibly, to secure the financial effects caused by them.

Atmospheric phenomena are difficult to predict, which does not mean that it is impossible (in the further part of the publication it will be possible to prove it to a large extent). If it were possible to accurately forecast the weather over a longer time horizon, then there would be no weather risk and at the same time the whole problem. Despite significant developments in synoptic meteorology, it is basically possible to make a weather forecast with a satisfactory accuracy for a few or several days ahead, which in the case of most companies means that the period of time is far too short.

In addition, the phenomena under consideration require measurement of a very extensive infrastructure. It is worth noting that in the sphere of the risk itself, there is no correlation with other categories of risk associated with running a business.

An important feature of weather risk also seems to be its impact effect. It can be both positive when weather conditions are favorable for the entity and negative when they are unfavorable.

In addition to the local nature of this type of risk, it is important to underline its characteristic feature, i.e., its size, which changes during the year, despite the fact that the aforementioned seasonality features are observed over the years.

As indicated by Szopa (2012), a very important feature of weather risk is the lack of cash market financial instruments. Therefore, in the financial sphere we are dealing only with derivative weather instruments.

The influence of nonextreme weather events on financial results of business entities

Surely each one of us is able to indicate a situation when he or she experienced the "tricks" that are sometimes played by the weather. Apart from relatively rare and extreme events such as earthquakes, hurricanes, tornadoes, or floods, other weather anomalies occur more often. As has been already mentioned, their possible deviations are within the range of normal distribution. Typical examples of such conditions are sudden cooling in summer or on the other hand, warm winters, heavy rainfall, or gusty winds. Indeed, each of these phenomena has an impact on people, however, it is not always the same. We need to agree with one thing, that in the case of the relation to economic activity it is quite significant, because according to some estimates almost 70% of the so-called world economic activity is susceptible to weather influences (Brabazon and Idowu, 2002). Analyzing situations where too mild a winter has a significant impact on the profits of the energy sector, and a cold summer has a similar effect on the producers of cooling drinks or ice cream or observing the possible costs of companies dealing with snow clearance, the fact that the weather has a significant impact on the activity of companies is clearly crystallized. The number of these examples can be multiplied, as the so-called weather impacts are nowadays extremely important.

As stated by Preś (2007), the notion of weather risk is used in literature to describe financial threats to the company due to certain weather events, such as heat, cold, or precipitation of all kinds. More generally, Corbally and Dang (2002) treat the problem according to which this risk includes a set of hazards in the form of catastrophic and noncatastrophic events. Cogen (1998) refers to the financial result in the definition. In his opinion, weather risk is identified with the uncertainty of cash flows and profits caused by weather variability over a given period of time.

Combining the weather risk with the conducted business activity, it seems justified to define it as the entity's financial exposure to weather events. According to Sokołowska (2009), the source of these events is immanent weather variability. Since the impact of changes in weather conditions on the financial results of entities can be expressed in a monetary measure, the weather risk is included in the financial risk catalog (Dziawgo, 2012). A similar tone is also expressed by WRMA, according to which weather risk should be understood as a financial gain or loss caused by the volatility of daily weather conditions.[1] Clemmons (2002) treats this issue as a possible description of the financial exposure of business activities to the weather events already mentioned before. Importantly, this exposure is usually of a noncatastrophic nature and affects the profitability of entities rather than their fixed assets. The representation of L. Clemmons has contributed to a twofold understanding of the problem. Small anomalies from average weather conditions have been identified with weather risk, whereas weather phenomena of rare frequency but significant harmfulness in terms of property have been attributed to the catastrophic risk (Table 6.1).

Analyzing the above, it becomes clear that the risk of nonextreme weather events concerns a larger number of entities. Moreover, the frequency of its implementation is much higher than in the case of the natural disaster risk, although the damage caused in this case is less severe. Therefore, the risk of nonextreme weather events will be the subject of further research, which, in a way, was already indicated much earlier.

Unfavorable weather conditions may have a significant impact on demand or, in extreme cases, may prevent the normal functioning of a particular entity. However, the profile of the business activity is of great importance here, the strength of the weather risk impact is not the same in every case.

The industries directly exposed to the risk of changes in weather conditions are *basically* (Cao et al., 2003; Malinow, 2002; Brix and Jewson, 2005):

---

[1]When estimating the values of individual structural parameters, the method of the least squares was used; because the model is linear in nature, the number of observations is greater than the number of estimated parameters and there is no collinearity between the explanatory variables of the model (Welfe, 2003).

TABLE 6.1 Weather risk and catastrophic risk—basic differences.

Differential criterion	Weather risk	Catastrophic risk
Impact on enterprises	Applicable to certain types of enterprises	Applicable to all enterprises
Nature of influence on the company	Influence on financial results	It causes losses in fixed assets
The size and duration of the impact on the enterprise	It causes small losses or additional profits in the short term (up to a few days), significant in the long term (months, seasons)	It causes great losses usually in a short period of time
Frequency of risk materialization	Small	Big
Type of weather-related phenomena identified that may affect the enterprise	Specific weather parameters, meteorological phenomena to a lesser extent	Specific catastrophic meteorological phenomena (high air temperatures, heavy rainfall, hurricane, hail, snow blizzards, etc.)

Source: Own elaboration.

- Energy and mining — these are companies producing or distributing electricity, heat or, for example, gas; moreover, entities such as hydroelectricity or wind power plants are also included here
- Agriculture — the industry most sensitive to changes in weather conditions; the effects of adverse effects in this case are visible with a certain delay due to the long production cycle
- Construction — mainly companies in the field of road and bridge construction
- Transport — mainly air and sea transport
- Tourism and recreation — in this case, the amount of precipitation becomes important as a factor determining possibly lower revenues
- Food industry — a branch characterized by significant seasonality, the most vulnerable are the producers of ice cream, cooling drinks, beer, etc.
- Cleaning companies are companies engaged in cleaning or, most importantly, clearing snow; in their case, the impact of changing weather conditions mainly relates to winter periods
- Cities and municipalities — these entities are substantial consumers of heat energy, which significantly increases costs in the event of significant temperature drops; moreover, during the same period they are exposed to increased costs related, for example, to snow clearance
- Health care activities
- Provision of insurance services.

The impact of atmospheric conditions on selected types of economic activity is presented in Table 6.2. The list is interesting as the potential impact is presented in a division into weather parameters that do not always have the same impact on a given industry. Some

TABLE 6.2　Impact of weather conditions on selected types of economic activity.

Type of activity	Weather parameters with an impact on the activity			Effects of adverse weather conditions
	Air temperature	Precipitation	Wind	
Agriculture	Strong impact	Strong impact	Strong impact	Lower yields or unfavorable crop prices
Construction	Significant impact	Significant impact	Significant impact	Delays in works requiring specific weather conditions
Entertaining activity	Significant impact	Significant impact	Significant impact	Cancellation of events or lower revenues from ticket sales
Production of beverages	Strong impact	Insignificant	Insignificant	Decrease in sales value (mainly in the summer)
Electricity providers	Strong impact	Strong impact	Strong impact	Increase in energy demand; decrease in production of energy from renewable sources
Gas or heating providers	Strong impact	Insignificant	Insignificant	Increasing the amount of heat needed for heating rooms
Skiing resorts	Insignificant	Significant impact	Insignificant	Lower revenues from ticket sales
Air transport	Significant impact	Significant impact	Significant impact	Delay or cancellation of flights

Source: Own elaboration.

of them sometimes do not even matter. There are also those that are susceptible to the impact of each of them.

What seems to be important here is recalling the results of American research in this area. In 2001, a unit of the US Department of Commerce conducted a study on the analysis of the impact and strength of the impact of noncatastrophic weather risks on the US economy.

The overall assessment shows that about 40% of US gross domestic product (GDP) is dependent on weather conditions. The division into sectors, apart from the general characteristics of the impact, shows that not all of them are equally susceptible to weather conditions. There are cases of a lack of direct reflection. The magnitude of exposure to weather risk depends to a large extent on the nature of the business. For comparison, 2003 data from the National Research Council (National Research Council, 2003) confirm only the results of earlier studies, suggesting that between 25% and 42% of US GDP is susceptible to weather. J. K. Lazo, on the other hand, in a 2011 study (Lazo, 2012) showed that US GDP changes by ±1.7% depending on weather conditions. In value terms, this amounts to 485 billion dollars per year.

Moving the analysis to the European market, Willams and Diplock (2003) discovered similar patterns in the relation of weather conditions to the GDP indicator for some European countries, as was the case in the United States. In addition, D. McWilliams (McWilliams, 2004) confirmed earlier observations already in 2004 by means of analyzing European data from 1970 to 1995; in quarterly periods he indicated a strong and significant impact of weather on the value of gross domestic product. In addition, he concluded that the impact of weather conditions on a country's economy depends to a large extent on the climate in that country. In countries where the climate is relatively warm and dry, the lack of precipitation and the increase in temperature are negatively correlated with the economy. In countries with humid and cooler climates, on the other hand, both a decrease in temperature and above-average precipitation have a negative impact. Generally speaking, D. McWilliams stated that the most important factor is the lack of precipitation, which has a negative impact on the EU economy. The 1 mm/day increase in annual precipitation contributes to an increase of more than €2.5 billion in the GDP of the member states, according to McWilliams.

With regard to Europe, Pollard et al. (2008) also conducted research within the scope being considered. Using the example of Great Britain, they showed that the weather can affect 70% of companies.

It is hard to find similar analyses in Polish conditions. There are no studies on the impact of changes in particular weather factors on sectors of the economy. This does not mean, however, that attempts in this respect are not conducted. An example is the study done by Michalak (2014). However, it refers only to the assessment of the impact of temperature and precipitation on the amount of energy sold by one of the Łódź companies in the years 2006—11 (monthly data). The study is limited to a single-equation linear econometric model in the form of[2]:

$$Y_t = \alpha_0 + \alpha_1 X_1 + \alpha_2 X_2 + \gamma_1 dm2 + \ldots + \gamma_{11} dm12 + \varepsilon_t, \tag{6.1}$$

where:

Y_t — the amount of energy sold, in TJ,
X_1 — temperature, in °C
X_2 — the amount of precipitation, in mm,
$dm2, \ldots, dm12$ — periodical zero-one variables,
$\alpha_0, \alpha_1, \alpha_2, \gamma_1, \ldots, \gamma_{11}$ — respectively: a free word and standing parameters for explanatory variables and for zero-one variables.[1]

Referring directly to the interpretation of assessments of structural parameters obtained as a result of model estimation (1), D. Michalak assesses both the accuracy of their estimates and substantive verification. Thus, referring to the assessment of the significance of the parameters obtained on the basis of the t-Student's statistics, with the assumed level of significance $\alpha = 0.05$, it was obtained that the "temperature" variable has a significant influence on the dependent variable. The influence of $X2$ variable turned out to be insignificant in this case. A substantive analysis is limited to the assessment of the directions of the influence of explanatory variables on the dependent variable. While the negative value of parameter $\alpha1$ is unquestionable—with the increase in temperature the amount of energy sold decreases—there is no logical justification for the negative value of $\alpha2$ parameter. This would mean that the increase in precipitation causes a decrease in energy sales.

---

[2]The figure shows changes in the average annual value of the correlation coefficient between the average monthly temperature and electricity demand in the years 2002—15. The calculations of monthly values are based on hourly observations.

In order to support the observations presented by D. Michalak, analogous research was carried out for companies from the energy sector, increasing their scope (range).[2] Taking into account data on energy sales for seven large Polish cities in the years 2012—16, as well as temperature and precipitation data corresponding to these locations, models analogous to the described formula (6.1) were estimated. The obtained estimates are to a large extent based on the theses formulated in D. Michalak's studies. Studies on the significance of structural parameter assessments in the case of weather variables indicate a significant influence of temperature variable and the lack of such influence in the case of the precipitation factor. A certain explanation for this state is the analysis made already at the initial stage of modeling, consisting in a correlation study between weather variables and the modeled variable, i.e., the amount of energy sold. At this stage a high, negative dependence of temperature and endogenous variable is noticeable, but there is no (significant) dependence for a similar correlation with the amount of precipitation.[2]

In a sense, the substantive assessment of the models received is also becoming analogous, as the negative values of structural parameter assessments for the *X1* variable are confirmed. However, there is no unanimity in the case of observations concerning the *α2* parameter. Here, the question of a negative or positive sign looks exceptionally different. There is no clear trend in this respect.

It should be clearly stated that each of the seven models estimated in this case was characterized by a significant adjustment to the empirical data. This may be evidenced by the values of determination coefficients exceeding 0.91.

As can be seen from the above examples, the impact of weather changes, generally speaking, on the condition of economic entities is sometimes really significant. Agriculture and energy are the most vulnerable in this group. With regard to the latter, an important regularity should be pointed out in this case—the lower the outside temperature (Dischel, 2002), the higher the amount of energy consumption. The relationship between temperature and power demand in Poland is shown in Fig. 6.4.

The influence of atmospheric phenomena on the change in temporary demand for electricity in the case of other weather factors is presented in Table 6.3.

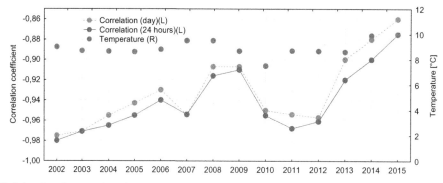

FIGURE 6.4   Correlation values between temperature and power demand, for a time step of 1 month.[2] *Source: Own elaboration.*

TABLE 6.3 Effects of atmospheric phenomena on changes in temporary electricity demand for England and Wales.

Phenomenon	A change in demand (MW)	Change in demand with respect to the total demand
Temperature change − decrease by 1°C at temperatures below 0°C	+400	1%
Change in wind force − increase in wind speed by 10 knots at temperatures below 0°C	+700	2%
Change in cloudiness − from cloudless sky to full cloudiness	+1500	4%
The occurrence of rainfall − change from lack of precipitation to heavy rainfall	+800	2%

Source: Own elaboration.

A similar analysis for agriculture is rather intuitive. The effect of temperature and precipitation on yields, quality, and size is well known and widely described in literature (Lerner, 2004; Skees, 2001; Gardner, 2003). Due to the length of production cycles, weather parameters interact with a certain delay. At first, they contribute to lower agricultural production, then they influence product prices, which is reflected in the manufacturers' income (Majewska, 2013). In the energy sector, on the other hand, the impact of weather changes is almost immediate.

The influence of particular weather factors on agricultural production is not clear, and moreover, it concerns plant cultivation and animal husbandry to a different extent. Considering only the temperature, it becomes noticeable that for each plant there are the highest and the lowest limits (maximum and minimum) for this factor. Beyond these limits, the plant cannot live and develop. In addition, there is a certain "optimum" temperature value for the plant, at which it grows best. Similar values also exist for animal organisms. This also applies to precipitation (rain, snow) or wind strength. Therefore, the course of meteorological phenomena cannot be ignored in agriculture.

Construction is another sector that is not very resistant to weather turbulences. Unfavorable changes in weather conditions can lead to significant losses in the industry. According to Connors (2003)), in the case of larger projects, the contracts are accompanied by penalty clauses for delays, which usually amount to compensating the contracting authority for any loss of profit. Negative temperatures or high levels of precipitation make it difficult or even impossible to carry out construction work. This results in penalties for all delays being paid to the client (Lyon, 2003).

Weather risks in the construction industry should be considered both in terms of their impact on the construction process itself and on other indirect processes, such as the production of building materials. The list of constraints that can result, for example, from the occurrence of chronic heat waves in temperate climates can be really long for these processes. It is enough to mention here, among others:

- transport and supply disruptions (e.g., temporary bans on heavy goods vehicles)
- disruption of technological processes
- restrictions on the use of certain building materials
- high and expensive demand for cooling
- shortening the working time of employees
- restriction of energy supply (related to problems with cooling of power units.

The effects of possible changes in weather conditions should also be considered in the cross-section of the individual stages of the construction process. While evaluating only their overall impact without specifying the individual weather factors, you can indicate the following consequences:

- in the design phase, the need to use materials adapted to changing weather conditions, the choice of location taking into account flooding and buildings subsiding,
- during the construction phase, it is necessary to organize the storage of materials without exposing them to the effects of weather events,
- in the maintenance phase of buildings, faster consumption of materials, increase in maintenance costs, and insurance costs of facilities.

The big impact of changing weather conditions is also visible in transport. The problem concerns both maritime and inland waterway transport, which is heavily dependent on wind, road and rail transport, as well as air transport. For the latter, two weather parameters, wind and precipitation, already have a significant impact. Heavy snowfall or strong winds can lead to significant delays or cancellations of flights (Saunderson, 2004). The cost of delays or, worst of all, cancellations of flights is a considerable burden[3] for airlines. In all categories of transport, climate vulnerability must be considered from the point of view of three basic elements: infrastructure, means of transport, and social comfort. Most climate factors affect all types of transport, but some of them are particularly relevant to a particular type of transport. The functioning of the transport sector (feasibility of providing a transport service) depends on its sensitivity to the impact of the agreed climate categories. The sensitivities of the different means of transport are shown in Table 6.4.

Tourism and recreation are also economic sectors that are dependent on the changing weather situation. However, the location plays an important role here. Mountain resorts are exposed to losses due to a lack of snow in winter or higher temperatures during this period (McIntyre, 2001). Coastal resorts, on the other hand, are susceptible to this type of risk in summer, but the decrease in tourist traffic is mainly caused by low temperatures and a significant amount of rainfall.

In order to illustrate the scale of potential threats, it is worth quoting the results of research conducted by Bigano et al. (2005). The aim of their analysis was to determine changes in the volume of tourist traffic caused only by temperature changes in Italy. Thus, an increase in temperature by 1°C in July was reflected in an increase in the number of overnight stays in coastal regions by 24,783. In August, an increase in temperature by 1°C resulted in an increase in the number of overnight stays by 62,294. In the Alpine regions, a corresponding increase in temperature during the winter months resulted in a decrease in the number of overnight stays by 30,368.

TABLE 6.4 The scope of the impact of UKK on various types of transport.

Agreed climate category	Infrastructure	Means of transport	Social comfort
Vulnerability of road transport elements			
Frost	2	2	2
Snow	3	1	2
Rain	3	1	1
Wind	3	2	1
Heat	2	1	2
fog	1	0	2
Vulnerability of rail transport elements			
Frost	3	1	1
Snow	3	1	1
Rain	3	0	1
Wind	3	0	0
Heat	1	0	1
fog	0	0	2
Vulnerability of inland waterway elements			
Frost	3	2	3
Snow	2	2	0
Rain	2	0	1
Wind	2	2	2
Heat	0	2	1
fog	0	2	2
Vulnerability of air transport elements			
Frost	2	2	1
Snow	3	1	1
Rain	1	1	1
Wind	2	2	2
Heat	1	2	1
fog	0	2	1
0 - neutral	1 — hindering	2 — restricting	3 - preventing

Source: Own elaboration.

Similar studies carried out in other European countries have shown that an increase in summer temperature by 1°C results in a 0.8%—4.5% increase in the number of tourists. Much higher increases are recorded in coastal regions than in landlocked regions. Studies carried out in the United Kingdom have shown that outbound tourism was more sensitive to climate change. In addition, trips for shorter stays are more sensitive to temperature changes.

In addition, research conducted in Italy has shown that in addition to the right temperature during the season (when some tourists are not able to change their plans anyway), temperature expectations are also important variables (extreme preseason weather situations, temperature in the same months of the previous year). Changes in tourist demand caused by higher temperature variability (or the impossibility of anticipating changes) may include, inter alia, a greater interest in last minute purchases, self-organization of the trip, purchase of travel cancellation insurance, etc. (Kachniewska et al., 2012).

When it comes to municipal entities, mainly municipal cleaning companies, the degree of possible financial risks may also be high. It concerns mainly the situation of heavy snowfall during the winter. The determinant of the potential danger is therefore the variable costs of clearing snow from roads (Biello, 2002). At present, local governments (slowly) are trying to limit such threats by concluding appropriate contracts for this type of service with external companies, thus ceding possible threats to the outside in a certain sense. There is also a risk on the part of companies providing such services. In the absence of snowfall, the interest in such entities drops to zero, obviously in the variant when contracts for the performance of the work are concluded.

Referring to studies by Dutton (2002) and Starr-McCluer (2000), and more specifically to the relationship between weather risk and retail trade, it should be stressed that this relationship refers to companies manufacturing or trading in clothing. This applies to both winter clothing and swimwear as well as footwear manufacturers (Nicholls, 2004).

It can therefore be concluded that the weather risk in economic activity is, in a sense, ubiquitous. Noncatastrophic events of this type have a strong impact on the economy, both on a micro and macro scale. They often decide to be or not to be. This impact is manifested by the impact on the following variables of the enterprise (McIntyre, 2001; Majewska, 2013; Michalski and Kupczyk, 2008):

- sales volume and prices of products, services, or goods (impact on operating income),
- purchase volume and factor prices (impact on operating costs),
- production, commercial or service capacity, and thus on sales volume and prices (impact on both revenues and operating costs),
- profit from financial operations in assets based on weather parameters (impact on revenues or financial costs).

Analyzing the above, it can be concluded that in a group of companies directly threatened by changing weather conditions, temperature, precipitation, or wind influence sales revenues, and this ultimately is reflected in the financial results of the companies. Therefore, it is often stated that noncatastrophic weather events, and more precisely the risk resulting from them, is associated with the volume risk (Preś, 2007). It should also be added that the impact of the weather parameters being considered is much greater on volume than on price (Cogen, 1998; Erich, 2003).

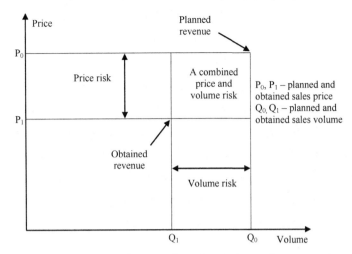

FIGURE 6.5   Relation of sales volume and price volume in relation to the company's revenues. *Source: Own elaboration.*

To sum up, according to Michalski and Kupczyk (2008), any weather risk analysis makes sense when weather parameters are also volume risk factors, as price risk management methods are well developed and commonly used. Of course, the volume of sales or purchases can be influenced by many other factors, such as the current trends in fashion or currency exchange rate in foreign trade, but in most cases the determining factors are weather conditions. Then the risk of nonextreme weather events can be identified with the volume risk (Fig. 6.5), and then it is possible to manage it effectively.

Temperature fluctuation and demand for electrical power in Poland

As it was shown a little earlier, changes in temperature have a significant impact on the demand for electrical power. This applies both to households and businesses. While in the case of gas, heating oil, or coal, the relationship between the increase in temperature and the demand for these energy carriers is inversely proportional, in the context of electricity, the impact of temperature on its demand is already nonlinear in nature. The observed ambiguity results from the fact that the potential demand for electricity may increase both at higher and lower temperatures. In the first case, there is a situation resulting from the need for cooling, and in the second case, when the heating of buildings is necessary.

This can be confirmed by global studies. For example, A. F. Colombo, D. Etkin, and B. W. Karney (1999) assessed the impact of climate variability, and thus weather events, on energy demand for nine Canadian locations. They showed that high temperatures are accompanied by an increase in electricity consumption. C. S. Chen, M. S. Kang, J. C. Hwang, and C. W. Huang. express a similar view. (2001). In a study carried out on the example of Taiwan, they showed that a 5°C rise in temperature is accompanied by a 22% rise in electricity demand. Studies carried out for exemplary Spanish weather stations (Valor et al., 2001) also

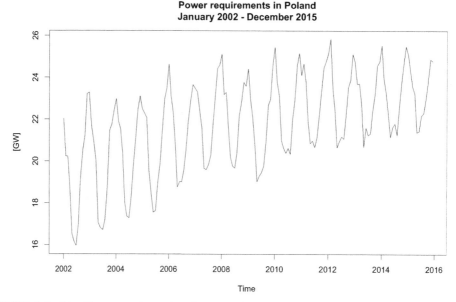

FIGURE 6.6   Monthly energy demand in the national energy system in Poland. *Source: Own elaboration.*

point to the increased electricity consumption in the summer. The reasons for this are mainly caused by the climate and the major role of refrigeration equipment. On the other hand, analyses for Ireland (Liu, Sweeney, 2008) show that household electricity demand is sensitive to climate change. The research indicates a 15%−23% decrease in electricity demand in winter compared to the corresponding summer period. In Polish conditions, the works of T. Popławski, E. Starczynowska, and B. Rusek (2012) and J. Łyp, T. Popławski, and E. Starczynowska (2011) deserve attention in this respect. These works comprehensively assess the impact of atmospheric factors on the situation on the energy market.

Going strictly to the description of the above, it should be emphasized that the demand for electricity shows daily, weekly, and annual seasonality. Some tendencies for Poland, in this respect, can be observed, for example, after the analysis of Fig. 6.6.

Referring to a significant extent to the results obtained by the team composed of J. Jurasz, J. Mikulik, and A. Piasecki (2016) coefficients of linear correlation were determined for the relation between temperature indication and electricity demand for different time periods. However, the analysis was limited to day, month, and year time periods and more detailed values were omitted. The values of temperature readings used in the study are averaged and come from the data of the Institute of Meteorology and Water Management (IMGW), and more precisely they are a resultant of all weather stations where these type of measurements are carried out with a time step of 1 day. On the other hand, the energy data come from the information available to Polskie Sieci Energetyczne/Polish Power Systems/. The results of the conducted research are presented in Table 6.5.

Correlative dependencies at a relatively high level, indicating the occurrence of such dependencies, were recorded for monthly periods. Then, regardless of the fact whether the analysis was carried out for 24-h data or only for the period of the day, the values of correlation

TABLE 6.5 The relationship between temperature and power in the period 2002—15.

Time step	Correlation coefficient	
	24 h	Day
1 day	−0.591	−0.527
1 month	−0.752	−0.736
1 year	0.243	0.312

Source: Own elaboration.

coefficients exceed 0.7, which indicates a strong correlation. Similarly, there are studies for daily periods, but the correlation levels are slightly lower and reach values in the range of 0.5—0.6. Negative indications in each given case should be emphasized here. This means that the drop in temperature is accompanied by a significant increase in power demand. The lack of such a confirmation occurs only in the case of annual periods.

In the study on the impact of weather conditions on electricity demand, it is worth distinguishing between periods when this energy can potentially be used for heating purposes as well as the time when its consumption for cooling purposes increases. The threshold value adopted for the calculation was set at 15°C. For such a division, the regression equations describing the temperature ratio of power demand for two temperature ranges were determined. The ranges and values of direction coefficients of the so estimated linear equations for the following years are shown in Fig. 6.7.

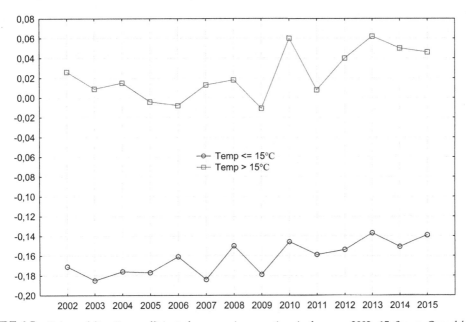

FIGURE 6.7 Values of direction coefficients for regression equations in the years 2002—15. *Source: Own elaboration.*

A much higher value of the structural parameter standing at variable temperatures in the temperature range below the level of 15°C is clearly visible. In this case, changes of 1°C result in a much higher increase in power demand than in the temperature range above 15°C. For example, for the last period, a temperature drop of 1°C in the first temperature range generates an increase in electricity consumption of around 140 MW. In the case of a similar analysis for the second range, any changes in power demand are not as strong. There is, however, a certain correlation in the fact that after 2009 the dynamics of changes in energy demand slightly weaken for the variant temp ≤15°C. Each subsequent degree above the base temperature triggers relatively smaller changes in the energy market after this period than was the case until 2009. If we consider the variant of temperature >15°C, the opposite trend is noticeable. Minor changes in the temperature range result in a much stronger response to energy market demand. This is an extremely unfavorable phenomenon. This type of case could be observed very clearly in August 2015, when high temperatures caused partial failure of the power system (blackout risk). This was partly due to the increase in water temperature in the cooling circuits of thermal power plants, which led to a decrease in their efficiency. If the rising temperature in the following years leads to an increase in electricity consumption, it will be necessary for the power system to use less heat-sensitive energy sources. Therefore, looking at the average temperature in Poland for annual periods between the years 1951 and 2014 (Fig. 6.8), one may have some concerns about the efficiency of the energy system in the so-called warm periods. There is a slight growing trend, which in a sense may indicate the

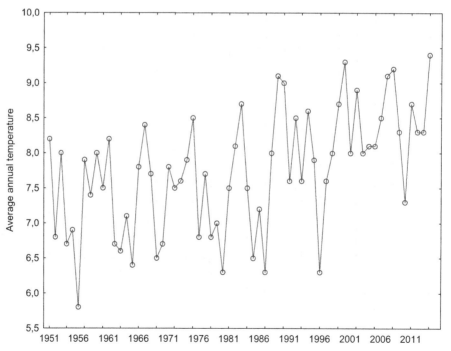

FIGURE 6.8 Evolution of the average annual air temperature in Poland in the years 1951–2014. *Source: Own elaboration.*

phenomenon of climate warming. Therefore, it becomes justified to rely on alternative sources of energy, such as photovoltaics. Firstly, such sources are less sensitive to high temperatures, and secondly, high temperatures are accompanied by high levels of sunshine, which then favors the generation of energy that could cover the demand generated, for example, by air conditioners.

References

Dziawgo, D., 2012. Credit Rating Na Miedzynarodowym Rynku Finansowym. PWE, Warszawa.

Banks, E., 2002. Weather Fundamentals. W E. Banks. Weather Risk Management, Palgrave.

Biello, D., November 2002. New Entrants Take Chill off Weather Market. Environmental Finance, pp. 14–15 strony.

Bigano, A., Goria, A., Hamilton, J., Tol, R.S., 2005. The effect of climate change and extreme weather events on tourism. In: Lanza, A., Markandya, A., Pigliaru, F. (Eds.), The Economics of Tourism and Sustainable Development, pp. 173–197 (Northampton).

Brabazon, T., Idowu, S.O., 2002. Weather derivatives. Accountancy Ireland 7–9.

Brix, A., Jewson, S., 2005. Weather Derivative Valuation. Cambridge Univeristy Press, Cambridge.

Cao, M., Li, A., Wei, J., 2003. Weather Derivatives: A New Class of Financial Instruments. Papers of University of Toronto, Toronto.

Chen, C.S., Kang, M.S., Hwang, J.C., Huang, C.W., 2001. Temperature effect to distribution system load profiles and feeder losses. Power Systems, IEEE Transactions on 16 (4), 916–921.

Clemmons, L., 2002. Introduction to Weather Risk Management. W E. Banks, Weather Risk Management: Markets, Products and Applications. Palgrave, New York.

Cogen, J., 1998. What is Weather Risk. PMA Online Magazine. No. 5.

Colombo, A.F., Etkin, D., Karney, B.W., 1999. Climate variability and the frequency of extreme temperature events for nine sites across Canada: implications for power usage. Journal of Climate 12 (8), 2490–2502.

Connors, R., 2003. Weather Derivatives Allow Construction to Hedge Weather Risk. Cost Engineering 45 (3), 21–24.

Corbally, M., Dang, P., 2002. Underlying Markets and Indexes. W E. Banks, Weather Risk Management: Markets, Products and Applications. Palgrave, New York, pp. 87–104.

Czekaj, Z., 2016. Derywaty pogodowe jako instrumenty zarzadzania ryzykiem pogodowym. Ruch Prawniczy, Ekonomiczny i Spocjologiczny 217–228.

Dischel, B., 1999. Weather Risk Management at the Frozen Falls Fuel Company. Chicago Mercantile Exchange, pp. 90–94.

Dischel, B., 2002. Introduction to the weather market: dawn to mid-morning. In: Dischel, W.B. (Ed.), Climate Risk at the Weather Market (Str. 4). Risk Books.

Dutton, J., 2002. The weather in weather risk. In: Dischel, W.B. (Ed.), Climate Risk and the Weather Market. Risk Books.

Erich, C., 2003. Weather Risk Management. Journal of Financial Regulation and Complience 12 (No.2), 164–168.

Gardner, L., 2003. New Options for Managing Agricultural Weather Risk. CPCU Journal. CPCU Society.

Hull, J., 2010. Riska Management and Financial Institutions. Pearson, Boston.

Kachniewska, M., Niezgoda, A., Nawrocka, E., Pawlicz, A., 2012. Rynek Turystyczny. Ekonomiczne Zagadnienia Turystyki. Wolters Kluwer Polska, Warszawa.

Lazo, J., 2012. Economic of weather. APEC Research Center of Typhoom and Society 2 (4), 27–29.

Lerner, J., 2004. Answering Agriculture's Needs. Environmental Finance.

Liu, X., Sweeney, J., January 2008. Impacts of climate change on energy demand in greater dublin region, Ireland. In: The 20th Conference on Climate Variability and Change. New Orleans, USA.

Lyon, P., 2003. Weathering the storm. Energy Risk 8 (2), 21–23.

Majewska, A., 2013. Instrumenty Pochodne Jako Narzedzia Wspomagajace Zarzadzanie Ryzykiem W Przedsiebiorstwie (Szczecin: Volumina.pl).

Malinow, M., 2002. End Users. W E. Banks, Weather Risk Management: Markets, Products and Applications. Palgrave, New York.

McIntyre, R., 2001. Weather Risk Management. Conference Materials. WRMA.

McWilliams, D., 2004. Does the Weather Affect the European Economy? Conference Materials. WRMA.

Michalak, D., 2014. Konstrukcja instrumentu zabezpieczajacego przed niekorzystnym wpływem niekatastroficznego ryzyka pogodowego. Ekonomia XXI wieku 3 (3), 94–112.

Michalski, G., Kupczyk, J., 2008. Wpływ ryzyka pogodowego na finansowa efektywność przedsiebiorstwa. In: Dudycz, T. (Ed.), Wartość Jako Kryterium Efektywności. Indygo Zahir Media, Wrocław, pp. 101–110.

National Research Council, 2003. Fair weather: effective partnerships in weather and climate services. The National Academies Press, Washington, DC.

Nicholls, M., October 2004. Confounding the Forecasts. Environmental Finance str. 5.

Pollard, J., Oldfield, J., Randalls, S., Thornes, J., 2008. Firm finances, weather derivatives and geography. Geoforum (39), 616–624.

Popławski, T., Starczynowska, E., Rusek, B., 2012. Wykorzystanie metody pojemności integralnych informacji Hellwiga do określenia wpływu czynników pogodowych na obciażenie w KSE. Przeglad Elektrotechniczny 88, 138–141.

Preś, J., 2007. Zarzadzanie Ryzykiem Pogodowym. CeDeWu, Warszawa.

Rip, A., Kemp, R., 1998. Technological Change. In: Rayner, S., Malone, E.L. (Eds.), Human Choice and Climate Change, 2. Battelle Press, Columbus, OH, USA, pp. 327–399.

Saunderson, E., 2004, March. Looking beyond the Energy Sector. Environmental Finance.

Skees, J., 2001. The potential role of weather markets for U.S. Agriculture. The Climate Report 2 (4).

Sokołowska, E., 2009. Pochodne instrumenty pogodowe w zarzadzaniu ryzykiem. In: Berneś, B., Berneś, B. (Eds.), Zarzadzanie Finansami Firm - Teoria I Praktyka. Wrocław: Wydawnictwo Uniwersytetu Ekonomiczego We Wrocławiu (Red.).

Starr-McCluer, M., 2000. The effects of weather on retail sales. Finance and Economics Discussion Series (8), 15.

Szopa, A., 2012. Właściwości Ryzyka Pogodowego. Annales Univeristattis Mariae Curie-Skłodowska, Sectio H, Lublin.

Valor, E., Meneu, V., Caselles, V., 2001. Daily air temperature and electricity load in Spain. Journal of Applied Meteorology 40 (8), 1413–1421.

Welfe, A., 2003. Ekonometria. PWE, Warszawa.

Willams, S., Diplock, T., 2003. Why do retailers continue to blame the weather for poor retail sales? Report PMC.

Łyp, J., Popławski, T., Starczynowska, E., 2011. Kompleksowa analiza wpływu czynników meteorologicznych na zmienność obciażeń Krajowego Systemu Elektroenergetycznego. Przeglad Elektrotechniczny 87, 97100.

Further reading

Brockett, J., Wang, M., Yang, C., 2005. Weather derivatives and weather risk management. Risk Management and Insurance Review 8 (1).

NAIC, 2003. Weather Financial Instruments (Temperature): Insurance or Capital Markets Products? Property and Casualty Insurance Committee draft White Paper, 9 Sept. 2003.

International conflicts related to energy: a case study of Lithuania's opposition to the construction of Russian nuclear power plants near its borders

Justinas Juozaitis

Research Centre, General Jonas Žemaitis Military Academy of the Republic of Lithuania, Vilnius, Lithuania

Abbreviations

BOO Build-Own-Operate
CNS Convention on Nuclear Safety
CEN Continental European Network
EU European Union
IAEA International Atomic Energy Agency
IPS/UPS Integrated Power System/Unified Power System
NATO North Atlantic Treaty Organization
NPP Nuclear Power Plant

OSCE Organization for Security and Co-Operation in Europe
PACE Parliamentary Assembly of the Council of Europe
SEED Site and external events design
TPP Thermal Power Plant
UN United Nations
WENRA Western Europe Nuclear Regulators Association

Introduction

With a different degree of success, Russia is implementing two nuclear power plant (NPP) projects close to Lithuanian borders—one near the town of Ostrovets, Belarus, and the other in Kaliningrad, close to the river Neman. Russian state-owned company Rosatom plans to finish the former in 2020, while it had frozen the construction of the latter in 2013 but hopes to resume it. In 2016, Lithuania declared its intention to prevent the construction of Ostrovets NPP and 1 year later it adopted a law that prohibits purchasing electricity from Belarus as well as specified other measures for achieving this goal (Government of the Republic of Lithuania, 2017; Parliament of the Republic of Lithuania, 2016a, 2017). Despite explicitly revealing its principal political objection against Ostrovets NPP only in 2016, Lithuania opposed both Russian NPPs indirectly by questioning their nuclear safety, highlighting environmental risks, accusing of noncompliance with international law, and raising these issues in international organizations since finding out about their locations in 2008.

Lithuanian opposition to the development of nuclear power close to its borders seems like an ordinary case in a broader context of international disputes over strategic energy infrastructure, and the second Nord Stream project is a case in point. Just as Lithuania opposes the construction of Ostrovets and Baltic NPPs, certain European countries together with the United States object to the building of a natural gas pipeline that will further strengthen the interconnection between Russian and German natural gas systems via the Baltic Sea (Riley, 2018a; 2018b). Moreover, just as Lithuania protests against Russian NPPs, Russia tries to hinder the development of Lithuanian strategic energy infrastructure that would reduce its dependence on Russian energy supply (Cesnakas, 2018), and similar comparisons are plentiful.

Handl helps to reinforce this point further by showing that attempts to disturb the construction or to stop the operation of the already-built energy infrastructure beyond one's territory are also frequent in the nuclear energy sector, especially in instances when NPPs are situated close to national borders (2015, p. 211, 212). For example, following its post-Fukushima phase-out policy, Germany pressured Belgium to close its aging Tihange and

Doel NPPs. On the basis of such policy, close geographic proximity and recent safety-related shutdowns, Germany also pressured France to put its Feseinheim NPP out of operation (Federal Ministry for the Environment, Nature Conservation and Nuclear Safety, 2015, 2018). Another long-time opponent of nuclear energy, Austria, even threatened to prevent the Czech Republic from joining the European Union (EU), if Prague would not install additional safety features in its Temelin NPP (Bock and Drabova, 2005, p. 13,14). While looking beyond Western Europe, one finds Azerbaijan and Turkey sharing a skeptical position toward nuclear safety of Armenian Metsamor NPP (Trend, 2018). Ironically, Lithuania belongs to the same category of examples because the EU forced it to choose between risking its future membership in the organization and committing itself to decommission Ignalina NPP (Elsuwege, 2008).

International disagreements over strategic energy infrastructure, however, are mostly based on either symmetrical relationship, when countries or groups of countries involved in a dispute are roughly equal regarding their capability to impact the processes in international politics or asymmetrical relationship when a stronger country tries to coerce its weaker neighbor. Whereas Lithuanian opposition to Russian NPPs illustrates a different kind of asymmetry—a fundamentally weaker country pressures a great power to stop the development of nuclear energy in its territory or within its sphere of influence.[1] Given these specific circumstances, to what extent can a small state disturb the great power in developing strategic energy infrastructure beyond its borders?

As a general rule, contemporary research on small states indicates that they can overcome the problem of power asymmetry if they advance their national interests through international organizations and appeal to international law (Archer et al., 2014). As far as this argument goes, highly institutionalized environment mitigates the effects of power asymmetry and allows small states to have a more ambitious foreign policy that can even challenge the conflicting preferences of more powerful countries (Galbreath and Lamoreaux, 2007; Panke, 2010; Thorhallsson, 2018). However, there are authors who dispute this claim by showing that great powers often choose to advance their national interests in an identical way (Lamoreaux, 2014), thus implicitly suggesting that international organizations and agreements are not necessarily an exclusive instruments of small states, rather they are arenas for states to act out power relationships between them (Mearsheimer, 1994). Hence, the effects of international law and organizations on small states' capabilities to advance their interests might be exaggerated as power asymmetry can manifest itself not only within bilateral relations but also to be present in a highly regulated environment.

---

[1]That is not to say that there are no cases when smaller countries object to the construction or functioning of strategic energy infrastructure in a territory of a stronger country. Luxembourg, for example, openly advocates for the closure of French Cattenom NPP, while Austria sued the European Commission to EU's General Court over its decision to approve of UK's aid to Hinkley point C NPP. There are fundamental differences, however, concerning institutional context, the intensity of opposition, and means of opposition. Luxembourg, for instance, offered financial compensation to France if Paris would agree to shut down its Cattenom NPP. Austria, on the other hand, limited its opposition within legal limits of the European Union. Hence, disagreements took place in a highly institutionalized context. Lithuania, on the contrary, objects to the processes that are taking place beyond Euro-Atlantic space, is openly hostile to Russia's NPPs, and objects to them not only by using legal instruments but also by exercising national power as Vilnius is banning electricity imports, refusing to construct new power lines, and to share its active-power reserve capacities.

Building upon the disagreement presented above, the chapter assumes that international law and membership in international organizations can indeed help the small states to overcome the problem of power asymmetry and tests it by studying Lithuania's attempts to prevent the construction of Russian NPPs. It aims to determine wherever relevant international agreements and memberships in international organizations helped Lithuania to oppose the construction of Ostrovets and Baltic NPPs. Since the research on international energy disputes mostly deals with disagreements involving relatively equal countries or stronger states that try to coerce weaker ones, analysis of Lithuania's attempts to oppose the construction of Russian NPPs contributes a new angle on studying the energy disputes.

As far as the methodology is concerned, the chapter is a classic case study. By examining Lithuania's official documents, statements, and interviewing officials, the analysis first outlines Lithuania's objectives in using relevant multilateral agreements, compliance review mechanisms, and memberships in international organizations to oppose Russian NPPs. Then it analyses the documents issued by the entities mentioned above and determines the extent to which Lithuania has managed to achieve its objectives. Having established that, the study concludes by discussing wherever the observed achievements had any impact on the development of Russian nuclear projects. In order to control the overlap between Lithuanian opposition on the national (direct actions vis-à-vis Belarus and Russia) and international levels (working through international organizations and international agreements), the study also examines Lithuanian national efforts to stop the construction of Russian NPPs in parallel.

The chapter is structured into three sections. The first section introduces the general features of Russian NPPs and highlights their interdependencies with Lithuanian electricity network. The second one explains why Lithuania opposes them. The last section discusses the effectiveness of Lithuania's efforts to stop their construction by utilizing international legal mechanisms, employing memberships in international organizations, and resorting to national actions.

Introducing Russian NPPs

Ostrovets NPP

An NPP project in Belarus consists of two Russian VVER-1200 reactors, having a total electricity generation capacity of 2400 MW. Belarusian special commission chose the construction site near the town of Ostrovets in December 2008 after considering two alternative sites (Krasnopolyansk and Kukshinovsk) in the Mogilev region, close to the Russian border, and Belarus began developing the supporting infrastructure the next year (Pedraza, 2015, p. 625). At the end of 2011, Belarus and Russia finalized their talks regarding the financing model and a Russian state-owned company, Atomstroyexport, a subsidiary of Rosatom, was officially chosen as the general contractor for the construction of Ostrovets NPP. In November 2013, the contractors started pouring concrete into the foundations of the first unit, while the construction of the foundations of the second unit began shortly after, in April 2014. If everything goes according to plan, the first reactor should begin its commercial operation by the end of 2019, while the second unit should do the same in late 2020. Two additional reactors might be built after finishing the first ones if the conditions prove to be favorable (Ministry of the Emergency Situations of the Republic of Belarus).

It is important to note that Ostrovets NPP is not the first, but the third, attempt to develop nuclear power in Belarusian territory. The first Soviet endeavor to construct two NPPs in the 1980s was abandoned due to the Chernobyl disaster, while the second Belarusian effort in the 1990s did not progress beyond conceptual planning and feasibility studies mainly because of lack of funds. The current NPP project, on the other hand, was not scrapped due to the Fukushima disaster or encounter any substantial financial difficulties, and most likely it will be finished by 2020 (Novikau, 2016, 2017).

The main reason behind the success of Ostrovets NPP is the substantial Russian involvement in the project. In order to illustrate this point, most researchers highlight that Russia is funding 90% of the total project estimate (11 billion U.S. dollars) by loaning the money to Belarus; and Russia is providing a state-owned general contractor as well as technology (Vlček, Jirušek, 2015). Russian involvement, however, does not end here. Moscow is facilitating this project by supporting it politically, organizing training for Belarusian civilian and military personnel, providing air defense systems (TOR) for Belarusian soldiers guarding the airspace around Ostrovets NPP and committing itself to exporting nuclear fuel for the entire life cycle of the plant (Kamynskaya, 2011; Gorin et al., 2015; Ioffe, 2018). Even though the Ostrovets project does not rely on the build-own-operate (BOO) model, when Russian state-owned companies would claim ownership of the Ostrovets NPP, Belarusian authorities would lack resources to construct an NPP (as the previous attempt during the 1990s clearly shows) and Russian thorough support, therefore, plays a vital role in the successful development of the project. Hence, Ostrovets NPP is as much a Russian project as it is Belarusian.

Fallacious strategic justification helps reinforce this point further. Belarusian official documents indicate that the strategic logic behind Ostrovets NPP is twofold. The first purpose of nuclear power is to diversify the Belarusian energy mix, which is dominated by Russian oil and natural gas, while the second interrelated aim is to reduce energy dependence on Russia (President of the Republic of Belarus, 2007a; 2007b; Council of Ministers of the Republic of Belarus, 2015). Considering the extent of Russian involvement in the construction of Ostrovets NPP and keeping in mind that introduction of nuclear power will not alter Russian dominance in Belarusian oil and natural gas sectors, Ostrovets NPP will not decrease Belarusian energy dependence on Russia—on the contrary, it will increase it (Cesnakas and Juozaitis, 2017).

From the economic perspective, the project also has two purposes: to change the natural gas—based Belarusian electricity generation with more competitive electricity production that relies on nuclear power and to export surplus electricity westward, mainly through Lithuania and to Lithuania itself (Dyner, 2018; Hussein, Kardas, and Klysinski, 2018). In principle, plans to export electricity aim to exploit favorable circumstances that emerged after Lithuanian Ignalina NPP was shut down in 2010. Belarusian and Lithuanian electricity systems are interconnected with five transmission lines (Litgrid, 2014, p. 34), having a total of 1300 MW transfer capacity (Masiulis, 2016, p. 8), and they were mostly used to export surplus Lithuanian electricity to Belarus when Ignalina NPP was working. After Lithuanian NPP ceased to be operational, an opportunity emerged to use the existing power lines for a different purpose—to export electricity from Belarus to Lithuania. From Lithuania, Belarusian electricity can reach the other two Baltic states, Poland and Sweden. Ignalina NPP was also backed by a hydro pump storage power plant near Kruonis, and Belarus assumed that it would also back up Ostrovets NPP.

In sum, the Lithuanian electricity system plays an important strategic role in Belarusian plans. Strong interconnectivity between Belarusian and Lithuanian electricity systems allows Belarus to export large quantities of surplus electricity westward, while the pump storage power plant in Kruonis could provide the plant with reserve capacity. However, underlying strategic importance also enables Lithuania to use it as a means for opposing the construction of Ostrovets NPP.

Baltic NPP

From the technical standpoint, Baltic NPP is similar to Ostrovets NPP as both NPPs consist of two VVER-1200 reactors, each having a generation capacity of 1200 MW (World Nuclear Association, 2018). The idea to construct a new NPP in Kaliningrad first appeared in a program of regional development in 2007 (Banktrack, 2018). Rosatom showed interest in such a proposal and on April 16, 2008, and as a result it signed the Agreement on Cooperation with the Government of the Kaliningrad Region. The Russian government officially approved the construction of Baltic NPP on September 25, 2009, and initial work began on February 25, 2010 (Nuclear Threat Initiative, 2010, p. 4). The concrete was poured into the foundations of the first unit in April 2012, and they were finished by December 2012. Despite Rosatom's plans to launch the first unit in 2016 and to begin operating the second in 2018, the construction of Baltic NPP was entirely frozen in June 2013 and has not resumed since then (World Nuclear Association, 2018).

Russian authorities argue that suspension is temporary and are trying to support this narrative by advertising Baltic NPP in international conferences (Rosatom, 2016a; 2016b) and drafting further studies (International Atomic Energy Agency, 2015a,b). However, current developments indicate that Rosatom will not de facto resume the construction of Baltic NPP as the major obstacle for the successful implementation of the project—insufficient electricity demand—remains unsolved. For any NPP to function, a stable and adequate level of electricity consumption must be ensured, and Russia fails to do that for its Baltic NPP (Stsiapanau, 2016). Since 2011, electricity generation in Kaliningrad already exceeds its needs, and the region exports its surplus electricity to Lithuania. Moreover, studies indicate that even if Russian authorities would shut down all local power plants, the capacity of just one unit of Baltic NPP (1200 MW) would substantially exceed peak electricity demand (700 MW) in Kaliningrad (Usanov and Kharin, 2014, p. 8; Varanavičius, 2018).

In order to export electricity from Kaliningrad, sandwiched between Lithuania and Poland, Russia logically needs to secure agreements with Vilnius and Warsaw for the construction of additional electricity interconnections. Kaliningrad is not interconnected with the Polish electricity system, while it has three cross-border electricity links with Lithuania (Kadisa et al., 2016, p. 55). However, only 600 MW is available for export from Kaliningrad to Lithuania (Masiulis, 2016, p. 6), which is not enough for securing electricity demand for the Baltic NPP. Hence, the construction can only be resumed if Russia finds markets for exporting electricity surplus.

In sum, the specific geographic conditions, the lack of interconnections, and insufficient domestic demand complicate Russian possibilities to construct Baltic NPP and provides Lithuania with opportunities to oppose the project.

Why is Lithuania against Russian NPPs?

Before explaining the reasons behind Lithuanian attempts to stop the construction of Ostrovets and Baltic NPPs, the section starts by introducing the peculiarities of the Lithuanian position first. From 2008 through 2015, Lithuania expressed a couple of different and sometimes conflicting positions regarding Russian NPPs.[2] Despite the different political declarations, a systematic study of the statements made by top Lithuanian decision-makers exposed two important patterns. First, they shied away from expressing straightforward protests against the construction of Ostrovets and Baltic NPPs and instead requested to build them transparently and constructively, ensure the highest nuclear safety standards, minimize environmental risks, and comply with international law. Second, key Lithuanian officials refrained from relating the aforementioned nuclear projects with Russian geopolitical objectives and limited themselves on voicing concerns over their proximity, nuclear safety, environmental risks, noncompliance with international law, and other suspected flaws. However, behind the prudent official positions lay the unspoken convictions that Ostrovets and Baltic NPPs are indeed tools of Russian political pressure against Lithuania and therefore it wanted to stop their construction (Interview, 2014a; 2014b; State Security Department, 2015; State Security Department and Second Investigation Department Under the Ministry of National Defense, 2017). Despite such latent point of view, Lithuanian opposition was very conscientious at that time—it focused on exposing the shortcomings of Russian NPPs and urged Russia and Belarus to fix them both directly and through various international organizations, but was not explicitly opposed to their construction (Juozaitis, 2016).

The Lithuanian position changed in December 2016, when the newly formed government explicitly underscored that it is: "Against the construction of nuclear power plants in our region." (Parliament of the Republic of Lithuania, 2016b, p. 50). Two months later, a written agreement of all Lithuanian political parties represented in the Parliament denounced Ostrovets NPP as a Russian geopolitical project, and political parties committed themselves to utilize all available means to prevent its construction (Parliamentary Political Parties of the Republic of Lithuania, 2017). In April 2017, a law prohibiting purchasing electricity from neighboring countries with unsafe NPPs reinforced this position even further.[3]

---

[2]For example, Lithuanian officials had explicitly stated that they are not against the development of nuclear power in Belarus per se, but they would like Belarus to be more transparent, cautious and respectful to international nuclear safety standards and law (Ministry of the Foreign Affairs of the Republic of Lithuania, 2011a). Similarly, Lithuania urged Russia to be transparent about the construction of Baltic NPP by answering to Lithuanian queries and implementing the standards of international law (Parliament of the Republic of Lithuania, 2012a) in order to reach a common goal of assuring nuclear safety in the region (Ministry of the Foreign Affairs of the Republic of Lithuania, 2013). Finally, Lithuania suggested relocating the construction site of NPP in Belarus further away from Lithuanian border. (President of the Republic of Lithuania, 2011).

[3]According to the law on Necessary Measures of Protection Against the Threats Posed by Unsafe Nuclear Power Plants in Third Countries article 2.1.: "Unsafe nuclear power plant shall mean a nuclear power plant under construction or operating in a neighboring third country, the designing, installation or operation of which has resulted in non-compliance with the requirements for environmental protection or nuclear safety and radiation protection and violation of international agreements and conventions and which, due to its geographical location or technological characteristics, poses a threat to the national security of the Republic of Lithuania, its environment and public health."

After key national institutions endorsed the principal political opposition against the construction of Russian NPPs, in September 2017 the Lithuanian government adopted a plan that specified concrete measures to be used against Ostrovets NPP. They included terminating the electricity trade with Belarus from the moment Ostrovets starts working, denying Minsk the ability to procure reserve services from Kruonis pump-storage power plant, and synchronizing its electricity grid with the Continental European Network by 2025 (Government of the Republic of Lithuania, 2017), thus completely cutting off the links between Belarusian and Lithuanian electricity systems.[4] After Rosatom suspended the construction of Baltic NPP in Kaliningrad in 2013, Ostrovets NPP received much more attention. However, Lithuania has a strategic approach to further oppose the construction of Baltic NPP, should it be resumed.

Having briefly introduced the Lithuanian official position and its changes from 2008 to 2018, the next step is to explain it. Due to the likely influence of cautious Lithuanian protests against Russian NPPs during 2008—15, some researches tend to associate Lithuanian opposition with the shortcomings of Russian NPPs (Backaitis, 2011; Handl, 2015, pp. 211, 212, 218), not with factors exposing how these projects challenge Lithuanian national interests. It is understandable to some extent as Lithuanian decision-makers chose not to reveal their political concerns and focused on the flaws of Russian NPPs, and this led some to believe that Lithuania opposes them only because of their shortcomings. This section, however, argues on the contrary. Lithuania first and foremost objects to the construction of Ostrovets and Baltic NPPs because it perceives them as being threats to its national security by conflicting with its fundamental foreign and energy policy priorities. Even though it is entirely reasonable to consider the flaws of the NPPs mentioned above as a source of concerns provoking Lithuanian resistance to some extent, they can also be considered as arguments to justify it.

In order to clarify the subtle differences between concerns and arguments, and to explain Lithuanian opposition in doing so, this section divides itself into three subsections. The first one introduces the conventional argument that stems from the earlier version of the Lithuanian official position. The second one deconstructs the conventional explanation by exposing a number of logical inconsistencies within it and establishing a subtle division between the shortcomings of Russian NPPs as issues that provoke Lithuania's negative response and as arguments that justify its opposition against them. The last one concludes the section by elaborating on specific Lithuanian political motives to oppose the construction of Ostrovets and Baltic NPPs.

Introducing the conventional explanation

As far as the conventional argument goes, Lithuanian objections against the construction of Russian NPPs in Kaliningrad and Belarus emanate from three national concerns. The first concern is over the geographic proximity of these objects. The construction site of Baltic NPP is located less than 20 km away from Lithuania's southwestern border, while Ostrovets NPP is being built 23 km away from Lithuania's eastern frontier (IAEA, 2010, p. 11) and 50 km

---

[4]Electricity exchange between asynchronous electricity systems is available if converter stations are built. Plans of Lithuanian TSO "Litgrid" (Litgrid, 2018) indicate that Lithuania will not build any converter stations on the Belarusian border.

away from Lithuania's capital city, Vilnius. As well, both NPPs are located near major rivers (Nemunas and Neris), essential sources of Lithuanian drinking water. The second concern is over the safety of Russian NPPs, emanating from subjective mistrust of Russian nuclear technologies (Ministry of the Foreign Affairs of the Republic of Lithuania, 2011) and work ethics (Parliament of the Republic of Lithuania, 2012a) as well as objective worries about multiple incidents that took place during the construction of Ostrovets NPP. The last concern is over legal issues of these projects—Lithuania criticizes Belarus for violating international agreements (for a detailed list, please see Table 7.1) and accuses Russia of selective application of international law. Other researchers sometimes present a longer list of Lithuanian concerns by adding environmental risks, human rights violations, or lack of safety equipment and structures (Hussein, Kardas and Klysinski, 2018, pp. 25—27), but one can position all of them under the umbrella of broader categories of Lithuanian concerns regarding proximity, nuclear safety, and international law.

Handl indicates that geographic proximity of planned or already operational NPPs facilitates controversy in relations between the owner and its neighbor (2015, p. 211, 212). The closeness of NPPs provokes scrutiny from the bordering countries as they are interested in foreign infrastructure objects that might affect their environment or could pose a risk to security and well-being of their citizens (Nuclear Energy Association, 2016). Hence, Lithuanian worries over the proximity of Ostrovets and Baltic NPPs are consistent with international practice, and that serves as another reason for researchers to assume that closeness is a concern provoking opposition rather an argument that justifies it.

Moving on to nuclear safety, one can reasonably argue that Lithuanian fears of Ostrovets NPP come from documented incidents during its construction and Belarus's reluctance to acknowledge them in a timely and transparent manner (Ministry of Energy of the Republic of Lithuania, 2017). Even though Belarusian Vice-Minister for Energy Michail Michadiuk once admitted the occurrence of 10 incidents that resulted in three fatalities (Charter 97, 2016a), in general, Belarusian authorities were reluctant to acknowledge incidents and to share information about them. For example, Belarus informed the public about the fall of 330 tons of a nuclear reactor pressure vessel on July 26, 2016. The incident took place on July 10, 2016, and the information about it was first published in the social media by a local activist, Mikalai Ulasevich, on July 25, 2016. Hence, only after the incident became public did Belarusian authorities inform the public about an uncommon situation in the construction site (Charter 97, 2016b). Since Belarusian institutions are systematically limiting the access of its citizens to information about Ostrovets NPP (Amnesty International, 2013), it is likely that some incidents were successfully concealed, once again adding to the plausibility of considering nuclear safety as a concern that prompts Lithuanian opposition.

In contrast to Ostrovets NPP, Lithuanian concerns over nuclear safety of Baltic NPP in Kaliningrad are not related to specific incidents during the construction of the plant itself as only the foundations of the first unit were completed before Russia postponed the project in 2013. Concerns emanate from a general subjective mistrust of Russian contractors and technology and observed accidents in other Russian NPPs. The collapse of a frame built of steel armature in Leningrad-2 NPP and crumbling of its containment building in 2011 are among the most frequently mentioned examples by Lithuanian officials. These incidents are also often juxtaposed with the disaster in Fukushima that took place in the same year, and the explosion in Chernobyl NPP 25 years earlier. Similarly, Lithuania voices concerns about lack of

TABLE 7.1   Lithuanian position regarding Belarus's compliance with international law and recommendations regarding Ostrovets NPP.

International instrument	Lithuanian official position
Nuclear Safety Convention	Belarus failed to assess possible negative impact on society and the environment (Art. 17.2), coordinate site selection with Lithuania (Art. 17.4), and give clear priority to nuclear safety (Art. 10).
IAEA's missions and recommendations on nuclear safety, 2013	Constructing an NPP too close to Vilnius; Belarus must invite the IAEA experts to review the nuclear safety of Ostrovets NPP, especially to conduct a full SEED mission.
Espoo Convention	Belarus failed to take necessary legal, administrative, and other measures for the establishment of an environmental impact assessment procedure (Art. 2.2), inform the Lithuanian public about Ostrovets NPP (Art. 2.6), to notify Lithuania properly (Art. 3.2) draft documentation of environmental impact assessment properly (Art. 4.1 and 4.2), hold expert consultations (Art. 5a), draft a decision to commence construction in a proper manner (Art. 6.1 and 6.2).
Aarhus Convention	Belarus failed to ensure access of the Lithuanian public to information about decisions made about Ostrovets NPP, translate relevant documents into Lithuanian (Art. 3.9), allocate reasonable time for the examination of EIA report, ensure the participation of Lithuanian public in the events informing about Ostrovets NPP (Art. 6.2), provide the opportunity to submit comments regarding the findings of state Ecological Expertiza (Art. 6.3), ensure the opportunity for the Lithuanian public of early participation in Ostrovets NPP project and to submit its position regarding the selection of alternative locations (Art. 6.4), ensure that in decisions due account was taken of the outcome of the public participation (Art. 6.8).
Helsinki Water Convention	Violations are not directly specified. Lithuania raised environmental concerns regarding the Neris river, a part of Nemunas river basin that is important source of Lithuanian drinking water and a water source for cooling Ostrovets NPP, during the meetings of the parties of convention.
WENRA recommendations, 2013	Failure to ensure that Ostrovets NPP would sustain a major airplane crash.
EU "stress tests"	"Stress tests" highlight the deficiencies of Ostrovets NPP. Recommendations provided in the EU stress test peer review report must be implemented before the Ostrovets NPP is launched.

Source: Ministry of the Foreign Affairs of the Republic of Lithuania (2018b). Statement by the Ministry of Foreign Affairs on Astravets Nuclear Power Plant under Construction in Belarus https://www.urm.lt/default/en/news/statement-by-the-ministry-of-foreign-affairs-on-astravets-nuclear-power-plant-under-construction-in-belarus-; Ministry of the Foreign Affairs of the Republic of Lithuania (2018c). L. Linkevičius: The EU Stress Tests Proved that the Ostrovets NPP Is Not Safe, https://www.urm.lt/default/en/news/l-linkevicius-the-eu-stress-tests-proved-that-the-ostrovets-npp-is-not-safe; Ministry of the Foreign Affairs of the Republic of Lithuania (2015). 2014 Metų Veiklos Ataskaita [Activity Report for 2014] http://www.urm.lt/uploads/default/documents/Ministerija/veikla/veiklos_ataskaita/URM_2014_veiklos%20ataskaita.pdf; United Nations Economic and Social Council, 2013.

transparency and accuses Russia of not sharing information about the project (Parliament of the Republic of Lithuania, 2012a).

Viewed from a legal standpoint, Lithuanian concerns regarding Ostrovets NPP are justified by Belarusian noncompliance with Espoo and Aarhus Conventions. In 2014, Belarus was declared to be in noncompliance with the Espoo Convention due to many violations regarding the Environmental Impact Assessment (EIA) report (United Nations Economic and Social Council). During the same year, Belarus was also ruled to be in noncompliance with the Aarhus convention as a consequence of restrictions on public access to information regarding the environmental impact of Ostrovets NPP and its participation in the decision-making process (United Nations Economic Commission for Europe, 2014). In 2017, the parties of the Aarhus Convention expanded the list of Belarusian noncompliances to the convention by adding, among other things, the arrests, harassment, penalization, and prosecution of Belarusian environmental activists (United Nations Economic Commission for Europe, 2017).

As far as the Baltic NPP is considered, Lithuania is concerned about Russian reluctance to initiate the procedure of environmental impact assessment in a transboundary context per Espoo Convention (Ministry of the Foreign Affairs of the Republic of Lithuania, 2013). Even if Russia is not bound to conduct such assessment as Moscow has not yet ratified the convention mentioned above, other Russian strategic energy projects undertook this procedure. For example, not only the provisions of Espoo Convention but also other treaties were considered by Russia when assessing the environmental impact of both Nord Stream natural gas pipeline projects due to their international status (Nord Stream, 2009; Nord Stream 2 2017).

In the end, it seems that there are plenty of arguments to support the conventional explanation of Lithuanian opposition to Russian NPPs. Lithuanian concerns over the location, nuclear safety, and legal aspects of Ostrovets and Baltic NPPs, despite several subjective fears, are mostly supported by objective evidence and do not diverge from the international practice to oppose the construction or operation of NPPs close to one's national borders. The question remains, however, does Lithuania oppose the construction of Russian NPPs mainly because it perceives them to be to close, unsafe, and illegal or there are more important forces in play?

Deconstructing the conventional explanation

Having outlined the essence of conventional explanation, this section shows why geographic proximity, nuclear safety, and legal problems cannot be defined as concerns that explain Lithuanian behavior, and they should also be perceived as arguments for justifying its protest against the construction of Russian NPPs and, sometimes, even as instruments helping to oppose them.

Taking the international practice into account, one cannot doubt that concerns regarding the closeness of Russian NPPs provoke Lithuanian opposition against them, but the intensity of Lithuanian protest suggests that closeness of Ostrovets and Baltic NPPs juxtaposed with the worries over their nuclear safety are not the only factors that incite it. For example, on the basis of close geographic proximity and nuclear safety issues, Germany pressures France to put its Feseinheim NPP out of operation. Due to the same reasons, Germany also presses Belgium to close its aging Tihange NPP (Federal Ministry for the Environment, Nature Conservation and Nuclear Safety, 2015, 2018). Despite having similar concerns to Lithuanian ones and far greater capability to enforce its will in international politics, Germany is not planning to limit electricity trade or utilize any other measures of such intensity if France or even Belgium disagree to do so. Therefore, even if close geographic proximity generally causes some level of opposition from the affected party, it does not reach such levels of intensity observed in Lithuania.

The main reason why nuclear safety is not merely a concern that provokes Lithuanian opposition to Ostrovets NPP is the precise time when doubts regarding its nuclear safety were first expressed. On January 8, 2009, in a meeting with EU's Commissioner for External Relations and European Neighborhood Policy, Benita Ferrero-Waldner, the former Lithuanian Foreign Minister Vygaudas Ušackas stated that: "Lithuanian politicians and society are mostly concerned about the safety and environmental impact of the planned [Ostrovets] nuclear power plant" (Ministry of Foreign Affairs of the Republic of Lithuania, 2009). Despite raising concerns about nuclear safety, Lithuania could not have possessed substantial evidence to support them at that time due to two main reasons. First, the aforementioned statement was made only 3 weeks after Belarusian special commission recommended constructing an NPP near Lithuanian borders on December 20, 2008 (Ministry of the Emergency Situations of the Republic of Belarus). Second, Lithuania formulated such a position despite having practically no information about the Ostrovets NPP as only very broad technical specifications without a specific reference to a possible construction site were shared by Belarusian authorities; Minsk distributed its preliminary environmental impact assessment to interested parties only on August 24, 2009 (Ministry of the Environment of the Republic of Lithuania, 2015). Hence, at the earliest stages of the Ostrovets NPP project, Lithuania used nuclear safety as an argument justifying its protest against it rather than opposing Ostrovets because of its knowledge over specific safety flaws as they just could not have been known at the moment the concerns were first expressed.

It is not to say, however, that nuclear safety is merely a fictional argument vindicating Lithuanian protests against Ostrovets NPP. As time passed by, Lithuanian officials found themselves increasingly anxious as they were dissatisfied by the quality of Ostrovets's environmental impact assessment, lack of transparency, and, most importantly, reoccurrence of incidents during the construction process. From the moment these issues became apparent, nuclear safety could be credibly considered as a concern that prompts Lithuanian opposition against Ostrovets NPP (Parliament of the Republic of Lithuania, 2016). It is to say, however, since Lithuania used nuclear safety as an argument to justify its protests against the Ostrovets NPP before knowing its deficiencies, nuclear safety by no means can explain Lithuanian hostility toward the aforementioned NPP by itself.

Baltic NPP reveals a similar tendency. Even though a public report by a Russian NGO Eco-defense exposes inadequacies of Baltic NPPs environmental impact assessment as early as 2009 (Slivyak et al., 2009), and there were other accusations regarding the lack of proper safety measures by local Russian residents, Lithuania fears were mostly grounded not on them but on subjective distrust of Russian nuclear technologies and work ethics. If one accepts an assumption that the roots of Russian working culture and technological progress lay in the Soviet era, Lithuanian perception is inconsistent with operating a Soviet-built Ignalina NPP 20 years after regaining independence in 1990 at the same time. On the one hand, Lithuania is skeptical about Russian ability to ensure the safety of its NPPs; on the other hand, Vilnius operated the Soviet-built Ignalina NPP and was keen on prolonging its operation despite commitments made to the EU to close it. Again, nuclear safety concerns do not seem to adequately explain Lithuanian opposition to the Baltic NPP.

Systematic efforts to question the compliance of Russian NPPs with international law can hardly be considered as a product of concerns regarding their legal issues. Instead, international agreements are instrumentalized for opposing them. Lithuania rigorously accused

Russia and Belarus of violating international law in various international organizations and other institutions dealing with infrastructure projects with a transboundary impact. For example, Lithuania accused Belarus of violating not only the already mentioned Espoo and Aarhus Conventions but also the Nuclear Safety Convention, Helsinki Water Convention, and IAEA's and Western Europe Nuclear Regulators Association's post-Fukushima recommendations on additional safety measures. Moreover, Lithuania raised its doubts regarding noncompliance of Russian NPPs with the aforementioned documents in the EU, NATO, OSCE, IAEA, UN, Council of Europe, and Baltic Council of Ministers. It also raised them in such multilateral formats as the Nuclear Security Summit, Nordic-Baltic Eight, Europe—Asia Meeting, as well as bilaterally (Ministry of the Foreign Affairs of the Republic of Lithuania, 2011a, p. 2, 2012a, p. 54, 2015, p. 3, 17, 18, 78, 2016, p. 27, 2017, p. 17; 2018a, p. 6, 7). As it was underscored in the introduction, such behavior reflects a bedrock assumption in the small state literature—Lithuania has limited capability to oppose the development of nuclear infrastructure beyond its borders, and therefore the country defends its national interests by utilizing international legal instruments and memberships in international institutions. Hence, questioning the compliance with international law is a means to oppose Russian NPPs and cannot be considered as a variable that explains Lithuanian opposition as of a consequence.

One can question this position by asking: if legal issues and nuclear safety flaws of Russian NPPs are not the principal reasons for Lithuanian opposition, while the underlying cause is the belief that these projects undermine Lithuanian foreign and energy policy, why did Lithuania focus on presenting the former and silencing the latter during 2008—15? Since Lithuania together with the other two Baltic states, Latvia and Estonia, was developing its own Visaginas NPP project in parallel, Lithuania's possibilities to raise political concerns over Ostrovets and Baltic NPPs were substantially constrained. Raising political questions until 2016, when Visaginas NPP was still under the development, would have been counterproductive as Belarus and Russia repeatedly accused Lithuania of using nuclear safety and international law as political tools in order to dispose of competition to its own NPP project, and that would have strengthened the validity of their claims. Only when the Peasants and Greens formed a ruling coalition at the end of 2016 did it become clear that Lithuania will de facto stop the development of Visaginas NPP.[5] Simultaneously, Lithuania changed its approach toward Baltic and Ostrovets NPPs by denouncing them as a part of a broader Russian geopolitical scheme to project power into the Baltic Sea Region, explicitly revealing its willingness to stop their construction.

All in all, even though geographic proximity of Baltic and Ostrovets NPPs, issues with their nuclear safety, and factual or perceived noncompliance with international law can eventually be conceptualized as concerns, they cannot explain why Lithuania is against them. The closeness of NPPs to national borders usually provokes suspicion and sometimes protests, but the opposition does not reach such levels of intensity when countries draft laws to block

---

[5]Visaginas NPP project progressed rapidly until an advisory referendum in 2012, in which Lithuanian citizens voted against the construction of the aforementioned NPP. After such outcome, Visaginas NPP remained a heavily debated subject in Lithuanian public discourse until the end of 2016. By that time a newly formed Lithuanian government led by the Peasants and the Greens explicitly stated that Lithuania is in principle against the development of nuclear power and Visaginas NPP is excluded from the updated Lithuanian energy strategy in 2018.

the electricity trade or prepare to desynchronize from the common electricity system. There are objective reasons to doubt the nuclear safety of Ostrovets NPP, but Lithuania raised its nuclear safety concerns before they became apparent. At the same time, Lithuania is skeptical about Russian capability to ensure nuclear safety of Baltic NPP despite operating the Soviet-built Ignalina NPP for 20 years and wanting to continue to do so. Finally, legal issues can also be considered as a source of concern, but Lithuania utilized international law first and foremost as an instrument to systematically oppose the construction of Russian NPPs.

Explaining Lithuanian opposition

After presenting the conventional explanation and exposing its logical inconsistencies, this section demonstrates how Ostrovets and Baltic NPPs pose a threat to Lithuanian national security by undermining its energy and foreign policies and therefore prompting opposition against them. When the race between Ostrovets, Baltic, and Visaginas NPPs began in 2006–07, the logic behind Lithuania's decision to construct an NPP of its own was twofold. First, Lithuania was under the obligation to the EU to put the second unit of its Ignalina NPP out of operation until 2010 and lacked competitive alternatives for electricity generation or import. Second, Lithuania perceived this project as a means of avoiding dependence on electricity imports from Russia as such dependence in the natural gas and oil sectors threatened its national security on many occasions (Cesnakas 2012, 2013).

Despite perceiving the reliance on energy imports from Russia as national security threats (Parliament of the Republic of Lithuania, 2005, 2007, 2012a, 2012b), Lithuania still found itself importing most electricity from Russia from the moment Ignalina NPP was shut down in 2010 and until electricity interconnections with Sweden and Poland were completed in 2015 (Molis et al., 2018). Lithuania perceived Ostrovets and Baltic NPPs as Russian instruments aimed at preventing Lithuania from constructing Visaginas NPP, thus ensuring its long-term dependence on Russian electricity import. For example, reports by State Security Department of Lithuania suggest that Russia utilized Ostrovets and Baltic NPPs to put Lithuanian strategic energy projects in a negative light and then instrumentalized the negative attitude of Lithuanian public to oppose against them (2015, 2016). Lithuanian public opinion did turn negative toward the development of nuclear power, and Visaginas NPP was eventually scrapped, but it remains unclear wherever these developments are the consequence of Russian attempts to form Lithuanian public opinion or if other forces drove it. For instance, Genys and Leonavicius do not list Russian involvement as relevant to this change in Lithuanian public attitude toward nuclear energy (2018).

Even though Lithuania failed to construct the Visaginas NPP, it eventually managed to complete the electricity interconnections with Poland and Sweden. Not only did these projects diversify Lithuanian electricity imports but also they were an essential advancement in accomplishing one of its fundamental foreign policy priorities, EuroAtlantic integration. However, if the plans presented earlier to utilize Lithuanian electricity network for exporting electricity generated in Russian NPPs to Western markets would be implemented, it would deny Lithuania the benefits mentioned above. Allowing for Russia and Belarus to utilize these interconnections for supplying electricity to Sweden and Poland would mean that Lithuania will not be able to import electricity from these countries at the same time and the country would be forced to rely heavily on the electricity supplies from Russia. Hence, it would coincide with Russian strategic interests to maintain its influence in Lithuania. Most importantly, however, such a scenario would be a significant setback in Lithuanian efforts to deepen its EuroAtlantic integration and strengthen its independence from Russia.

In the end, Russian foreign policy aimed to further integrating Lithuania into the Russian energy system and preventing to achieve closer ties with European Union is the most pressing concern, explaining Lithuanian opposition against the construction of Baltic and Ostrovets NPPs. Close geographic proximity, nuclear safety flaws, and legal issues can also be considered as Lithuanian concerns regarding Russian NPPs to some extent. However, Lithuania mostly uses them as arguments to justify its opposition against them or as a means to combat their construction, and the following chapter further elaborates on the latter argument.

Lithuanian attempts to prevent the construction of Russian NPPs

As it was briefly discussed in the opening paragraphs of this chapter, international organizations and legal instruments are considered as tools for small states that help them to overcome the problem of power asymmetry when dealing with more powerful countries. The following sections examine how Lithuania utilized them in order to stop the construction of Ostrovets and Baltic NPPs and discuss its effects on the outcome of the aforementioned Russian nuclear projects. The study proceeds in three levels of analysis. The first level explores how Lithuania utilized international legal instruments and various recommendations for combating Russian NPPs. The second level examines Lithuanian attempts to oppose the construction of Ostrovets and Baltic NPPs by working through international organizations. The third level goes beyond the traditional instruments conceptualized within the realm of small state studies and overviews Lithuanian national efforts in protesting against the construction of NPPs close to its borders. Having presented the findings on all levels of analysis, the study discusses their impact on the implementation of Russian NPP projects.

Utilizing international legal instruments and recommendations

Since Russia has not ratified the Espoo and Aarhus Conventions and because construction of the Baltic NPP was frozen in 2013, Lithuania mostly used international legal instruments and recommendations to combat the construction of Ostrovets NPP in Belarus. Vilnius presented the most extensive lists of suspected violations to the implementation committees of Espoo and Aarhus Conventions. Starting from the former, Lithuania officially complained in 2011 that Belarus made many violations in relation to the EIA procedure of Ostrovets NPP. Moreover, Lithuania accused Minsk of failing to notify its authorities and inform the Lithuanian public per the requirements of the convention, to draft a decision to commence construction and to hold expert consultations in a proper manner (United Nations Economic and Social Council, 2013). Moving on to the latter, Lithuania complained that Belarus is in noncompliance with the Aarhus Convention because it limits the access of the Lithuanian public to information about environmental aspects of Ostrovets NPP in many ways (Ministry of the Environment of the Republic of Lithuania, 2015).

A position statement by the Lithuanian MFA indicates that the country also perceives Belarus to be in violation with the Convention on Nuclear Safety (CNS) and the Helsinki Water Convention. Regarding the CNS, Vilnius claims that Belarus failed to assess possible negative impact on society and the environment, coordinate site selection with Lithuanian authorities, and give clear priority to nuclear safety. As far as the Helsinki Convention is concerned, Lithuania does not specify concrete violations, rather Vilnius uses it as a platform to voice environmental concerns regarding the river Neris that is an important source of Lithuanian drinking water. Not only did Lithuania accuse Belarus of not complying with international law but also of ignoring post-Fukushima recommendations on nuclear safety by the IAEA not to construct new NPPs close to large population centers and the Western Europe Nuclear Regulators Association (WENRA) to ensure the protection of newly built NPPs against the possibility of a major airplane crash (2018b).

Lithuania was also active in using the EU's and IAEA's instruments as it pressured Belarus to conduct the stress tests designed by the former and invite on-site missions offered by the latter. Especially, Lithuania wanted Belarus to invite IAEA's site and external events design review service (SEED). Lithuania's attempts to take advantage of international legal instruments and recommendations in opposing the construction of Ostrovets NPP are summarized in Table 7.1.

Hence, by defending its interests through international legal instruments, Lithuania behaved in line with expectations of small state literature, but how successful were these efforts? Lithuania's major but sole success in this regard was the decision adopted by the Sixth Meeting of the Parties to the Espoo Convention in 2014 declaring Belarus to be in noncompliance with the convention, thus validating the credibility of Lithuania's position that Ostrovets NPP has legal issues (for a detailed list of violations, please see Table 7.2). In some instances, responsible international authorities declared Belarus to be in noncompliance with the provisions of international law not because of Lithuanian complaints but as a result of charges brought up by other entities. For example, Lithuanian accusations regarding noncompliance with the Aarhus Convention are still being investigated, but the Fifth and Sixth Meetings of the Parties to the Aarhus Convention have already declared Belarus in noncompliance with

TABLE 7.2 IAEA missions to ostrovets NPP.

Date	IAEA missions in Belarus
August 5, 2019	Preoperational safety review
October 8, 2018	Emergency preparedness review
January 16, 2017	Site and external events design review
October 2, 2016	Integrated regulatory review
June 18, 2012	Integrated nuclear infrastructure review no. 2
June 18, 2012	Integrated nuclear infrastructure review no. 1
May 30, 2011	Education and training appraisal
October 4, 2010	Emergency preparedness review

Source. IAEA.

the convention due to the complaint of Belarusian NGO Ecohome (United Nations Economic Commission for Europe, 2017). Therefore, these decisions yet again helped Lithuania to reinforce its positions regarding legal issues of Ostrovets NPP. In other instances, however, Lithuania has not yet managed to secure favorable decisions from the responsible authorities overseeing the compliance of the instruments mentioned above.

Lithuania was notably unsuccessful in utilizing IAEA's instruments to denounce the nuclear safety of Ostrovets NPP. In hopes that IAEA would expose fundamental safety flaws, Lithuania pressured Belarus to invite its experts for conducting peer review missions. Belarus did establish close cooperation with the IAEA and invited its experts to overview various safety aspects of Ostrovets NPP eight times (please see Table 7.2). However, instead of criticizing Belarus, IAEA strictly limited itself to suggestions on how to improve the nuclear safety of Ostrovets NPP. Moreover, it praised Belarus for its willingness to cooperate and sometimes, contrary to Lithuania's expectations, complimented Belarus's commitment to nuclear safety. For example, one IAEA's official stated after conducting an integrated regulatory review in Belarus: "They [Belarusian authorities] are committed to providing effective regulatory oversight of the nuclear program as well as a diverse range of activities with radiation sources" (IAEA, 2016). A similar statement was made after the IAEA examined Belarus's emergency preparedness and response framework: "The team found that Belarus has solid arrangements in place for emergency preparedness and response" (IAEA, 2018).

The so-called SEED mission is another interesting example, exposing how Belarus managed to use the cooperation with the IAEA to its advantage. According to the IAEA, SEED mission consists of six modules and its member states can request one or more depending on their needs (2018). Belarus invited IAEA's experts to review the safety of Ostrovets NPP against site-specific external hazards and to assess how the lessons learned from the disaster in Fukushima were implemented, thus choosing a limited SEED mission. In contrast to Lithuania's expectations, the results of this review (IAEA, 2017a) and statements by the IAEA's officials that preceded it were favorable to Belarus: "This mission demonstrated that appropriate steps have been taken to establish the design parameters of the nuclear power plant to protect it against the worst credible external event" (IAEA, 2017b). After learning the results of the SEED mission, Lithuanian authorities blamed Belarus for having a selective approach to nuclear safety due to not selecting all possible mission modules. Despite such a position, the fact remains that Belarus managed to use the IAEA to its advantage in legitimizing the nuclear safety of Ostrovets NPP, while Lithuania failed to do the opposite.

As far as the EU's stress tests are concerned, it is complicated to access their significance at this point. One the one hand, the fact that their preliminary phase took place at all during 2017 and 2018 is beneficial to Lithuania by itself because they locked EU's attention on the Belarusian project as European Commission wants to continue the stress-testing process and asked Belarus to prepare additional documents and to subject them to another review. On the other hand, however, the conclusions of the first review in 2018 were also limited to recommendations on the technical level (ENSREG, 2018), while the press statement by the European Commission highlighted Belarusian willingness to cooperate, even if they were not obliged to do so (2018b). After the report and EU's statements were published, Lithuania focused on the technical level and maintained that recommendations in the report expose the safety flaws of the Ostrovets NPP (Ministry of the Foreign Affairs of the Republic of Lithuania, 2018c). On the contrary, Belarus focused on the political level and interpreted the EU's position as a political endorsement of its commitment to nuclear safety (Belta, 2018). Looking from an objective standpoint, the stress tests themselves have so far not helped Lithuania in justifying its claims over nuclear safety flaws of Ostrovets NPP; however, neither they helped Belarus to justify theirs.

Moving on to Kaliningrad, Lithuania made only several attempts to use various legal instruments against Baltic NPP. Despite the absence of ratification, Lithuania still urged Russia to comply with the requirements of Espoo Convention, especially with the ones related to the EIA procedure. Moreover, Lithuania wanted Russia to undergo EU's stress test procedure and invite international experts from the IAEA to examine the construction site of Ostrovets NPP (Ministry of the Foreign Affairs of the Republic of Lithuania, 2012b). Russia replied by asking the IAEA to conduct a peer review of its EIA report of the first unit of the Baltic NPP against IAEA's nuclear safety standards and the requirements of the Espoo Convention (IAEA, 2015a). However, Russia has not invited EU's experts to conduct the so-called stress tests in Baltic NPP, nor has it requested any of IAEA's mission teams to inspect the aforementioned object.

In sum, Lithuanian success in utilizing international instruments was limited. Vilnius managed to prove that Belarus is in noncompliance with the Espoo Convention and persuade Minsk to use the instruments offered by the IAEA (various peer review missions) and the EU (stress tests). Notwithstanding, Belarus was not found to be in noncompliance with the CNS or the Helsinki Water Convention, while IAEA's missions helped the country to put Ostrovets NPP in a positive light and thus created a counterargument against Lithuanian criticism.

Employing membership in international organizations and cooperation mechanisms

Lithuania achieved far greater success in securing favorable political declarations from international organizations rather than using international legal mechanisms against Ostrovets NPP, while they were seldom used against Baltic NPP due to its early conservation in 2013. The first measurement of success is general positions adopted by international organizations. The fiercest statement was made in a resolution by the Parliamentary Assembly of the Council of Europe (PACE) that practically mirrors the Lithuanian official position: "suspend the construction of the Astravets Nuclear Power Plant (NPP) because of numerous violations, the lack of respect for international standards for nuclear safety and serious safety violations and major incidents during the construction of this plant" (Parliamentary Assembly of the Council of Europe, 2017: Art. 5.6). Nordic-Baltic Eight issued another favorable statement that, among other things, explicitly recognizes the necessity of creating a level playing field in electricity trade vis-à-vis third countries (2016), thus somewhat strengthening the legitimacy of Lithuania's plans to limit the electricity trade with Belarus. Naturally, such position is also shared by the Baltic Council of Ministers, which: "Acknowledged the importance of the application of the uniform principles related to the entrance to the electricity market from third countries < ... >" (2018, p. 3). Should the construction of the Baltic NPP resume unexpectedly, agreements on electricity trade with third countries would apply to it as well.

EU's key institutions have also expressed a number of positions regarding the construction of Ostrovets NPP that favor Lithuania's efforts in stopping it to some extent. For example, the European Commission does not support the idea of limiting the electricity trade between

Lithuania and Belarus, even though it acknowledges that nuclear safety of Ostrovets NPP is a common problem of all the EU, instead of downplaying its significance to a Lithuanian national issue. It is also involved in the process of stress testing, therefore, not limiting itself to expressing a supporting political position but engaging in genuine cooperation both with Lithuanian and Belarusian authorities (European Commission, 2018). The European Parliament assumed an even stricter stance by advocating: "to ensure that progress in EU-Belarus relations is conditional on increased openness and cooperation, and on full compliance with international nuclear and environmental safety standards, on the part of Belarus" (2018). However, these standards are not explicitly defined, and they could relate to findings by IAEA's and EU's peer review teams or Belarusian compliance with Espoo and Aarhus Conventions.

On the other hand, there were instances when attempts to employ the membership in international organizations to oppose the construction of Ostrovets NPP did not necessarily play in Lithuania's favor. For example, the European Commission allocated funding for strengthening the capabilities of the Belarusian nuclear regulator (2016b), a move that sparked controversy in Lithuania. Furthermore, Lithuania failed to secure a similar political statement to PACE in the Parliamentary Assembly of the OSCE. In general, however, Lithuania was successful in employing memberships in international organizations and formats for cooperation in order to secure favorable positions toward Ostrovets NPP.

Not only did Lithuania secure such positions but it also managed to convince a number of international organizations to support its specific claims regarding Belarusian noncompliance to international law and recommendations (detailed in Table 7.1). For instance, PACE maintains that Belarusian authorities should not issue an operating license to Ostrovets NPP until it complies with the Espoo and Aarhus Conventions, completes all the available modules of IAEA's SEED mission, and ensures that recommendations provided by WENRA and EU's stress test peer review team are fully implemented. The Baltic Council of Ministers also urges Belarus to comply with the requirements of environmental conventions and highlights the importance of implementing the recommendations made in the report summarizing the outcome of the EU's initial stress tests in 2017—18. The argument is summarized in Table 7.3.

As it can be observed, Lithuania was active in opposing the construction of Ostrovets NPP through international organizations, and it made some important strides in that regard. In most instances, Lithuania managed to strengthen the credibility of its position regarding Belarusian noncompliance with international law or recommendations by securing favorable declarations from PACE and the Baltic Council of Ministers. PACE even supported Lithuania's determination to prevent the construction of Ostrovets NPP. However, international organizations and cooperation mechanisms were not successfully exploited in supporting Lithuanian efforts to prevent the construction of Baltic NPP.

Using national means

As it was observed while introducing Ostrovets and Baltic NPPs, Lithuania had the opportunity to put pressure on Russia and Belarus by exploiting the advantageous geographic circumstances and favorable interdependencies in the electricity grid, and it utilized both of them. Regarding the Baltic NPP, Lithuania refused to negotiate with Russia about the

TABLE 7.3   Lithuania's position regarding Belarus, its validation and political support.

Lithuanian position regarding Belarus	Validation by a responsible authority or a declaration of political support
Espoo Convention. Violations of Art. 2.2, 2.6, 3.2, 4.1, 4.2, 5a, 6.1 and 6.2).	Meeting of the parties to the Espoo Convention at its sixth session, June 2—5, 2014 - Belarus is in noncompliance with: Art. 2.6 — failure to inform Lithuanian public about Ostrovets NPP; Art. 4.2 — failure to draft documentation of EIA properly; Art. 5a — failure to hold expert consultations; Art. 6.1 and 6.2 — failure to draft a decision to commence construction in a proper manner. (Parliamentary Assembly of the Council of Europe, 2018) — not to issue an operational license for the Ostrovets NPP before it complies with the convention. (Baltic Council of Ministers, 2018) — Belarus must comply with the requirements of Espoo Convention.
Aarhus Convention. Violations of Art. 3.9, 6.2, 6.3, 6.4 and 6.8.	Lithuanian submission regarding Belarusian noncompliance with Aarhus Convention is being considered by the compliance committee. Belarus was declared to be in noncompliance to a number of articles by the sixth and fifth meetings of the parties to the Aarhus Convention as of a consequence of a complaint by a Belarusian NGO Ecohome. (Parliamentary Assembly of the Council of Europe, 2018) — not to issue an operational license for the Ostrovets NPP before it complies with the convention. (Baltic Council of Ministers, 2018) — Belarus must comply with the requirements of environmental conventions.
IAEA SEED mission. Belarus must invite the IAEA experts to conduct all six modules of SEED mission.	(Parliamentary Assembly of the Council of Europe, 2018) — not to issue an operational license for the Ostrovets NPP before a full scope of SEED mission is completed.
WENRA recommendations, 2013. Failure to ensure that Ostrovets NPP would sustain a major airplane crash.	(Parliamentary Assembly of the Council of Europe, 2018) — not to issue an operational license for the Ostrovets NPP before reactors' protection against a crash of a major commercial airplane is enhanced.
EU "stress tests". Recommendations provided in the EU stress test peer review report must be implemented before the Ostrovets NPP is launched.	(Parliamentary Assembly of the Council of Europe, 2018) — not to issue an operational license for the Ostrovets NPP before the recommendations of the peer review report on the Belarus stress test are implemented in full. (Baltic Council of Ministers, 2018) — underlines the importance to fully implement the recommendations of the peer review report. European Commission, 2018 — recommendations by the peer review require should be implemented and the status of their implementation should be a subject of another independent review.

Sources: United Nations Economic Commission for Europe. ACCC/S/2015/2 Belarus, https://www.unece.org/environmental-policy/conventions/public-participation/aarhus-convention/tfwg/envppcc/submissions/acccs20152-belarus.html; United Nations Economic and Social Council (2013). Report of the Implementation Committee on its Twenty-Seventh Session https://www.unece.org/fileadmin/DAM/env/documents/2013/eia/ic/ece.mp.eia.ic.2013.2e.pdf; United Nations Economic and Social Council (2014). Decision VI/2, https://www.unece.org/fileadmin/DAM/env/eia/meetings/Decision_VI.2.pdf; United Nations Economic and Social Council (2017). Report of the Compliance Committee. Compliance by Belarus with its Obligations under the Convention. https://www.unece.org/fileadmin/DAM/env/pp/mop6/English/ECE_MP.PP_2017_35_E.pdf; Parliamentary Assembly of the Council of Europe (2018). Nuclear Safety and Security in Europe http://assembly.coe.int/nw/xml/XRef/Xref-XML2HTML-EN.asp?fileid=25175&lang=en; Baltic Council of Ministers (2018). Joint Statement https://lrv.lt/uploads/main/documents/files/JOINT%20STATEMENT%202018.pdf

possibilities to expand the interconnection between Sovietsk and Klaipeda to a capacity of 1500 MW (Nagevičius, 2013), thus playing its hand to the full extent. Nevertheless, Russia could circumvent Lithuania's refusal to strengthen its interconnections on the border with Kaliningrad by persuading Poland to link its electricity system with the Russian enclave and convincing other countries to interconnect their electricity grids via the Baltic Sea. However, it failed to do so. Russia proposed that Poland build an asynchronous interconnection with a transmission capacity of 1000 MW and link Mamonowo with Olsztyn (Ea Energy Analyses, 2010). Even though Poland did consider such a possibility, Warsaw was not interested in the end. Russia also suggested to Germany to construct a submerged interconnection, having a transmission capacity between 800 and 1000 MW, under the Baltic Sea but did not manage to capture Berlin's interest (Menkiszak, 2013).

Moreover, Russia failed to expand the Kaliningrad's cross-border electricity interconnections despite trying to lure external investments into Baltic NPP, something not seen before in NPP projects developed on Russian soil. To increase chances of finding customers for the Baltic NPP and making agreements regarding additional cross-border electricity infrastructure, Russia offered potential investors to acquire up to 49% equity in the project (Usanov and Kharin, 2014, p. 8). According to Menkiszak, Rosatom negotiated with such Western companies as Enel, E.ON, EdF, GDF Suez, Fortum, CEZ, and Iberdrola but they failed to produce any results (2013). With no export markets available, it is safe to assume that the Baltic NPP project in Kaliningrad will not resume until they are found.

Apart from the Russian failure to secure the market for the Baltic NPP, other signs indicate that the construction will remain frozen. First, in 2017 the reactor pressure vessel initially designed for the first unit Baltic NPP was sent to Ostrovets NPP, replacing the one that was accidently dropped from a height of 2−4 m (World Nuclear Association, 2018). Second, in 2018 European Commission, Poland, Lithuania, Latvia, and Estonia finally reached an agreement to synchronize the electricity systems of the Baltic states with the Continental European Network via Poland (European Commission, 2018a), thus further reinforcing the claim that Baltic and Polish export markets will remain closed.

Third, Russia is reacting to the desynchronization of the Baltic states electricity systems from the Integrated Power System/Unified Power System (IPS/UPS) by constructing additional electricity generation capacities in Kaliningrad, and that magnifies the issue of electricity surplus even further. On March 2, 2018, Russia commissioned two new thermal power plants (TPPs), and the other two TPPs are under construction. The recently completed TPPs (Mayakovskaya and Talakhovskaya) consist of four gas turbines, each adding generation capacity of 78 MW. Four additional natural gas turbines of Pregolskaya TPP are expected to be finished in 2019 and will contribute 454 MW generation capacity (Inter RAO, 2018), while the three coal-burning units of Primorskaya TPP are expected to be completed by 2020 and will increase Kaliningrad's generation capacity by 195 MW (Inter RAO). Providing that the remaining two projects will be finished successfully, these four new power plants will add additional 951 MW to Kaliningrad's already surplus generation capacity of 950 MW (President of the Russian Federation, 2018) that has a peak load only of 700 MW (Varanavičius, 2018, p. 16). These projects clearly show that Russia is preparing for isolating Kaliningrad's electricity system—not for resuming the construction of Baltic NPP.

Due to the strong interconnectivity (1300 MW) along the Lithuanian−Belarusian border, Lithuania had to resort to more decisive response measures if it wanted to oppose the construction of Ostrovets NPP and to hinder Belarusian plans to export its electricity to Lithuania. It was already mentioned that Lithuania decided to limit electricity trade with Belarus and to forbid it to use the services of Kruonis pump-storage power plant at first and then completely cut the links with the Belarusian electricity system after synchronizing its electricity grid with CEN. Poland assisted Lithuania in its efforts by declaring that it will not import the electricity generated in Ostrovets NPP (Astapenia, 2018), thus denying Belarus direct access to Western markets. Despite that these national measures were far stronger than the ones used against Baltic NPP in Kaliningrad and will entail certain financial costs for Belarus, they have failed to persuade Russia and Belarus to discontinue the construction of Ostrovets NPP.

The Belarusian electricity system can withstand such Lithuanian and Polish pressure and still accommodate Ostrovets NPP as it has four cross-border electricity interconnections with Russia and two with Ukraine (IAEA, 2015b) and it also has a far larger electricity demand then Kaliningrad does (Belta, 2017). Even though these advantages give Belarus more room to maneuver, it still faces the problems of ensuring the smooth operation of Ostrovets NPP during night hours (using the Krounis pump-storage power plant would have solved this issue) and finding new ways to generate revenue after being banned from exporting its electricity westward. The magnitude of these issues created by Poland and Lithuania together with its technological and economic impact on Belarus should be studied in greater detail as it is not possible to somehow specify them in a couple of sentences. It is reasonable to assume, however, that these decisions are making an impact because Belarus is trying to provide the industry with incentives to consume more electricity during the nighttime and is removing the subsidies for the electricity tariffs for the household consumers (Charter 97, 2017; Belaeas, 2018; Belta, 2015).

In sum, the case study shows that Lithuania did not limit itself to utilizing various international legal mechanisms and memberships in international organizations for opposing the construction of Russian NPPs—it also relied heavily on the response measures at the national level. To what extent did Lithuanian efforts hinder the construction of Russian NPPs and which instruments were most effective in doing so?

Discussing the outcome of Lithuania's attempts to oppose the Russian NPPs

Despite securing favorable political declarations from international organizations and proving that Belarus is in noncompliance with the Espoo Convention, these breakthroughs had only a minor impact on the construction of Ostrovets NPP. For example, Rosatom decided to replace the fallen nuclear reactor pressure vessel not because it considered being damaged but in an attempt to protect Belarus from Lithuanian accusations, and Froggatt and Schneider estimates that it delayed the construction of Ostrovets NPP by 8 months (2018, p. 155). On the other hand, it is impossible to verify wherever such decision was indeed connected with focused Lithuanian efforts to denounce the nuclear safety and expose the legal flaws of Ostrovets NPP by working through international organizations. Lithuania also persuaded Russia to undergo a couple of additional legal procedures regarding the Baltic NPP, but they did not constrain the development of the NPP in any way.

Lithuania managed to achieve greater impact by a national effort. Its refusal to expand the capacities of power lines linking Kaliningrad to Lithuania substantially contributed to freezing the construction of Baltic NPP in 2013 as Moscow failed to solve the problems of insufficient demand. Lithuanian law banning electricity import from Belarus will negatively impact the economic rationale of Ostrovets NPP, while denying Minsk the possibility to use the services of Kruonis pump-storage power plant forces Belarus to look for alternative ways to ensure the smooth operation of its NPP. Even though these Lithuanian measures will impose certain costs that are not precisely clear at the moment, they were not sufficient for stopping the project.

In between the lines of Lithuanian national effort and EU's support lay another important achievement—agreement to synchronize the Baltic state's electricity systems with CEN by 2025—which is already imposing financial costs for Russia and Belarus. Synchronization will expand Lithuania's options in regulating electricity flows with Belarus and Kaliningrad because it allows for disconnecting the existing power lines with them, thus enabling Lithuania to stop the physical electricity flow between the national electricity systems. The regional infrastructural developments imply that the widening of Lithuania's strategic options was taken seriously by Russia and Belarus. In order to prepare for the breakup of the BRELL ring, they began upgrading their transmission grids (Varanavicius, 2018), while the geopolitical importance of Kaliningrad enclave provoked Russia to double its already surplus generation capacity and to construct an alternative natural gas supply route to the existing one through Lithuania by developing a regasification terminal of liquefied natural gas. Once again, Lithuania was successful in imposing additional financial costs for Russia and Belarus, but that has not stopped the construction of Ostrovets NPP. As far as the Baltic NPP is concerned, the agreement on synchronization was reached 5 years after its construction was frozen, so it cannot be considered as a factor that helped to do so. However, synchronization contributes to the other conditions that ensure this project will not be revitalized.

Even though it seems that Lithuania made the most impact by engaging Russian NPPs directly as opposed to approaching them through international legal mechanisms and international organizations, it was the membership in these organizations that allowed to employ the already discussed national measures in the first place. For instance, Lithuania would not be capable of synchronizing with CEN in the absence of EU's political support and financial assistance. Moreover, Lithuania would have to be far more careful in its bilateral relations with Russia if NATO would not protect it against Russian military countermeasures or the EU would not safeguard the country against any hostile economic activity.

Conclusion

This chapter revealed that Lithuania opposes the construction of Russian NPPs not so much on the basis of their shortcomings but mostly because it perceives them as threats to its national security or, more precisely, instruments of Russian foreign policy that seeks to hinder the development of Lithuanian strategic energy infrastructure and to constrain its European integration in doing so. The closeness of Ostrovets and Baltic NPPs to the national borders and concerns over their legal issues and nuclear safety flaws have also played a part in provoking Lithuania's opposition, even though significantly lesser than national security considerations. Lithuania used the above mentioned factors as arguments that justify its opposition against Russian NPPs and sometimes even as instruments that help to protest against them.

The Lithuanian case distinguishes from the broader context of disputes over strategic energy infrastructure that usually involves more or less equal countries or great powers and its weaker neighbors because opposition against Russian NPPs does not fit in either of these categories. In line with the bedrock assumptions of small state studies, Lithuania actively worked through international organizations and tried to take advantage of international legal instruments in order to stop the construction of Ostrovets NPP and, to a lesser extent, to hinder the development of the Baltic NPP project as it failed to progress beyond the foundations of the first unit and was frozen in 2013. Lithuania also engaged both Russia and Belarus directly. To prevent the construction of Baltic NPP, Lithuania refused to expand its electricity interconnections with Kaliningrad. In order to stop the development of Ostrovets NPP, Lithuania adopted a law that bans the electricity trade with Belarus after the NPP becomes operational and prohibits it from procuring the reserve capacities from the Kruonis pump-storage power plant. Moreover, it linked the opposition against the construction of Russian NPPs with its strategic priority to synchronize the electricity system with CEN.

Lithuania was least successful in using international legal instruments against Russian NPPs as only Ostrovets NPP was declared to be in noncompliance with the Espoo Convention due to Lithuanian complaint. The country was far more successful in exploiting its membership in international organizations because it managed to secure many favorable political declarations that either endorse its general position toward Ostrovets NPP or support Lithuania's specific accusations of Belarusian noncompliance to international law and nuclear safety-related recommendations. However, neither favorable decisions by the responsible review bodies of international conventions nor the political pressure from international organizations translated into a significant impact that somehow slowed the development of Russian NPPs.

Only national measures made a notable impact on the development of Russian NPPs. The Lithuanian decision not to expand the cross-border electricity transmission links with Kaliningrad contributed to Russia's failure to secure the demand for the Baltic NPP and eventually forced Moscow to freeze the project. On the other hand, Lithuania's law that prohibits purchasing electricity from Belarus will entail certain financial costs, while the refusal to allow Belarus to procure the reserve capacities of Kruonis hydro-pump storage plant creates technological issues.

Notwithstanding that Lithuania managed to make the most significant impact on the development of Russian NPPs by national means this does not contradict the bedrock assumptions of small state literature. In the absence of membership in NATO and the EU, Lithuania would find itself vulnerable to Russian countermeasures and could not oppose the construction of Ostrovets and Baltic NPPs on the national level in the first place.

Author biography

Justinas Juozaitis is a junior researcher at General Jonas Zemaitis Military Academy of Lithuania and a lecturer at Vytautas Magnus University, where he is also a Ph.D. candidate in political science. Before joining the Lithuanian Military Academy in 2018, he worked at

Energy Security Research Center and was a member of the Future Energy Leaders program (FEL-100) at the World Energy Council. In 2015 and 2017, he was awarded scholarships from the Lithuanian Academy of Sciences for academic achievements in a category of young scientists. Justinas Juozaitis is interested in energy, small states, intergovernmental organizations, and international relations theory.

References

Amnesty International, 2013. What Is Not Permitted Is Prohibited. Silencing Civil Society in Belarus. https://www.refworld.org/docid/51767bb64.html.

Archer, C., Bailes, J.K.L., Wivel, A., 2014. Small States in International Security. Europe and beyond. Routledge, New York.

Astapenia, R., 2018. Will the West Join the Lithuania's Crusade against Belarus NPP? https://udf.by/english/featured-stories/168719-will-the-west-join-the-lithuanias-crusade-against-belarus-npp.html.

Backaitis, S., 2011. Dangers from Proposed Belarus and Russian Nuclear Power Plants to Lithuania. Lithuanian American Council, Chicago.

Baltic Council of Ministers, 2018. Joint Statement. https://lrv.lt/uploads/main/documents/files/JOINT%20STATEMENT%202018.pdf.

Banktrack, 2018. Baltic Nuclear Power Plant — Kaliningrad. https://www.banktrack.org/project/baltic_nuclear_power_plant_kaliningrad/pdf.

Belaes, 2018. Construction of the Power Distribution System of Belarusian NPP Is Completed — the Ministry of Energy. http://www.belaes.by/en/news/item/2200-raboty-na-belaes-idut-po-grafiku-minenergo.html.

Belta, 2015. Belarusian Power Grid's Readiness for Nuclear Power Plant Launch Reviewed. https://eng.belta.by/president/view/belarusian-power-grids-readiness-for-nuclear-power-plant-launch-reviewed-86044-2015.

Belta, 2017. Electricity Consumption in Belarus to Grow by 3bn kWh by 2020. https://eng.belta.by/economics/view/electricity-consumption-in-belarus-to-grow-by-3bn-kwh-by-2020-107690-2017/.

Belta, 2018. National Report on Belarusian Nuclear Power Plant Stress Tests Compliant with EU Requirements. https://www.belarus.by/en/press-center/speeches-and-interviews/national-report-on-belarusian-nuclear-power-plant-stress-tests-compliant-with-eu-requirements_i_0000083878.html.

Bock, H., Drabova, D., 2005. Transboundary Risks the Temelin Case. https://inis.iaea.org/collection/NCLCollectionStore/_Public/41/109/41109418.pdf.

Cesnakas, G., 2012. Energy security in the Baltic-Black Sea Region: energy insecurity sources and their impact upon states. Lithuanian Annual Strategic Review 10, 155—198.

Cesnakas, G., 2013. Energy security challenges, concepts and the controversy of energy nationalism in Lithuanian energy politics. Baltic Journal of Law & Politics 6 (1), 106—139.

Cesnakas, G., 2018. Baltic states. In: Butler, E., Ostrowski, W. (Eds.), Understanding Energy Security in Central and Eastern Europe: Russia, Transition and National Interests. Routledge, London and New York.

Cesnakas, G., Juozaitis, J., 2017. Nuclear Geopolitics in the Baltic Sea Region. Exposing Russian Strategic Interests behind Ostrovets NPP. Atlantic Council, Washington, D.C.

Charter 97, 2016a. Deputy Minister of Energy: 10 Accidents Happened at NPP, 3 Dead. https://charter97.org/en/news/2016/9/15/222688/.

Charter 97, 2016b. How Did Reactor Fall at the Belarusian NPP? https://charter97.org/en/news/2016/7/27/215441/.

Charter 97, 2017. Electricity Tariff "trap" Prepared for Belarusians. https://charter97.org/en/news/2017/10/31/267736/.

Council of Ministers of the Republic of Belarus, 2015. Концеﻪция Лнерﻪетической Безоﻪасности Ресﻪублики Беларусь (Concept of Energy Security of the Republic of Belarus). http://www.government.by/upload/docs/file5a034ca617dc35eb.PDF.

Dyner, A.M., 2018. The importance of the energy and petroleum industries for Belarus. PISM Bulletin 110, 1181. http://www.pism.pl/publications/bulletin/no-110-1181.

EA Energy Analyses, 2010. Energy Perspectives for the Kaliningrad Region as an Integrated Part of the Baltic Sea Region. https://www.bdforum.org/wp-content/uploads/2010/12/thematic_reports_energy_persp_kalinin-grad_2010.pdf.

Elsuwege, P., 2008. The uncertain future of the Ignalina nuclear power plant: Lithuania's obligations under the Treaty of Accession to the EU. Teise 68, 155—162.

ENSREG, 2018. EU Peer Review Report of the Belarus Stress Test. http://www.ensreg.eu/sites/default/files/attachments/hlg_p2018-36_155_belarus_stress_test_peer_review_report_0.pdf.

European Commission, 2016. Annex 2 of the Commission Decision on the Annual Action Programme 2016 for Nuclear Safety. https://ec.europa.eu/europeaid/sites/devco/files/aap-financing-belarus-action_fiche-20160411_en.pdf.

European Commission, 2018a. Synchronization of the Baltic States' Electricity Grid with the Continental European System. https://ec.europa.eu/info/news/synchronisation-baltic-states-electricity-grid-continental-european-system-2018-sep-14_en.

European Commission, 2018b. Comprehensive Risk and Safety Assessments of the Belarus Nuclear Power Plant Completed. http://europa.eu/rapid/press-release_IP-18-4347_en.htm.

European Parliament, 2018. European Parliament Resolution of 19 April 2018 on Belarus. http://www.europarl.europa.eu/cmsdata/142472/EP_resolution_BY_April2018.pdf.

Federal Ministry for the Environment, 2015. Nature Conservation and Nuclear Safety France Remains Committed to Closure of Fessenheim. https://www.bmu.de/en/pressrelease/france-remains-committed-to-closure-of-fessen-heim/.

Federal Ministry for the Environment, 2018. Nature Conservation and Nuclear Safety Federal Environment Minister Svenja Schulze on Inaugural Trip to Brussels. https://www.bmu.de/en/pressrelease/bundesumweltministerin-svenja-schulze-zum-antrittsbesuch-in-bruessel/.

Galbreath, D.J., Lamoreaux, J.W., 2007. Bastion, beacon or bridge? Conceptualizing the Baltic logic of the EU's neighborhood. Geopolitics 12 (1), 109—132.

Gorin, V., Maruda, N., Polyukhovich, V., 2015. Personnel for Belarus' Nuclear Power Engineering: Training, Popularity, Prestige. Belta. https://atom.belta.by/en/conf_en/view/personnel-for-belarus-nuclear-power-engineering-training-popularity-prestige-6655.

Government of the Republic of Lithuania, 2017. On the Approval of the Action Plan for the Prevention of Unsafe Nuclear Power Plant Which Is Being Built in the Republic of Belarus and Threatening National Security, the Environment and Public Health of the Republic of Lithuania. https://e-seimas.lrs.lt/portal/legalAct/lt/TAD/deba7a6199db11e78d46b68e19efc509?jfwid=-2icx98646.

Handl, G., 2015. Preventing transboundary nuclear pollution: a post-Fukushima legal perspective. In: Beckman, R., Jayakumar, S., Koh, T., Phan, H.D. (Eds.), Transboundary Pollution— Evolving Issues of International Law and Policy. Cheltenhan & Northampton: Edward Elgar Publishing, pp. 190—232.

Hussein, J.H., Kardaś, S., Kłysiński, K., 2018. Troublesome Investment. The Belarusian Nuclear Power Plant in Astravyets. Centre for Eastern Studies, Warsaw.

Inter RAO. Primorskaya TPP http://irao-engineering.ru/en/projects/primorskaya-tpp/.

International Atomic Energy Agency. Peer review and advisory services calendar https://www.iaea.org/services/review-missions/calendar?type=All&year%5Bvalue%5D%5Byear%5D=&location=3510&status=All.

International Atomic Energy Agency, 2010. Peer Appraisal of the Arrangements in the Republic of Belarus Regarding the Preparedness for Responding to a Radiation Emergency. International Atomic Energy Agency, Minsk.

International Atomic Energy Agency, 2015a. International Peer Review of the Environmental Impact Assessment Performed for the Licence Application of the Baltic-1 Nuclear Power Plant, Kaliningrad, Russian Federation. https://www-pub.iaea.org/MTCD/Publications/PDF/ENV_KGD_web.pdf.

International Atomic Energy Agency, 2015b. Belarus. https://www-pub.iaea.org/MTCD/publications/PDF/CNPP2015_CD/countryprofiles/Belarus/Belarus.htm.

International Atomic Energy Agency, 2016. IAEA Mission Says Belarus Committed to Nuclear Safety; Further Regulatory Strengthening Needed Ahead of Reactor Start. https://www.iaea.org/newscenter/pressreleases/iaea-mission-says-belarus-committed-to-nuclear-safety-further-regulatory-strengthening-needed-ahead-of-reactor-start.

International Atomic Energy Agency, 2017a. Safety of the Belarusian NPP against Site Specific External Hazards. https://www.iaea.org/sites/default/files/documents/review-missions/seed_mission_report_belarus_2017.pdf.

International Atomic Energy Agency, 2017b. IAEA Mission Concludes Site and External Events Design (SEED) Review in Belarus. https://www.iaea.org/newscenter/pressreleases/iaea-mission-concludes-site-and-external-events-design-seed-review-in-belarus.

International Atomic Energy Agency, 2018. IAEA Reviews Belarus's Emergency Preparedness and Response Framework. https://www.iaea.org/newscenter/pressreleases/iaea-reviews-belaruss-emergency-preparedness-and-response-framework.

Inter RAO, 2018. The First Power Unit of the Pregolskaya TPP in Kaliningrad was Commissioned. http://irao-engineering.ru/en/pressroom/the-first-power-unit-of-the-pregolskaya-tpp-in-kaliningrad-was-commissioned/.

Interview with Ex-Lithuanian Official. Vilnius, February 13, 2014.

Interview with Lithuanian Official. Vilnius, April 27, 2014.

Ioffe, G., 2018. Belarusian nuclear power plant proceeding full speed ahead. Eurasia Daily Monitor 15 (91). In: https://jamestown.org/program/belarusian-nuclear-power-plant-proceeding-full-speed-ahead/.

Jirušek, M., Vlček, T., 2015. Energy Security in Central and Eastern Europe and the Operations of Russian State-Owned Energy Enterprises. Masaryk University & Prague Security Studies Institute, Prague.

Juozaitis, J., 2016. Lithuanian foreign policy vis-à-vis Belarusian nuclear power plant in Ostrovets. Lithuanian Foreign Policy Review 35 (1), 41−46.

Kadisa, S., Klementavicius, A., Radziukynas, V., Radziukyniene, N., 2016. Challenges for the baltic power system connecting synchronously to continental European network. Electric Power Systems Research 140, 54−64.

Kaminskaya, M., 2011. Minsk's Cooperation Agreement with Moscow on Building Ostrovets NPP Ratified in Closed-Door Parliament Hearing. Bellona. http://bellona.org/news/ukategorisert/2011-11-minsks-cooperation-agreement-with-moscow-on-building-ostrovets-npp-ratified-in-closed-door-parliament-hearing.

Lamoreaux, J.W., 2014. Acting small in a large state's world: Russia and the Baltic states. European Security 23 (4), 565−582.

Litgrid, 2014. Development of the Lithuanian Electric Power System and Transmission Grids. http://www.leea.lt/wp-content/uploads/2015/05/Network-development-plan-2015.pdf.

Litgrid, 2018. Lietuvos elektros Energetikos Tinklų 400-110 kV Tinklų Plėtros Planas 2018−2027, (Lithuanian Electricity Energy System 400-110 kV Network Development Plan 2018−2027). http://www.litgrid.eu/uploads/files/dir435/dir21/dir1/0_0.php.

Masiulis, R., 2016. Energetikos strategijos Kryptys (Energy Strategy Vectors).

Mearsheimer, J.J., 1994. The False Promise of international institutions. International Security 19 (3), 5−49.

Menkiszak, M., 2013. Russia Freezes the Construction of the Nuclear Power Plant in Kaliningrad. https://www.osw.waw.pl/en/publikacje/analyses/2013-06-12/russia-freezes-construction-nuclear-power-plant-kaliningrad.

Ministry of Energy of the Republic of Lithuania, 2017. The Nuclear Power Plant in Astravyets Poses a Threat for Lithuania's National Security. https://enmin.lrv.lt/en/news/the-nuclear-power-plant-in-astravyets-poses-a-threat-for-lithuania-s-national-security.

Ministry of the Emergency Situations of the Republic of Belarus. General Information about the Construction of the Belarusian Nuclear Power Plant https://gosatomnadzor.mchs.gov.by/en/bezopasnost-belorusskoy-aes/obshchaya-informatsiya-o-stroitelstve-belorusskoy-aes/.

Ministry of the Environment of the Republic of Lithuania, 2015. Submission of the Republic of Lithuania Requesting to Investigate the Compliance of the Republic of Belarus with the Provisions of the Aarhus Convention in the Course of the Implementation of the Project for the Construction of a Nuclear Power Plant in Belarus.

Ministry of the Foreign Affairs of the Republic of Lithuania, 2009. Susitikime su ES išorinių ryšių komisare aptarti Lietuvos energetinio saugumo klausimai (Lithuania's energy security aspects discussed in a meeting with EU's Commissioner for External Relations). http://www.urm.lt/default/lt/naujienos/susitikime-su-es-isoriniu-rysiu-komisare-aptarti-lietuvos-energetinio-saugumo-klausimai.

Ministry of the Foreign Affairs of the Republic of Lithuania, 2011a. 2010 Metų Veiklos Ataskaita [Activity Report for 2010]. http://www.urm.lt/uploads/default/documents/Ministerija/veikla/veiklos_ataskaita/2010m_URMataskaita.doc.

Ministry of the Foreign Affairs of the Republic of Lithuania, 2011b. A. Ažubalis to Belarusians: Experiment in the Wasteland, Not Near Our Border. http://www.urm.lt/default/lt/naujienos/aazubalisbaltarusiams-eksperimentuokite-dykvieteje-o-ne-salia-musu-delfilt-2011-m-balandzio-28-d.

Ministry of the Foreign Affairs of the Republic of Lithuania, 2012a. 2011 Metų Veiklos Ataskaita [Activity Report for 2011]. http://www.urm.lt/uploads/default/documents/Ministerija/veikla/veiklos_ataskaita/2011.pdf.

Ministry of the Foreign Affairs of the Republic of Lithuania, 2012b. Lietuva Siekia, Kad Kaliningrado AE Poveikio Aplinkai Vertinimas Būtų Atliekamas Pagal Tarptautinius Reikalavimus (Lithuania Aims that the Environmental Impact Assessment of Kaliningrad NPP Would Correspond to International Requirements). http://www.urm.lt/default/lt/naujienos/lietuva-siekia-kad-kaliningrado-ae-poveikio-aplinkai-vertinimas-butu-atliekamas-pagal-tarptautinius-reikalavimus.

Ministry of the Foreign Affairs of the Republic of Lithuania, 2013. Kaliningrado Atominės Elektrinės Projektas Privalo Atitikti Tarptautinius Saugos Reikalavimus (A Nuclear Power Plant Project in Kaliningrad Must Comply with International Safety Regulations). http://www.urm.lt/default/lt/naujienos/kaliningrado-atomines-elektrines-projektas-privalo-atitikti-tarptautinius-saugos-reikalavimus.

Ministry of the Foreign Affairs of the Republic of Lithuania, 2015. *2014 Metų Veiklos Ataskaita* [Activity Report for 2014]. http://www.urm.lt/uploads/default/documents/Ministerija/veikla/veiklos_ataskaita/URM_2014_veiklos%20ataskaita.pdf.

Ministry of the Foreign Affairs of the Republic of Lithuania, 2016. *2015 Metų Veiklos Ataskaita* [Activity Report for 2015]. http://www.urm.lt/uploads/default/documents/Ministerija/veikla/veiklos_ataskaita/URM_2014_veiklos%20ataskaita.pdf.

Ministry of the Foreign Affairs of the Republic of Lithuania, 2017. *2016 Metų Veiklos Ataskaita* [Activity Report for 2016]. http://www.urm.lt/uploads/default/documents/Ministerija/veikla/veiklos_ataskaita/URM%202016%20m_%20veiklos%20ataskaita%20(final).pdf.

Ministry of the Foreign Affairs of the Republic of Lithuania, 2018a. *2017 Metų Veiklos Ataskaita* [Activity Report for 2017]. http://www.urm.lt/uploads/default/documents/2017%20m_%20URM%20veiklos%20ataskaita.pdf.

Ministry of the Foreign Affairs of the Republic of Lithuania, 2018b. Statement by the Ministry of Foreign Affairs on Astravets Nuclear Power Plant under Construction in Belarus. https://www.urm.lt/default/en/news/statement-by-the-ministry-of-foreign-affairs-on-astravets-nuclear-power-plant-under-construction-in-belarus-.

Ministry of the Foreign Affairs of the Republic of Lithuania, 2018c. L. Linkevičius: The EU Stress Tests Proved that the Ostrovets NPP Is Not Safe. https://www.urm.lt/default/en/news/l-linkevicius-the-eu-stress-tests-proved-that-the-ostrovets-npp-is-not-safe.

Molis, A., Cesnakas, G., Juozaitis, J., 2018. Russia coerces, but the Baltic States persist: the importance of initiatives for integration and cooperation. Politologija 91 (3), 3—47.

Nagevičius, M., 2013. Ir ka mums daryti su tomis prakeiktomis rusiškomis atominėmis elektrinėmis? [And what should we do with these damned Russian nuclear power plants? https://lietuvosdiena.lrytas.lt/aktualijos/2013/02/28/news/ir-ka-mums-daryti-su-tomis-prakeiktomis-rusiskomis-atominemis-elektrinemis-5089871/.

Nord Stream 2, 2017. Espoo Report. https://www.envir.ee/sites/default/files/ns2_aruanne_en.pdf.

Nord Stream, 2009. Nord Stream Environmental Impact Assessment Documentation for Consultation under the Espoo Convention. https://www.nord-stream.com/download/document/69/?language=en.

Nordic-Baltic Eight, 2016. Nordic-Baltic Cooperation Progress Report 2016. http://club.bruxelles2.eu/wp-content/uploads/2016/12/rapportannuelbaltesnordiques@161220.pdf.

Novikau, A., 2017a. Nuclear power debate and public opinion in Belarus: from Chernobyl to Ostrovets. Public Understanding of Science 26 (3), 275—288.

Novikau, A., 2017b. What is "Chernobyl Syndrome?" the use of radiophobia in nuclear communications. Environmental Communication 11 (6), 800—809.

Nuclear Energy Association, 2016. The application of the Espoo convention on environmental impact assessment in a transboundary context to nuclear energy-related activities. Nuclear Law Bulletin 97, 63—69.

Nuclear Threat Initiative, 2010. Russian Nuclear Chronology. https://www.nti.org/media/pdfs/russia_nuclear.pdf?_=1316466791.

Panke, D., 2010. Small States in the European Union. Coping with Structural Disadvantages. Routledge, London and New York.

Parliament of the Republic of Lithuania, 2005. Resolution on the Approval of the National Security Strategy. https://e-seimas.lrs.lt/rs/legalact/TAD/TAIS.262943/format/OO3_ODT/.

Parliament of the Republic of Lithuania, 2007. *Nutarimas Dėl Nacionalinės Energetikos Strategijos Patvirtinimo* [Resolution on the Approval of the National Energy Strategy. https://www.e-tar.lt/portal/lt/legalAct/TAR.498E8E1207CE.

Parliament of the Republic of Lithuania, 2012a. Seimo Nario Pauliaus Saudargo Pranešimas: Ar Tikrai Rusija Gamins "švaria" elektra?[Address by a Member of the Seimas, Paulius Saudargas: Is Russia Really Going to Produce "Clean" Energy?". https://www.lrs.lt/sip/portal.show?p_r=15371&p_k=1&p_t=124767.

Parliament of the Republic of Lithuania, 2012b. Resolution on the Approval of the National Energy Independence Strategy. https://e-seimas.lrs.lt/portal/legalAct/lt/TAD/TAIS.432271.

Parliament of the Republic of Lithuania, 2012c. Resolution on the Approval of the National Security Strategy. https://www.bbn.gov.pl/ftp/dok/07/LTU_National_Security_Strategy_2012.pdf.

Parliament of the Republic of Lithuania, 2016a. Resolution Concerning the 30th Anniversary of Nuclear Disaster in Chernobyl and the Danger of Ostrovets Nuclear Power to Lithuanian and Proposal to the Government to Take Action to Mitigate the Threat. http://www.lrs.lt/sip/getFile?guid=ca8fead1-4de2-4dd0-80cb-b0d1f1a4697d.

Parliament of the Republic of Lithuania, 2016b. Nutarimas Dėl Lietuvos Respublikos Vyriausybės Programos. [Ruling Regarding the Programme of Government of the Republic of Lithuania]. https://e-seimas.lrs.lt/portal/legalAct/lt/TAD/886c7282c12811e682539852a4b72dd4.

Parliament of the Republic of Lithuania, 2017. Law on Necessary Measures of Protection against the Threats Posed by Unsafe Nuclear Power Plants in Third Countries. https://e-seimas.lrs.lt/portal/legalAct/lt/TAD/74d3ebb07bf-c11e7aefae747e4b63286?jfwid=10a2n9kodt.

Parliamentary Assembly of the Council of Europe, 2017. The Situation in Belarus. http://assembly.coe.int/nw/xml/XRef/Xref-XML2HTML-en.asp?fileid=23935&lang=en.

Parliamentary Assembly of the Council of Europe, 2018. Nuclear Safety and Security in Europe. http://assembly.coe.int/nw/xml/XRef/Xref-XML2HTML-EN.asp?fileid=25175&lang=en.

Parliamentary Political Parties of the Republic of Lithuania, 2017. Susitarimas Dėl Bendrų Veiksmų Dėl Nesaugios Astravo AE, (Agreement Regarding Common Measures Regarding the Unsafe Ostrovets NPP). http://www.lrs.lt/sip/getFile?guid=511b8a2b-8f85-4262-8b7a-de784e01d616.

Pedraza, J.M., 2015. Electrical Generation in Europe. The Current Situation and Perspectives in the Use of Renewable Energy Sources and Nuclear Power for Regional Electricity Generation. Springer, New York.

President of the Republic of Belarus, 2007a. Concept of Energy Security of the Republic of Belarus (Decree of the President of the Republic of Belarus No.433. http://www.lse.ac.uk/GranthamInstitute/law/concept-of-energy-security-of-the-republic-of-belarus-decree-of-the-president-of-the-republic-of-belarus-no-433/.

President of the Republic of Belarus, 2007. О Приоритетныч наЦравленияч укреЦления Лкономической БезоЦасности Государства (Economy and Thrift — the Main Factors of the Economic Security of the State). http://president.gov.by/ru/official_documents_ru/view/direktiva-3-ot-14-ijunja-2007-g-1399/.

President of the Republic of Lithuania, 2011. The U.S. Is Lithuania's Partner in Energy Independence. https://www.lrp.lt/lt/jav-lietuvos-partnere-siekiant-energetinesnepriklausomybes/pranesimai-spaudai/11500.

President of the Russian Federation, 2018. The President Launched Two Thermal Power Plants in Kaliningrad Region. http://en.kremlin.ru/events/president/news/56968.

Riley, A., 2018a. A pipeline too far? EU law obstacles to Nordstream 2. International Energy Law Review 1—25. https://ssrn.com/abstract=3114202.

Riley, A., 2018b. Nord Stream 2: Understanding the Potential Consequences. Atlantic Council, Washington, D.C.

Rosatom, 2016a. ROSATOM Representatives Took Part at Platts European Power Summit. https://www.rosatom.ru/en/press-centre/news/rosatom-representatives-took-part-at-platts-european-power-summit-/?sphrase_id=574484.

Rosatom, 2016b. Viktor Riedel: Baltic NPP Is an Interlink between Central and Eastern Europe's Energy Markets. https://www.rosatom.ru/en/press-centre/news/viktor-riedel-baltic-npp-is-an-interlink-between-central-and-eastern-europe-s-energy-markets/?sphrase_id=574486.

Froggatt, A., Schneider, M., 2018. The World Nuclear Industry Status Report 2018. A Mycle Schneider Consulting Project, Paris and London.

Slivyak, V., Koroleva, A., Raguzina, G., 2009. Critical review of the environmental impact assessment (the EIA) of the baltic NPP project. Kaliningrad: Ecodefense.

State Security Department, 2015. Annual Threat Assessment 2014. Vilnius.

State Security Department and Second Investigation Department under the Ministry of National Defence, 2017. National Security Threat Assessment 2016.

Stsiapanau, A., 2016. Nuclear exceptionalism in the former Soviet union after Chernobyl and Fukushima. In: Hindmarsh, R., Priestley, R. (Eds.), The Fukushima Effect— A New Geopolitical Terrain. Routledge, New York.

Thorhallsson, B., 2018. Studying small states: a review. Small States & Territories 1 (1), 17—34.

Trend, 2018. Armenia's Metsamor NPP — Bomb Waiting to Explode, Azerbaijani MP Says. https://en.trend.az/azerbaijan/politics/2963892.html.

United Nations Economic Commission for Europe. ACCC/S/2015/2 Belarus, https://www.unece.org/environmental-policy/conventions/public-participation/aarhus-convention/tfwg/envppcc/submissions/acccs20152-belarus.html.

United Nations Economic and Social Council, 2013. Report of the Implementation Committee on its Twenty-Seventh Session. https://www.unece.org/fileadmin/DAM/env/documents/2013/eia/ic/ece.mp.eia.ic.2013.2e.pdf.

United Nations Economic and Social Council, 2014. Decision VI/2. https://www.unece.org/fileadmin/DAM/env/eia/meetings/Decision_VI.2.pdf.

United Nations Economic and Social Council, 2017. Report of the Compliance Committee. Compliance by Belarus with its Obligations under the Convention. https://www.unece.org/fileadmin/DAM/env/pp/mop6/English/ECE_MP.PP_2017_35_E.pdf.

United Nations Economic Commission for Europe, 2014. Decision V/9c on Compliance by Belarus. https://www.unece.org/fileadmin/DAM/env/pp/mop5/Documents/Post_session_docs/Decision_excerpts_in_English/Decision_V_9c_on_compliance_by_Belarus.pdf.

United Nations Economic Commission for Europe, 2017. Decision VI/8c. Compliance by Belarus with its Obligations under the Convention. https://www.unece.org/fileadmin/DAM/env/pp/compliance/MoP6decisions/Compliance_by_Belarus_VI-8c.pdf.

Usanov, A., Kharin, A., 2014. Energy security in Kaliningrad and geopolitics. Baltic Sea Policy Briefing 2, 1—18.

Varanavičius, L., 2018. The Necessity of Synchronization of the Baltic States' Electricity Network with the European System. Litgrid, 28 June 2018.

World Nuclear Association, 2018. Nuclear Power in Russia. http://www.world-nuclear.org/information-library/country-profiles/countries-o-s/russia-nuclear-power.aspx.

Case study: free public transport as instrument for energy savings and urban sustainable development—the case of the city of Tallinn

Gunnar Prause, Tarmo Tuisk

Department of Business Administration, Tallinn University of Technology, Tallinn, Estonia

Introduction

As the world is changing, humankind is always faced with several new paradigms. Rapid urbanization that has taken place throughout the 20th century has caused a new reality where majority of the world's inhabitants are already settled in cities. In coming decades

Energy Transformation towards Sustainability
https://doi.org/10.1016/B978-0-12-817688-7.00008-2

the share of urban population will reach over 70%. First, the cities need to find possibilities to manage this growth. Second, they also need to look for solutions for a plethora of other challenges at the same time. These challenges are the outcomes of economic and cultural globalization, environmental problems, and competition between these cities themselves—in order to attract more and more investments. Third, the "architects of the future" and first at all the researchers need to focus on and more thoroughly investigate the ways these current changes influence people on how they live, work, study, consume goods and services, spend their leisure time, and how this relates to urban sustainable development and energy saving while the world has been forecasted to run out of classical (fossil) resources. On the one hand, the implementation of alternative energies can be costly and dependent on geographical location and access possibilities to these new resources; on the other hand, the implementation of green supply chain concepts in the mobility sector can lead to competitive advantages (Hunke and Prause, 2014). All these aspects lead the authors to the understanding that contemporary cities are already facing a new reality.

The integration of digital technologies into different urban systems (e.g., transportation and intracity logistics) and cities' infrastructure in general has been increasingly shaping inhabitants' everyday experiences in respect to their mobility. At the same time we know that in all contemporary cities their administrations, environmental planners, and transportation organizers have been seeking possibilities on how to minimize the share of driving inside their cities to lower the impact of public space usage, decrease pollution levels and the number of fuel-consuming cars. Overall, unfortunately the studies show that in spite of all the measures applied, the number of cars is still continuously rising. Only small-scale improvements have been achieved by incremental policy changes, but in reality much more and profound interventions should be implemented to enable major changes to occur in existing urban transportation to force any modal shift, which has been seen as a key to granting sustainability of the cities at large. These environmentally oriented changes should be considered in combination with socioeconomic sustainability as well.

This chapter addresses the research question to estimate the influence of socioeconomic aspects for sustainable development in the context of free urban transportation. It also tries to measure the impact of fare-free public transportation (FFPT) to economic sustainability. This research is based on the case study of Tallinn free public transport using secondary data sources and statistical analysis. This chapter includes the following sections. First, is the introduction, the theoretical part—containing the history of current socioeconomic reality with respect to specific settlements in the city of Tallinn. Also, a theoretical framework for understanding the pricing of urban public transport is presented. Then, the system of different methods (expert interviews and secondary data analysis) is discussed. Research results will be described and the discussion and the findings are presented. The results are summarized in the conclusions.

Theoretical background

Taking into account socioeconomic aspect of urbanization, we know that growing inequalities in all European countries, including also the most egalitarian ones, form major challenges endangering stability and sustainability of their cities together with their competitiveness. Overall, the spatial gap when comparing the richest and the poorest increases in all major European cities leading to the loss of social stability among the communities residing there. Thereby, different measures that have been applied to overcome the gap have been analyzed by Tammaru et al. (2015). Their analysis, based on the example of Tallinn, shows that Estonia (and also Latvia and Lithuania) are unique due to their radical institutional transition from a state-controlled socialist system to one of the most liberal market-oriented systems in Eastern Europe. Tammaru and other authors name this historical era as "market experiment" unfolding in Tallinn. According to their findings, a significant part of the population—labeled as the Russian-speaking minority—represents a residential pattern that is still determined by the new housing construction and central housing allocation regime that was initiated and implemented during the Soviet period (1944—91). The researchers conclude that Tallinn's history from the Soviet period exemplifies the reality about what happens with long-established minorities when no significant new immigration is taking place and markets determine residential sorting. Tammaru et al. (2015) report that their main finding demonstrates a considerable increase in occupational residential segregation as socioeconomic inequalities become manifested in urban space, with increasing overlap between socioeconomic and ethnic segregation. The socioeconomic and spatial gap that existed during the Soviet period has drastically increased between the two communities despite national policies and municipal attempts together with Estonian and foreign resources that have been used to facilitate cultural and social integration in Tallinn, and in Estonia in general. One more recent study of Vilnius (capital city of Lithuania) shows also that the effect of political preferences among different ethnic groups could not simply be related to their ethnic origin, but rather to socioeconomic status (Burneika et al., 2017), where additionally the wealthiest and the poorest groups tend to become settled more and more separately. In order to lower this kind of segregation level in society, and to avoid accumulation of many negative effects, some efforts can be made by increasing the mobility for these groups or communities. Societal integration can be achieved by enhancing mobility for those who have no sufficient resources to pay transportation fare; the municipalities can provide their support here. One of the measures has been providing FFPT to broaden possibilities and help residents save their financial resources. For a sustainable city, a modal shift from heavier polluting and street space consuming car usage toward economic public transportation (economic buses, electric trams, and trolleybuses) will be the target solution.

Thereby, the pricing of public transportation has been quite often considered as a policy instrument that can be devised to influence or make the modal shift a reality. European Union cities and countries at large have been targets of several studies. We'll have a look into a special Eurobarometer that was carried out in May—June 2013 by the European Commission (2013). The results showed that the Europeans are convinced about the two measures that can facilitate urban transport. These are lower fares (supported by 59%) and "better" public transport (56%). These measures were pointed out by all travel mode users. These two options were high in particular for those who "considered road congestion to be an important problem." For 50% of European Union (EU) member countries participating in the study, "lowering the fares" was the most often supported instrument. At the same time respondents from Hungary (28%), Czechia (27%), Estonia (23%), Latvia (22%), and

Romania (22%) reported higher levels of daily public transport use than the EU average (16%). In Germany the same indicator was 13% and in the Netherlands just 8%.

Despite the fact that the most public transport systems need subsidies for operation, a typical financing formula includes two components: the cost per rider and passenger paid fee. The sum that all passengers have to pay in order to cover his/her cost of ride is called the passenger fare, and it is usually received from the rider by using one of three possibilities. First, the transport user pays the total fare, which seldom occurs in modern countries, and covers all operating costs of his/her ride. Second, the line or service operator i.e., the public transport, pays the full passenger operating cost of each passenger. This kind of free public transport system is also not common today (De Witte et al., 2006). Third, most of the transport user paid fares are partially subsidized, which means that a certain share of this cost is paid by the rider, called the "partial cost recovery" and the rest of the fare is subsidized (i.e., mixture of the first and the second option). In addition, a third party can pay the total cost of the ride, based on an agreement between the transport system and the external party. This has been applied usually in the cases when employers pay for their employees, also when social security institutions cover tickets for their clients, and when schools or universities cover costs for their students (Brown et al., 2001).

Historically, FFPT is not a totally new idea. For decades there has existed public debate that for various reasons public goods and services should be free, granting universal access for everybody, also handling this way road traffic jams, and decreasing negative environmental impact caused by various modes of urban transport. The idea of fare-free public services and goods addresses the principle that access to services like schools, libraries, museums, roads, green areas, and Wi-Fi should be free for use by everybody. Implementing the same approach for public transport underlines that mobility is a "service" and should be fare-free as the rapid growth of cities forces people to settle further away from the city center, further from the locations of schools and work places. Thereby, when the transport costs are high, this can be an obstacle for employees in approaching and participating in the labor market. In addition, an important argument that supports introducing FFPT is improving social inclusion within the society. Here the examples of the Baltic states' capitals serve as appropriate examples. When one has to consider the modal shift on the personal level, as an analysis showed, in this case an urban dweller is more likely to replace using his/her private car with using a public transport vehicle only when the cost of car maintenance or usage increases visibly and very much less likely only due to the decrease of the cost of using the public transport (that the person in question does not use anyway) (Cats et al., 2017).

The case of Tallinn

Tallinn, the capital of Estonia, with a population of 441,101 (January 4, 2019), is today the largest city worldwide that provides FFPT to its citizens. Since 1991 when Estonia regained its independence, after the breakdown of the USSR, there have been several dramatic adjustments in returning to a "new" societal order—a market economy. Use of public transport (PT) had lessened by about 30% between 1991 and 2012. As salaries rose and quality of life improved after the collapse of the USSR, people who had means acquired cars and stopped using public transport, which had been the only mode available for most of the population during Soviet rule (Hess, 2018). For 2012, automobile ownership (425 per 1000 people) had doubled since 1991 (Cats et al., 2014). Comparative data from 2003 to 2015 reveals that car ownership and driving in general rose fast during the first decade of the 21st century. While in 2003 just 24% were driving and PT was used by 41%, then these percentages exchanged places 12 years later, for 2015 (Hess, 2017).

Just how was today's fare-free transportation implemented in Tallinn? The municipality carried out a plebiscite for city residents during March 19–25, 2012. The results showed that 75.5% of the voters supported fare-free public transport for the residents while 24.5% of the participants were against it. By informing their population through this voting possibility—as the municipality leaders later admitted (Alaküla, 2017)—the community was involved in decision-making and the political decision was locked. The fact that the voter turnout was just 20% was considered as unimportant as the decision that was expressed by the voters was not legally binding for the municipality. The City Council implemented FFPT on January 1, 2013. Already during the first year of FFPT it was extended to national train traffic within the borders of the city of Tallinn. The bus lines were prolonged 19–23 km after this change, but more importantly, new lines were implemented which crossed the city center where there was a lot of traffic congestion. Thereby, in 2013 Tallinn became the first capital in the EU providing free public transport to its citizens.

Since the FFPT in Tallinn only applies to citizens registered in Tallinn, a mechanism had to be found to determine who could ride for free and who had to pay. The implementation of this task was realized with the use of the highly developed Estonian e-governmental system that is managed by the long-term project e-Estonia (E-Estonia, 2018), which coordinates the Estonian way into the digital society in cooperation with the state, business, and citizens. In the meantime, "e-Estonia" has become an important driver of innovation for the whole of Estonia, including the transport and logistics sector. This development had already begun in the 1990s, when a radical restart of Estonian administration toward an "invisible, paperless and "24/7" administration" was initiated (Prause, 2016). This radical new beginning also gave rise to the possibility of building a new and advanced IT infrastructure, consisting of the Estonian identity card, the eID card, that is equipped with a chip for digital signature and the installation and implementation of the IT backbone "X-TEE" integrating all public and private service providers that are important for e-services. This integrated concept facilitates also the FFPT in Tallinn by allowing the ability to check the identity and the place of residence of a person using the public transport in Tallinn.

The solution in the Tallinn case consists of a green RFID card each client has to buy for 2€ at public selling places. This RFID card possesses an ID number that has to be linked via an internet portal with the personal eID card of the Estonian resident. When entering a public transport device, the green RFID card has to be scanned by a reader so it can be validated if the person has the right to travel fare free or not. People who are registered outside Tallinn and who wanted to use the same card need to load money to that card. A monthly ticket for them costs 23€. Overall, the realization of the FFPT systems would be more complicated and much more expensive without the use of the underlying e-services.

The readers (validating machines) in the transport vehicles are linked via WLAN with the X-TEE system allowing identifying the person and their place of residence. Furthermore, the checked data are collected to optimize the public transport system in Tallinn. Hence, the goal of e-Estonia, i.e., to pave a sustainable way to digital society, facilitates in our case also the way to FFPT.

After having started the FFPT in Tallinn, according to studies there was an increase in the use of public transport of +10%, as well as a slight decrease of car traffic in the city center (−6%), and a slight increase of automobile traffic around the city center (+4%). Increasing of parking tariffs in the city center (currently six euros per hour in Tallinn Old Town, 4.80 euros in city center surrounding the old town) took place as well.

When comparing increases in private cars ownership, the statistics show in Tallinn it dropped to 0.6%. But in neighboring municipalities, it continued 10% a year as it was before also in Tallinn (Alaküla, 2017).

The Estonian energy dilemma

As with all political decisions, the reasons for establishing the FFPT in Tallinn were manifold. Besides social and political reasons, also environmental aspects played an important role (Estonian Environmental Monitoring, 2013). Thus, three important aspects of the restructuring of the public transport system were related to energy savings, decarbonization of traffic, and to the reduction of pollution, especially air pollution, in Tallinn. These three targets were to be achieved by increasing the utilization of public transport by realizing a modal shift away from cars to more environmentally friendly transport modes as well as to reduce emissions from urban transport.

A closer look at the Tallinn public transport system reveals that three types of fuels are used for transportation, namely diesel, gas, and electricity. The bus services are mainly using diesel fuel but there are also some buses that are fueled with gas. But the most important lines of the urban transport system are represented by four trolleybus lines and four tram lines since they are serving the most populated connections. All these eight lines are using electricity, which seems to be the most favorable for Tallinn citizens concerning air pollution.

By considering the total primary energy supply, i.e., the total amount of primary energy that a country has at their disposal comprising imported energy minus exported energy and energy extracted from natural resources (energy production), Estonia is ranked at the end of EU28, at place no. 27 with 0.147 TOE/1000USD of GDP. The reason for this situation lies in the high percentage of fossil resources for electricity production. An analysis of the Estonian electricity production highlights a national energy dilemma because the main source of energy production in Estonia is based on shale oil, which is mined and produced to the greatest extent in northeastern Estonia in Ida-Virumaa County. Gavrilova et al. (2006) studied the environmental impact of oil shale industry in Estonia and pointed out the environmental impacts and risks related to its use. Besides the emissions of greenhouse gases (GHGs), sulphur oxides (SO_X), nitrogen oxides (NO_X), and particulate matter (PM), the use of oil shale is related with waste management issues.

Siirde et al. (2013) analyzed the GHG emissions from Estonian oil shale as a result of energy production and they calculated for electricity production an emission value of about 1 kg CO_2 per kWh, which is around four times the emission value of the other fossil fuels used in public transportation in Tallinn. A closer look at the other emission values including

SO_X, NO_X and PM also reveals that in these areas oil shale is much more environmentally unfriendly like diesel or gas so that the apparently cleaner lines operated by electric trolley-buses and tramways are only shifting the pollution from Tallinn to Eastern Estonia where the environmental impacts are worse than they would be in Tallinn by using fossil fuel in urban transport in Tallinn.

In the Organisation for Economic Co-operation and Development (OECD's) first Environmental Performance Review of Estonia, independent assessments of the country's progress toward their environmental policy objectives were made (OECD, 2016). The result of this assessment was that Estonia has to speed up its activities to lower its oil shale dependency in order to move toward a greener economy by reducing air pollution. The OECD further recommended improving energy efficiency, to broaden the import of clean energy from other European countries, and to increase investments into renewable energy. With its focus on continuous oil shale mining and use for energy generation, Estonia represents the most carbon intensive economy among the OECD countries, being the third most energy intensive economy as well. Currently, oil shale provides about 70% of the energy supply in Estonia, and the Estonian CO_2 emissions are 533 kg per USD 1000 of GDP in 2014, about twice as high as the OECD average.

Thus, a reduction of oil shale use would not only green the Estonian economy but it would also improve the health and well-being of its population. Despite the fact that Estonia had reached its 2020 EU target of 25% of renewable energy in gross final energy consumption already by year 2011, through use of biomass for heating and wind power generation, the country is still among those that are the lowest in renewable electricity generation and the lowest within the OECD.

But the situation is not so easy from the Estonian viewpoint since oil shale plays an important role in the country's security and social coherence strategy. Prause et al. (2019) pointed out the importance of oil shale industry for Estonian socioeconomic stability since Estonian oil shale industry contributes up to 5% to the national economy and plays an important role in a weakly developed Ida-Virumaa County, which borders the Russian Federation and is populated mostly of Russian speakers and where about half of this regional workforce is employed either directly or indirectly in the oil shale industry. Also, national security issues are related to oil shale industry, e.g., the question of national energy security and national social coherence.

The use of oil shale also has economic advantages because compared to other EU28 countries the Estonian energy prices for households and industry lie below the European average in the range of 0.135 €; /kWh (2017). Another important political issue is dedicated to the national dependency rate showing the extent to which an economy relies upon imports in order to meet its energy needs, i.e., it measures the share of net imports in gross inland energy consumption. Whereas the dependency rate of EU28 in 2016 was equal to 54%, i.e., about half of the EU's energy needs were met by net imports, the same figure for Estonia was only 6.8%, i.e., Estonia represented the member state that was least dependent on imported energy. This energy dependence rate plays an important role in national security policy due to its history as a former Soviet republic located directly at the Russian border.

Thus, Prause et al. (2019) conclude that in spite of the continuous discussions about the ecological impact of oil shale mining, processing, and consumption, there is a need for integrating this industry into Estonian smart specialization strategy. Their research results show that currently the active support to the oil shale industry in Estonia produces many more advantages than disadvantages. Hence, the cleaner air in Tallinn urban transport is paid by high environmental costs in the neighborhoods of the energy production plants in Eastern Estonia, and despite the need to decarbonize the Estonian economy the existing political arguments are slowing down necessary changes.

According to the Central Lab of Estonian Environmental Research Center (EERC, 2013), SO_2 concentrations in 2013 did not exceed European norms in any of three (Center, North-Tallinn, Õismäe) measuring stations of Tallinn. Essential decrease of pollution levels was noted in north Tallinn and Õismäe, while in the city center it remained at the level of 2012. The concentration of SO_2 was measured also in Kohtla-Järve (center of northeast industrial area) where the annual average exceeded ~ 5 times (=6.5 $\mu g/m^3$) the level of Tallinn. The experts of the EERC note that the low level of SO_2 1−2 $\mu g/m^3$ in Tallinn is based on the strict norms of sulfur for liquid fuels (gas, diesel) used in urban transport. The above mentioned slight modal shift from replacing the automobiles by using PT in Tallinn contributes to this SO_2 decrease as well. Emissions related to oil shale industry cause drastic ecologic differences between Kohtla-Järve and Tallinn. In Kohtla-Järve, SO_2 maximum hourly concentration in 2013 was 202.4 $\mu g/m^3$ while for Tallinn the same indicator was also 5 times less, at 48.8 $\mu g/m^3$.

The main source of NO_2 emission is transport, despite the fact that requirements for vehicles have become strict. Catalyzers are used on vehicles, which should lead to low levels of nitrogen dioxide. Although the vehicles have become much more environmentally friendly, still their number is increasing and thereby total emissions of NO_2 is increasing as well. When compared to 2012, unfortunately the NO_2 level increased, but on average remained below the lowest norm (26 $\mu g/m^3$) in Tallinn. The measured data showed the highest NO_2 concentration peaks during morning and evening rush hours on business days in the city center while two other measurement areas (north Tallinn and Õismäe) remained significantly less influenced.

The concentration of small particles PM_{10} had increased in 2013 (EERC, 2014) when compared to previous year in all air monitoring stations: in the city center, 17.5 $\mu g/m^3$; north Tallinn 11.7 $\mu g/m^3$; and Õismäe 13.2 $\mu g/m^3$. Still, the daily norm was exceeded only 5 times a year. These PM_{10} levels increased daily, based on the traffic load, and in particular during rush hours and business days.

The latest annually available data for Tallinn from 2017 showed that the SO_2 and NO_2 levels have remained on the same levels, i.e., where they were in 2012 and 2013, and the concentration of small particles PM_{10} was: in the city center 16.3 $\mu g/m^3$; North-Tallinn 10.1 $\mu g/m^3$; and Õismäe 9.4 $\mu g/m^3$. Thus, with respect to emissions of SO_2, NO_2, and PM_{10}, the changes in air composition have not been easily noticeable immediately after implementation of FFPT and even 6 years after this change. On the other hand, increasing numbers of cars in the traffic of Tallinn could be even larger and pollution higher if the FFPT was not implemented at all.

Measuring the impact

Allan Alaküla, representative of Tallinn Municipality to the EU, has explained several times in public that the impact can be understood in terms of indirect indicators. According to him there is no proper technology to estimate and measure the impact. He has claimed that people's mobility increased, not only for unemployed but also for those looking for better jobs used public transport more often. People started to go out more in the evenings and weekends, for consumption of local goods and services. They save from paying taxi services, emission of CO_2 is decreasing. Families go out more on weekends, leaving their cars at home (Alaküla, 2017). This politics-dominated narrative about the success story of the implementation of FFPT in Tallinn has been presented by Alaküla and other city officers continuously throughout the last 6 years. A clearer explanation about the FFPT implementation can be understood when we look at the budget changes related to the change from 2012 to 2013.

The municipality's calculations show that during 2013–18 the number of Tallinners increased by ∼30,000 residents (from ∼410,000 to ∼440,000). Every 1000 new residents brought about one million euros to the city budget (as a part of the municipality's resident's income tax will be relocated to the municipality budget). The public transport budget has grown from 53 million euros in 2012 to 63 million euros in 2017 and by now is about 2.5% of the yearly municipal budget. In 2012 ticket revenue was 17 million euros (12 million euros from residents and five million euros from nonresidents using public transport of Tallinn). By making the city transport fare-free, the government lost 12 million euros per year, but by attracting new taxpayers to register themselves in Tallinn the win was about 20 million euros a year. How did this increase in the number of taxpayers occur? Actually, most of these people who registered themselves where already living in Tallinn. When they registered themselves in Tallinn they transformed themselves also to taxpayers. The timing of the implementation of the FFPT was also important to the municipality, led by the Center Party over 20 years. The municipal government promoted FFPT before municipal elections, which took place 1.5 years after implementation of FFPT. In October 2013 the Center Party received 58.2% of the votes from Tallinners. This way the promise for FFPT was understood by many as a populist strategic statement just to be re-elected with a majority of the seats in the city council. Critics of this change have also pointed out that municipal leaders had artificially created this demand for FFPT— a demand that previously had even not existed (Hess, 2017).

Neighboring counties and municipalities of Tallinn were not impressed with Tallinn's FFPT. The tax revenue (to the municipalities) of these ∼30,000 people who in 2013 had registered themselves in Tallinn has been taken from their budget. To support the residents of these neighboring counties and their mobility in Tallinn, special agreements for discounts have been made between these municipalities and the municipality of Tallinn.

Findings and discussion

Tallinn has eight administrative districts, which have different numbers of Russian speakers. The increase of support among residents of above mentioned segregated Russian-speaking areas, which represent the most loyal voters of the dominating party, was expected after implementation of FFPT. The results of the authors' two-way ANOVA analysis (Tuisk and Prause, 2019) confirmed that in the districts with a higher share of foreign population, the Center Party got significantly more votes than in districts with a lower percentage of this minority population. This confirmed that the voters' profile had not changed when 2013 elections were compared with pre-FFPT implementation results in 2009 across all the districts (i.e., when taking into account the ethnic composition). In addition, it also became evident that there was no significant change in the share of votes across all the eight voting districts that the Center Party got before and after the start of FFPT. In short, FFPT did not significantly influence the share of votes on the political level, neither in districts with a high share of foreigners nor in those with a low share. The data from the same study also showed that the implementation of FFPT positively impacted the use of public transport in Tallinn as the bus usage went up from 53% to 60%. The same ANOVA approach for the change in car use before and after the FFPT did not show a significant change, i.e., the car usage rate dropped slightly, but there was no significant statistical evidence for the decrease after FFPT. Then we wanted to know what were the variables explaining most of the modal change that had taken place. Regression analysis showed that the distance to city center represented the most important variable, explaining 60% to the change in bus usage, followed by the possibility to use the public transport freely with 19%, and the share of foreigners in the district with 14%. The share of votes of the Center Party in the district only explained 7% of the changes in bus usage.

Tallinn city municipality orders annual studies from different survey companies in order to learn about satisfaction with city-provided services. The study of Eesti Uuringukeskus (Estonian Survey Center) (2017) reports that across all publicly available transport services regular buses and trolleybuses are most used (80%) when compared to others (tram and train). Almost 60% of those who use public transport to their main destination (i.e., to work or school) do not make any changes (stops) to reach their destination, 25% make one change, 10% make two, and only 3% have to do three changes. Despite these results that show that the transportation is quite smooth, about 50% of Tallinn residents find that there have been no changes in the transportation during last year, less than 40% claim that the situation has improved, and 3% said it has got worse. The biggest problem has been overloaded transport and insufficient cleanliness, as reported by 20% of Tallinn residents, while one-fourth were satisfied with public transport and could not point out any problems. The respondents also admitted that during the last 4 weeks 90% of them had used public transportation, and about one-third of them use it daily, more than one-fifth use it several days a week, less than one-fifth use it 1—2 days a week, and 14% even less. Ninety-six percent of Tallinners own the contact-free card, but only 75% of these owners always validate the card when entering public transport. This behavior shows that there is overall trust in this organization of public transport and satisfaction is quite high. Also, this finding can be taken as an indicator of successful e-governance (Prause et al., 2012), i.e., electronic data gathered from validated usage can and will be used to optimize these bus, tram, and trolleybus routes, which leads to sustainable management and exploitation of public transport vehicles, and adds to higher quality of this public service. According to mayor Taavi Aas, this electronic validation data in the public transport system shows that in 2012 there was a minimum 134 million rides in the public transport, then in 2013, it was already 142 million, and this number remained 142—143 million for next 4 years (Aas, 2018).

In 2017 just 2% of respondents said they or their household members had to give up a job that was located within Tallinn because of transport-related problems; these were mostly too high a number of changes or inconveniences. Also, two-thirds of residents positively supported the reorganization of road lanes during recent years, which now give higher priority in traffic to public transport vehicles. About half of residents have one (private or official) car in use by their household, while about 25% have two cars, and more than two cars are used by 6% of households. When compared to cars, bikes were used by household members in two-thirds of the households. For children aged 7–14, public transportation was the main way for commuting to school and back, the next mode was going by foot. In addition, there are families or households where children are taken to school or hobby classes by car. In these cases the main reason for that was big amount of time (admitted by 30%), but also an opinion that child's independent trajectory to the site is not safe (27%), and lack of suitable public transportation was 20% (Eesti Uuringukeskus, 2017).

When combining the outcomes of the authors' earlier analysis and secondary data that reflect satisfaction levels of different residents/users groups with the current public transportation system in Tallinn, we have to note that here we see that we have to consider "sustainability" not merely in environmental terms. In addition, all these "quality of life" related aspects in the reorganization of this transport cannot be ignored. Namely, the main expectation should be that improvements in transport routes and frequency besides the FFPT should not be underestimated. Currently, 6 years after the implementation of FFPT, the expectations are still high, and all efforts for improvements are still anticipated by the customers of this public service. To create sustainable cities requires much more than simply optimizing transportation routes, adding special lanes, and lowering the transport fares. Because of increasing population densities, for the urban quality of life it is crucial that all the residents belong to integrated, stable, and vibrant communities (Jenks et al., 2000).

Impact: FFPT spreading over Estonia

In autumn 2016 the same political party (Center Party) that has been ruling Tallinn for last 20 years was asked to form government as a lead coalition partner. After some preparation the Ministry of Economics and Communication implemented FFPT in almost all 15 counties of Estonia for local residents within the borders of the county. During earlier years only one-third (e.g., 11 million € in 2017) of public (bus) transport tickets was covered by collecting ticket money. The rest, 21 million €, was contributed from the state budget to cover bus operation costs. According to Minister Kadri Simson, from July 1, 2018, annual costs of this county-level transportation will be covered totally from the state budget. In each county public transport centers were established, which have the authority to decide on how to organize the schedules of county transport. Counties also have the right to opt out from this FFPT offer if they prefer. In addition, the government assigned up to 3.3 million € to support rescheduling and tightening these local schedules of FFPT (Riigikogu Stenogramm, 2018). Moreover, the mayor of Tallinn is considering also the possibilities for providing FFPT in Tallinn for all residents of Estonia. For this move he sees an obstacle in the unbalanced financial contribution that comes from the state. Namely, while the state supports all other counties by covering their FFPT costs, in Tallinn only its taxpayers have contributed to covering the costs. Aas proposes FFPT in Tallinn for all Estonian residents as of 2020, but in this case only if certain support from the national budget will be also received (Gnadenteich, 2018).

Again, like in the case of Tallinn, the implementation of the Estonian-wide public transport system is linked to the highly developed e-governmental system. The entrance card for participating in the All-Estonian public transport is linked to the ownership of a personal RFID card, which is in this case red, and can be purchased for 2€ at public selling places all over Estonia. This red RFID card possesses also an ID number and has to be linked via an internet portal with the personal eID card of the Estonian resident. When entering a public transport device, the red RFID card has to be scanned by a reader so it can be checked if the person has the right to travel fare free or has to pay a travel fee. In accordance with the specific transportation rules all over Estonia, the system decides if it is necessary to charge a fee or not. Since transport fees may appear, the public transport users have to pay a deposit on their red cards so that the administrational burden has to be minimized. So again like in the Tallinn case the realization of the FFPT system is based on the existing of the highly developed Estonian e-governance system. The administrative savings related to the use in the transport system are not fully analyzed as of now, but from investigations of other sectors in Estonia we know that the administrational costs are cut up to 2% (Prause, 2016). RFID cards, validators in transport vehicles, and all the information system for FFPT has been provided by an Estonian private company, AS Ridango (Ridango Ltd.). Since the beginning of this FFPT implementation in Tallinn, the same company has been providing not just the all the validation infrastructure locally but has also integrated all transport card owners from other cities, e.g., Tartu or Pärnu. By validating their card, Tallinn residents are recognized, and charged via the same card if they have preloaded money for that. Also, the four largest Estonian intercity bus companies have been integrated by the same system, enabling the use of the RFID card in many cases already as a way of authentication for door-to-door transport between the cities of Estonia (Ridango, 2016). Moreover, the company has succeeded to in providing their services, know-how, and experience from Tallinn and Estonia also internationally, as the Lithuanian cities of Klaipeda and Palanga have also implemented Ridango's technology for validation of passengers in their public transportation systems. A common PT ticket system will be applied in 2020 where residents of Tallinn, Tartu, and Helsinki will be united to the same validating system.

By summing up the previous discussion, it turned out that an assessment of FFPT in Tallinn leads to different results due to the multiple character of the decision. More than 6 years after the start of Tallinn's fare-free public transportation, the concept is still vital and there is no evidence of major setbacks. Meanwhile, as described above, other Estonian municipalities have also already implemented or are going to implement their own FFPT systems, so it seems that the Tallinn idea has started to expand over all of Estonia. This development may jeopardize the economic fundamentals of the Tallinn approach because here the financing of FFPT came with the additional tax income for the city due to increasing the number of registered citizens after enforcement of free urban transportation. Such an effect is impossible by expanding FFPT over the whole country since the total number of Estonian citizens is constant. Hence, new financing models with state budgets are discussed.

Considering the political impact, it turns out that the initiating Center Party in Tallinn was not able to benefit substantially from the FFPT when taking into account the shares of votes in municipal elections. Also the figures for mode change in urban transportation were rather moderate so that not all citizens are fully convinced to cross over from private cars to FFPT. Also the environmental effects were twofold. On the one hand, the rising users of FFPT in Tallinn contributed to fuel savings, reduced congestions, and reduced the pressure on urban public space; on the other hand, the high carbonization level of the Estonian economy and energy production blocks the way toward a green electrification of urban transport in Tallinn. The underlying political and socioeconomic specifics of Estonia making quick changes is complicated.

Finally, the research highlighted the complexity of organizing and administering efficient public transport in Tallinn as well as in the whole of Estonia. Since the current landscape of fares in public transport in Estonia resembles a patchwork, the organization and administration of ticketing requires a sophisticated IT infrastructure, which is realized on the basis of the powerful Estonian e-governmental system driven by e-Estonia. Thus the Estonian experience is only partly possible to be transferred to other regions in Europe.

Conclusions

The socioeconomic impact of FFPT in Tallinn and Estonia is multifaceted. For Tallinn as the starting point in Estonia, the first consequences were positive with increasing numbers of citizens together with additional tax income for the city, which compensated for the opportunity costs lost due to missing ticket fees. Also a change of transport mode became obvious after the start of FFPT in Tallinn, but the change rates were only moderate; the data show that the ridership of public transportation in Tallinn has increased 14% after implementation of FFPT and modal shift in the city toward using public transportation has increased from 55% to 63%. At the same time this shift is largely based on increase from walking mode to PT (as average passengers decreased by 10%).

The environmental impacts are twofold. On the one hand, the higher utilization of FFPT decreased fuel consumption, congestions, and occupied space in the city center, whereas on the other hand, a closer view of the energy production in Estonia reveals that clean energy in Estonia is until now not linked to electricity due to the Estonian specialty of using oil shale for energy production. This high carbonization level of Estonian industry and energy sector has political and strategic reasons and slows down a change toward greener economy. The research also showed that the political impact for the leading party in Tallinn that implemented FFPT was negligible.

Despite these above-presented contradictory data and intertwining of municipal decisions and politics, Tallinn's fare-free public transportation has been vital for over 6 years, and there is no evidence of major setbacks. Meanwhile other Estonian municipalities also implemented FFPT so that it seems that the Tallinn concept is expanding over the whole country. Since the current landscape of fares in public transport in Estonia resembles a patchwork, the organization and administration of ticketing requires a sophisticated IT infrastructure, which is based on the powerful Estonian e-governmental system driven by e-Estonia. Thus, the Estonian experience is only partly possible to be transferred to other regions in Europe.

Biographies

Gunnar PRAUSE is a Professor Business Development at Tallinn University of Technology. His research interests are e-governance, public transport, entrepreneurship, sustainable supply chain management and innovation. ORCID ID: orcid.org/0000-0002-3293-1331.

Tarmo TUISK is researcher at the Department of Business Administration at Tallinn University of Technology. His research interests are fare-free public transport, innovation, environmental sustainability and logistics. ORCID ID: orcid.org/0000-0001-5073-4772.

References

Aas, T., 2018. Taavi Aas: Tasuta Ühistransport Jääb, Ennast Igaveseks Nimetav Seltskond Võib-Olla Mitte. Accessed 11.01.2019. www.tallinn.ee/est/tasutauhistransport/Uudis-Taavi-Aas-tasuta-uhistransport-jaab-ennast-igaveseks-nimetav-seltskond-voib-olla-mitte?filter_otsing_uudis_rubriik_id=331.

Alaküla, A., 2017. Forum Smart City du Grant Paris. www.tallinn.ee/eng/freepublictransport/Forum-Smart-City-du-Grand-Paris.

Brown, J., Hess, D.B., Shoup, D., 2001. Unlimited access. Transportation 28 (3), 233−267. https://doi.org/10.1023/A:1010307801490.

Burneika, D., Baranauskaitė, A., Ubarevičienė, R., 2017. Social segregation and spatial differentiation of electoral alignment in Vilnius metropolitan area. Geographia Polonica 90 (2), 87−110. https://doi.org/10.7163/GPol.0089.

Cats, O., Reimal, T., Susilo, Y., 2014. Public transport pricing policy: empirical evidence from a fare-free scheme in Tallinn, Estonia. Transportation Research Record 2415, 89−96. https://doi.org/10.3141/2415-10.

Cats, O., Susilo, Y.O., Reimal, T., 2017. The prospects of fare-free public transport: evidence from Tallinn. Transportation 44 (5), 1083−1104. https://doi.org/10.1007/s11116-016-9695-5.

De Witte, A., Macharis, C., Lannoy, P., Polain, C., Steenberghen, T., Van de Walle, S., 2006. The impact of "free" public transport: the case of Brussels. Transportation Research Part A 40, 671−689. https://doi.org/10.1016/j.tra.2005.12.008.

e-Estonia (2018). E-Estonia. Public transport made smarter as Estonian companies team up abroad. https://e-estonia.com/public-transport-made-smarter-as-estonian-companies-team-up-abroad/.

Eesti Uuringukeskus, 2017. Satisfaction of Inhabitants with Public Services in Tallinn in 2017. https://aktal.tallinnlv.ee/static/Koosolekud/Dokumendid/pkldok18990.pdf.

Estonian Environmental Research Centre, 2013. Välisõhu Seire Linnades. [Monitoring of outside Air in Cities]. http://eelis.ic.envir.ee/seireveeb/aruanded/14068_aasta_linnaohk_2012_final.doc.

Estonian Environmental Research Centre, Välisõhu Seire Linnades. [Monitoring of outside Air in Cities], 2014. http://seire.keskkonnainfo.ee/attachments/article/2963/aasta_v%C3%A4lis%C3%B5hu%20seire_2013.doc

European Commission, 2013. Directorate-general for mobility and transport: Attitudes of Europeans towards urban mobility. Special Eurobarometer 406. http://ec.europa.eu/public_opinion/archives/ebs/ebs_406_en.pdf.

Gavrilova, O., Randla, T., Vallner, L., Strandberg, M., Vilu, R., 2006. Life Cycle Analysis of the Estonian Oil Shale Industry. Tallinn. www.elfond.ee(electronic_publication).

Gnadenteich, U., 2018. Taavi Aas: Ehk Aastal 2020 Võivad Kõik Tallinna Ühistranspordis Tasuta Sõita. www.postimees.ee/6484858/taavi-aas-ehk-aastal-2020-voivad-koik-tallinna-uhistranspordis-tasuta-soita.

Hess, D.B., 2017. Decrypting fare-free public transport in Tallinn, Estonia. Case Studies on Transport Policy 5 (4), 690−698. https://doi.org/10.1016/j.cstp.2017.10.002.

Hess, D.B., 2018. Transport in mikrorayons: accessibility and proximity to centrally planned residential districts during the socialist era, 1957−1989. Journal of Planning History 17 (3), 184−204. https://doi.org/10.1177/1538513217707082.

Hunke, K., Prause, G., 2014. Sustainable supply chain management in German automotive industry: experiences and success factors. Journal of Security and Sustainability Issues 3 (3), 15−22. https://doi.org/10.9770/jssi.2014.3.3(2.

Jenks, M., Williams, K., Burton, E., 2000. Urban consolidation and the benefits of intensification. In: Roo, G.D., Miller, D. (Eds.), Compact Cities and Sustainable Urban Development: A Critical Assessment of Policies and Plans from an International Perspective. Ashgate, Aldershot, pp. 17−30.

OECD, 2016. OECD Environmental Performance Reviews: Estonia 2017. OECD Publishings, Paris. https://doi.org/10.1787/9789264268241-en.

Prause, G., 2016. E-Residency: a business platform for industry 4.0? Journal of Entrepreneurship and Sustainability Issues 3 (3), 216—227. https://doi.org/10.9770/jesi.2016.3.3(1).

Prause, G., Thessel, F., Hunke, K., 2012. Review of current status of the use of E-government services by SME in the Baltic Sea region. In: Prause, G, G., Hunke, K., Thessel, F. (Eds.), Transnational Aspects of End-User Oriented E-Services in the Baltic Sea Region. Berliner Wissenschafts-Verlag, Berlin, pp. 4—41 (Regional Business and Socio-Economic Development; 7).

Prause, G., Tuisk, T., Olaniyi, E.O., 2019. Between sustainability, social cohesion and security. Regional development in North-Eastern Estonia. Journal of Entrepreneurship and Sustainability Issues 6 (3), 1235—1254. https://doi.org/10.9770/jesi.2019.6.3(13.

Ridango (2016). Ridango web page, Ridango signed a contract for delivering a new ticketing system for public transportation in Pärnu city and county. https://ridango.com/blog/ridango-signed-a-contract-for-delivering-a-new-ticketing-system-for-public-transportation-in-parnu-city-and-county/.

Riigikogu Stenogramm, 2018. http://stenogrammid.riigikogu.ee/et/201806061300#PKP-22954.

Siirde, A., Eldermann, M., Rohumaa, P., Gusca, J., 2013. Analysis of greenhouse gas emission from Estonian oil shale based energy production processes. Life cycle energy analysis perspective. Oil Shale 30 (2S), 268—282. https://doi.org/10.3176/oil.2013.2S.07.

Tammaru, T., Kährik, A., Mägi, K., Novak, J., 2015. 'Market experiment': increasing socio-economic segregation in the inherited bi-ethnic context of Tallinn. In: Tammaru, T, T., Marcińczak, S., Van Ham, M., Musterd, S. (Eds.), Socio-economic Segregation in European Capital Cities: East Meets West. https://doi.org/10.4324/9781315758879.

Tuisk, T., Prause, G., 2019. Socio-economic aspects of free public transport. In: Kabashkin, I., Yatskiv, I., Prentrovskis, O. (Eds.), Reliability and Statistics in Transportation and Communication. Springer, Berlin, pp. 3—13. https://doi.org/10.1007/978-3-030-12450-2_1 (Lecture Notes in Networks and Systems 68).

Carsharing concept implementation in relation to sustainability — Evidence from Poland

Sebastian Kot

North-West University, Faculty of Economic and Management Sciences, South Africa and
Faculty of Management, Czestochowa University of Technology, Czestochowa, Poland

Introduction

Today stationary and straight using goods are a relic. Such a state of affairs is caused by an increasingly evident tendency to the ceaseless change of place of residence, called also human mobility. These behaviors have different reasons; however, the most important of them is the desire for improvement of living conditions. These changes next influence human mentality, which manifests itself in abandoning the tendency of accumulation of a lot of goods.

The study of Jamrozik and Kisielewski (2018) demonstrates 88% of respondents declared that they have at least one car in the household or they are able to use the vehicle of the close person. Taking into consideration that nearly 40% of all respondents use the car in order to reach work, school or college, and an average time of this travel on foot may take out only 20 min (Jamrozik and Kisielewski, 2018), it is possible to acknowledge that in their case take advantage of carsharing services will be a very good solution. It will allow using vehicles in case of such a need without any trouble and at the same time will save users from incurring fixed costs. Thanks to such decisions exhaust emissions would be considerably reduced (Popp et al., 2018a,b).

No wonder that people more and more often consider also brief lending transport means, including bicycles, motorbikes, and car rentals, which are attracting growing attention.

Carsharing is an international notion referring to the system of sharing passenger cars by many users. Vehicles are being delivered to users by fleet operators, which are collecting charges according to publicly accessible toll systems for using them. It is possible to name fleet operators for most enterprises, public agencies, cooperatives, as well as associations or groups of individuals. Carsharing can be "defined as a service where members of shared-use vehicle organizations get access to a fleet of vehicles" (Nobis, 2006). "Carsharing" is the most frequently used term referring to the collective and alternating utilization of cars. It is as a service that enables a group of individuals to share cars with other persons (Katzev, 2003).

These services can be provided with a few communication channels, which are: carsharing enterprise websites, mobile applications, or dedicated SMS systems (Czym jest car sharing, 2018).

Although many researchers and scientists describe an influence of carsharing on the economic system, economy, or environment, it should be mentioned that this notion does not have an appropriately specified definition, which would be explicit and repeatable. However, there is a group of scientists who constantly make attempts to define this phenomenon. An example of such an attempt is a definition of Józefowicz (2007), who determines carsharing as a simple and user-friendly system of shared use of widely available cars by many users. An idea of this service is lending vehicles for a short time in order to move within the urban agglomeration.

Larisch (2014) in his work shows a little bit different approach. He thinks that carsharing is a mobile service enabling the sharing of vehicles, and its main objective is an efficient use of vehicles and environmental protection. He simultaneously emphasizes that this service has three variants of functioning:

- Stationary carsharing — The contractor appropriately determines the location where customers can pick up and return the vehicle.
- Nonstationary carsharing — It appears when the operator outlines a zone, within which there is a possibility of lending and leaving the car without determining specific stations of the receipt.
- Private carsharing — Very rarely applied solution consisting of hiring private owners' cars, who currently do not use them.

Contrary to appearances, carsharing services can be directed not only at individual customers but also for entire companies, using cars belonging to the operator or granting one's own car to other users at the moment, when it is not used. Such a solution allows the company to use the service in the form of the repurchase of the monthly pass, for an appropriately specified amount.

Year to year the carsharing concept has increased its number of supporters. At present, the interest and the support for this type of service come from many social groups, among which it is certainly possible to count users and environmentalists. In recent times, the idea of hiring cars per minutes has gained great interest on the part of the government and also self-government bodies, which actively act for its popularization in the society. However, this concept is not near perfection nor are all trips necessarily easy to replace; carsharing tends to be used for irregular trips rather than for routine travel such as commuting (Schmöller et al., 2015; Keller et al., 2018). The positive and negative benefits resulting from use of services of this type need to be carefully analyzed.

After our analysis of the carsharing services market, it is possible to certainly enumerate such benefits as (Opulski, 2018):

- Ease of use — applications and platforms dedicated for services of this type are intuitive, thanks to clients who can fully make use of services after making a few straight moves. As can be seen on the picture (Fig. 9.1) every platform enables to check in a fast and simple way where the closest vehicle is, letting also to book or lend it using one's smartphone. A few clicks are usually enough to open and start the vehicle. The situation is alike in case of completing the ride. Originators also thought about facilitating the payment for services by automatic downloading of the essential amount from customer's credit card or from funds available on a virtual account.
- Lack of problems and costs accompanying having an owned vehicle — assuming that the system user will entirely resign from having their own vehicle, it is possible to

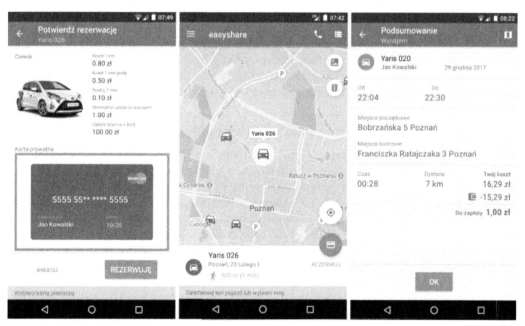

FIGURE 9.1   Image visible in easyshare application while using the service. *Source: Opulski P., Carsharing, czyli auto na minuty, Accessible in World Wide Web: https://www.pcworld.pl/news/Carsharing-czyli-auto-na-minuty410034.html.*

acknowledge that this eliminates costs connected with repair, maintenance, and use of vehicles. Based on straight calculations of costs of maintenance and exploitation of a 5-year vehicle and on the assumption that the car annually covers a distance of about 15,000 km, it is possible to spare altogether about 12,370 PLN/year. This price contains service charges of all kinds such as repairs, tires, fuel, and operating costs such as insurance and inspections (Dobroczyński, 2018). One should also consider the fact that every year the car loses part of its market value.

- You choose what you are driving in — one of the assumptions accompanying the idea of carsharing is that the user has a freedom of choice of the brand or a model of the car out of available stock. One can check preferences in relation to conveniences of cars, as well as choose parameters for the engine capacity, or the size, depending on what the customer wants to use the car for. For example, when going shopping with a the child, the customer will be considering a bigger vehicle; it will be totally different in case of the need for reaching the social round in the city center, where the size of the vehicle does not have such meaning. From a point of view of the transport and ecological policy, it is essential whether the user will really consider the trip by vehicle or not. Statistics show that persons owning the car are using cars even if it is not profitable from an economic point of view. An argument is usually that incurring running costs will still be necessary. It is worthwhile emphasizing that the driver moving their own car will have to cover the same distance twice. Meanwhile the driver renting a car has a freedom in choice of the vehicle and the place from which he will set off and in which he will finish his journey, thanks to that the probability exists that the return way will be shorter, and consequently will generate less pollution and this will affect the fall in the traffic intensity at certain sections.
- Lack of costs and problems with finding parking spaces — residents of major cities, who will often leave vehicles in the city center, will feel this benefit. Operators have specially selected parking spaces, which can be occupied only by cars of the specific supplier. Thanks to such a solution, the customer saves time and finances associated with traveling by car in order to find parking space and to pay a fee to occupy it.

Unfortunately at present vehicles services are not perfect and it is possible to see some disadvantages for carsharing, as there are (10 wad i zalet carsharingu, 2018):

- Costs incurred on short distances — from examinations of WysokieNapiecie.pl editors result that using the service on very short distances usually generates twice bigger costs than incurred with using one's own vehicle. In their calculations journalists took into account 30 factors affecting the price, for example, average speed of driving around the city, average transfer time through the city center. Calculation results show that using one's own Clio costs an average of 1.43 PLN/km and lending this model in a TrafiCar enterprise will be about 40% higher and will be about 2 PLN/km. So that the service will be advantageous, the driver would also have to consider costs incurred for paying parking fees, and as such this rate will increase even more.
- Payments for time spent in traffic jams and searching for place to park — Examinations prove that the average speed of the car in the center of Warsaw or Wrocław is about 25 km/h. It was found that 1 km of the route will take as many as 2—5 min for which the driver will have to pay additionally. It gets worse when there is congestion, when

the time of covering 1 km can grow even to 5 min. Assuming that the user covers a distance of 10 km, fares will be as follows depending on the situation on roads:

It appears from Fig. 9.2 that fares for the drive at little traffic intensity will remove only 17 PLN, however, in the moment of the heightened rush, it will increase to 34 PLN, which constitutes double of the basic quota. This situation will deteriorate if the driver is forced to look for a parking space for a long time because it can considerably extend the travel time, and costs could rise even to 40 PLN.

At this time the user can reach places with public transport at smaller costs or take advantage of services of a taxi with far greater travel comfort and far less stress.

- Huge fuel consumption — cars offered by operators have high fuel and electric energy consumption. For electric vehicles, this situation means more frequent need of charging by operators and increasing the number of charging stations; in case of cars with internal combustion engines, this fact will greatly influence costs of fuels and considerable air pollution.

Journalists checked the combustion of three chosen models of cars (Fig. 9.3), which showed that the level of combustion for individual brands could markedly differ. Hiring a hybrid Toyota Yaris is the best solution under the economic and ecological account. Whereas the worst answer would be the selection of Hyundai i30 because with such a high combustion level it will emit many harmful substances.

Such a situation results from driving along roads about high intensity and with a desire for shortening the transfer time, which many times means increasing speed. More and more often increased combustion is caused by users, who want to test cars, which means high speeds, high turnover of the engines, and fast acceleration.

FIGURE 9.2   Average operating cost of the vehicle on the distance of 10 km from taking into account surrounding changes. *Own study based on: 10 wad i zalet carsharingu. Jak sie jeździ "autem na minuty?", Accessible in World Wide Web: https://wysokienapiecie.pl/7693carharing_wynajem_auta_na_minuty_test_vozilla_traficar_panek_4mobility/#dalej.*

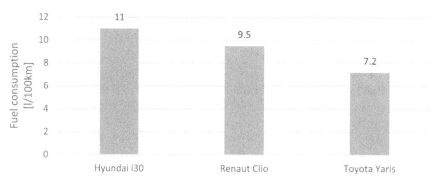

FIGURE 9.3   Average combustion of chosen cars models available in Carsharing services in Poland. *Own study based on: 10 wad i zalet carsharingu. Jak sie jeździ "autem na minuty"?, Accessible in World Wide Web: https://wysokienapiecie. pl/7693carharing_wynajem_auta_na_minuty_test_vozilla_traficar_panek_4mobility/#dalej.*

Development of carsharing concept

Carsharing is an essential international trend in the automotive sector. Although this notion seems to be modern, it is actually about 70 years old. The first project of this type appeared in the 1940s in Zurich, initiated by Safage cooperative (Shaheen et al., 2001). The cause of its establishment was desire for economic benefits. However, this plan did not get the favor of potential users and it collapsed in short time. However this idea appealed to social activists; in 1971, a Procotip initiative arose, which started in Montpellier in France, but it also quickly failed (Shaheen et al., 1999).

Only in 1974 did a concept of sharing as a foundation for the current form of the service arise. Its founder was a politician, Luud Schimmelpennink. The ideas constituted only a small element of a larger concept called the "White Plan" whose task was to encourage the population to share goods through an entire community (Brzeziński, 2018). An aim of this plan was to encourage living in accordance with one's surroundings. This plan, apart from transport, also included a lot of ideas for other aspects of living such as parenthood, uniformed services, and the construction industry.

In most every country, and also in Poland, the carsharing idea begins with the sharing of two-wheeled vehicles. However, the "White Plan" was very innovative and risky for this historical period. For example, it assumed abolishing traffic in the center of Amsterdam for cars and motorbikes and using only bicycle transport, as well as services supplied by taxi corporations (Historia CarSharingu, 2018). However, for that to be real and feasible in the city, they would have to invest in 20,000 bicycles. As it turned out, city authorities were not able to finance sources, and the project became a fiasco.

The political group Provo, who were called anarchists, wanted to provoke city authorities and the society to make attitudinal changes to environmentally friendly ideas, like the initiative of sharing. At first the plan implemented in 1967 seemed unrealistic and crazy, after all, it revolutionized the public transport worldwide (Story of cities, 2018). Preparing for the plan lasted as long as 6 years and was implemented gradually. In the first phase, a prototype of the first vehicle was shown. The first car used in the system aroused great interest.

The Witkar, a Dutch car, also called the "flying bathtub" was a two-man white electric vehicle. Its maximum speed was only 20 km/h. The car did not have a large battery capacity; therefore it drove only through the city center between strategically located stations, where users could charge batteries up in 5 min (Modern Living: The Witkars of Amsterdam, 2018). The vehicle was not too comfortable or elegant, however, very quickly it won the hearts of the Dutch. The Witkar was the first, single electric vehicle within a carsharing project. Such a state of affairs was caused by an insufficient amount of cash. The program was officially launched in 1974. Then 35 Witkar electric vehicles and five charging stations were put into public service. In the end, there were 100 vehicles and as many as 15 charging stations. The effect however was not enough to fulfill main aims of the program. According to current estimates of the contemporary situation, the number of vehicles should amount of 1000 minimum, i.e., ten times the size of the fleet and 100 charging stations. However, the idea did not include expansion since it was not supported by the Dutch authorities. After all, this activity was completed in 1986. There were many reasons for failure, among them it was possible to distinguish the mentality of the society, which aspired to have and not to use cars. Another reason was much smaller traffic in the city, and consequently, traffic jams achieved smaller scale. Low environment-friendly awareness was the third reason. The ignorance of authorities and society about effects of the exaggerated production of exhaust fumes for the ecosystem caused people not to notice the need for change (Brzezinski, 2018). Every concept has a few defects because engineers base their system on the docking of cars in determined stations. Not predicting that vehicles will be collected in most often resorts, where their number has often been too high, and will not reach places, where remaining stations are located. However, the greatest mistake of this concept was sluggishness and an indifference of those responsible for the project because they took no action in order to prevent such incidents. Nowadays contractors have already mastered certain systems for preventing phenomena of this type, for instance, the relocation of cars.

However, the action of authors of this system was not wasted, because this concept produced great interest by other state authorities so that in the 1980s and 1990s a number of projects started to appear in such countries as Germany, Canada, Switzerland, Sweden, and the United States.

The real possibilities of the concept of sharing vehicles and the base for widely understood services of the lease per minutes appeared at the end of the 1990s. Findings of the US Department of Transportation, demonstrat that 90% of business trips and over 50% of private trips include one person in a car. Moreover, in most cases the car is being used for the hour within the entire 24 hours. It means that vehicles are not being used for about 96% of the time. Therefore it is not profitable to incur the costs of purchase and maintenance of the vehicle. It is better to rent a car for a nominal amount in order to fulfill one's purpose.

Soon after publication of these studies the first companies conducting commercial carsharing activities appeared. In the United States they were FlexCar and ZipCar (Lane, 2005), and then City Club Car in London and Edinburgh. However, a sudden and considerable increase of interest for services of this type fell in 2012, when carsharing had over 1.7 million satisfied users on its scorecard. One should, however, emphasize that inhabitants of the United States constituted a considerable part of this population.

Even though services for sharing vehicles quickly won the hearts of users worldwide, and the number of companies that appeared on the market had dynamic growth, Poles had to wait for their first contractors. In September 2016, "4 Mobility company" presented its offer as the first enterprise to take a risk in this area. This offer included car rental in Warsaw. The next one was Cracow Traficar, also pioneering in the market. However, Panek company was the greatest surprise. In the first year of activity, this company made the greatest jmpact with Toyota Yaris hybrid cars. They had 300 units and this proved to be the largest order made by one entity in the history of this car model (Kubera, 2018).

Carsharing in Poland

Nowadays Warsaw and surroundings are the most developed market for carsharing services in Poland. Nearly a half, from the total 3000 to 5000 vehicles function in the form of lease for minutes carsharing. At present, there are three major players in this sector, which include Panek, 4 Mobility, and TrafiCar. There is carsharing on the streets of Cracow, Wrocław, Poznań, Gdansk, Łódź, and Silesia (Opulski, 2018).

The list of cities in which these services function is much shorter than that one, which it has reached. Contrary to appearances this activity has not been well accepted in many cities. The most frequent cause is an attachment to convenience and comfort of possession of one's own vehicle and fear of the consequences of using vehicles in systems of this type. Social anxieties have also appeared, concerning hiring vehicles by nonadults, people without permissions, or by people who are drunk and may pose a threat on the roads. A group of critics thought that vehicles would be very quickly vandalized by thieves or also young vandals. Other critics have claimed that people will not use such vehicles because they are not sanitary and could be sources of spreading illness (Car sharing w Polsce, 2018).

Currently, on the Polish market, there are six main carsharing entities and a few smaller private companies operating in small towns, which offer 3082 cars. The interesting fact is that a year ago there were eight market leaders, and the number of vehicles was 898 units. It means that in 2018 the size of transport increased by 243% in comparison to the previous year. One should, however, note that as many as five startup companies were forced to close or suspend their activity, and three new companies were launched (Raport Instytutu Keralla Research, 2018). To understand these market changes more precisely, see the two tables below presenting the size of the carsharing market in Poland in 2017 (Table 9.1) and 2018 (Table 9.2).

Here it is easy to notice the biggest market player, Express (TrafiCar) company. At the beginning, it "forced her way by storm" to the market and increases its majority and reach year to year. At present TrafiCar service includes 1650 vehicles, and delivery cars constitute 50. It is possible to notice in Fig. 9.4, the fleet of this enterprise constitutes 60% of all vehicles in carsharing services in Poland, while remaining participants altogether constitute only 40%. It means that this company leads the field in its sector once again; what is more, it is still developing, thanks to which it can become a leader also to the European scale.

One should pay attention to the fact that companies operate on the same market and their visions concerning provided services widely differ between themselves. A table that presents

TABLE 9.1 Size of carsharing market in Poland in 2017.

Firm and service	Number of cars
Experss (Traficar)	500
PANEK SA (Panek Car Sharing)	300
4Mobility (4Mobility)	50
Autolux (Lublin City Car)	15
VW Group Poland (Omni)	15
Energa Group (Carsharing Energa)	10
GoGet.pl (GoGet)	5
LeasePlan (SwopCar)	3
Total:	898

Own study based on: Raport Instytutu Keralla Research z czerwca 2017r., dot. Carsharingu w Polsce, Accessible at: https://www.keralla.pl/res/files/SYGNALNE/SYGN_28.06.17_453.pdf.

TABLE 9.2 Size of carsharing market in Poland in 2018.

Firm and service	Number of cars
Experss (Traficar)	1650 (including 50 vans)
PANEK SA (Panek Car Sharing)	600
4Mobility (4Mobility)	330
EasyShare (easyshare.pl)	200
Vozilla (Enigma)	200 (including 10 vans)
Click2Go	102
Total:	3082 (including 60 vans)

Own study based on: Raport Instytutu Keralla Research z października 2018r. dot. Rynku Carsharingu w Polsce, Accessible at: https://www.keralla.pl/res/files/SYGNALNE/SYGN_12_10_18_455.pdf.

a comparison of enterprises, including the most important aspects of the service of this type, follows.

Based on Table 9.3, one can notice that although enterprises operate in various regions, their services are very similar and include similar aspects. These services are in similar price ranges, some firms also implement additional conveniences in order to encourage their users. Among them, one may notice, for example, opened mode of use, which lets a person take and return the vehicle in any place in the catchment area. However, Panek Company gives extra benefits in the form of bonuses for washing or fueling (charging) cars. This has been a very good move for the company and has gained popularity among a substantial number of fans.

What is more, clients have a chance to save by devoting only a small amount of time to for reaching such places like car washes. The next change that recently took place offered every

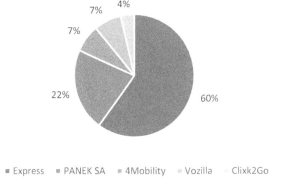

FIGURE 9.4   Percentage share of companies in the market in 2018. *Own study based on: Raport Instytutu Keralla Research z października 2018r. dot. Rynku Carsharingu w Polsce, Accessible in World Wide Web: https://www.keralla.pl/res/ files/SYGNALNE/SYGN_12_10_18_455.pdf.*

operator the possibility of hiring the vehicle and temporarily using it outside the area of normal activity. However everyone is obliged to return the vehicle in the zone of activity, otherwise, the operator of the vehicle calculates a financial penalty, which among others includes putting the car back into the right place.

To sum up, both the time of ride and layover and the distance a vehicle covers during its use influence the transport rate in every company. This price in some cities can also be dependent on a chosen car model. Simultaneously, diversification of the fleet at operators is an added value, which speaks in their favor since requirements and preferences of customers are greatly diversified. Based on the above statistics, it is possible to notice that these services are more and more often directed at companies that provide delivery trucks in offers.

Polish carsharing market in comparison to the world and Europe

Carsharing services have already existed on the market for almost 70 years, so it is nothing strange that the number of its supporters is rising considerably. Despite the first projects of the service appearing in the 1980s, its greatest spread took place after 2006. Such a state was caused by a fast flow of information and desire for seeking new, niche services markets. Analyzing Fig. 9.5, it is possible to state firmly that from 2006 to 2016 the number of vehicles and users have shown a great upturn. In carsharing history, 2012 turned out to be crucial when the meaning and functionality of mobile devices grew. A lot of companies were able to offer the possibility of searching for the closest vehicle, resulting in increased interest in services. On the basis of data from previous years, it is possible to state boldly that there is a high probability in this market and it will be growing even faster in the future.

Taking into consideration the population and the area of regions of the world, Europe is at present an indisputable leader of the carsharing market. In 2016, its residents constituted 29% of world users and offered fleets by companies constituted 37% of the worldwide number of vehicles on this market. Economic stimuli and growing environment-friendly awareness certainly affects such a state of affairs.

TABLE 9.3 Review of Carsharing Services in Poland.

Company's name	Express	Panek SA	4Mobility	Click2Go	Vozilla	EasyShare
Extent of activity	Warszawa. Wrocław. Poznań. Kraków. Piaseczno. Swarzedz. Trójmiasto. GOP	Warszawa. Proszków. Legionowo. Grodzisk Mazowiecki. Jabłonna. Piaseczno i Rembertów. Modlin (airport)	Warszawa. Poznań	Poznań	Wrocław	Poznań
Type of cars	Combustion	Hybrid	Combustion. Electric	Hybrid	Electric	Hybrid
Car brands in the fleet	Renault Clio. Opel Corsa. Renault Kangoo	Toyota Yaris	Hyundai i30. BMW3. Mini BMW 1. BMW i3	Toyota Yaris	Nissan Leaf. Nissan eNV200	Toyota Yaris
Initial bonus	no	20 PLN for registration	20 PLN for registration	no	no	no
Minimal fee per min (PLN)	0.5[a]	0.5	0.48	0.5	no	0.5
Minimal fee per km (PLN)	0.8[a]	0.65	0.8	0.8	0.9	0.8
Minimal fee or layover (PLN)	0.1[a]	0.1	0.12	0.1	0.1	0.1
Maximal fee for booking the car (PLN)	0.1	0.1	0.12	0.1	0.1	0.1
Maximal fee per 24 hours (PLN)	nd	120	120[b]	nd	nd	nd
Lease cost depends on the vehicle	no	no	yes	no	yes	no
Free parking in the paid parking zone	yes[c]	yes	yes	yes	yes	yes
Time of booking the car (min)	15	15	15	15	15	15

(Continued)

TABLE 9.3 Review of Carsharing Services in Poland.—cont'd

Company's name	Express	Panek SA	4Mobility	Click2Go	Vozilla	EasyShare
Free time for the preparation for the route (min)	2	3	2	1	2	no
Registration fee	no	1	1	1	no	1
Payment activation	Registration of the credit card	Transfer. Visit to the branch. Video conversation	Transfer. Transfer pay-link	Registration of the credit card. Prepayment	Registration of the credit card. Prepayment of 10 PLN	Registration of the credit card
Permanent place of pick up and return of cars	no	no	Optional	no	no	yes[d]
Possibility of using and temporary parking outside the zone	yes[e]	yes[e]	yes[e]	yes[e]	yes[e]	yes[e]
Access option without application (access card)	no	no	no	yes	no	no
Bonuses for washing car up	no	10 PLN (car wash). 25 zł (hand washing)	no	no	no	no
Bonuses for fueling/charging cars	no	10 zł (Orlen or Bliska)	no	no	no	no

bd, no data; nd, not applicable

[a] Payment dependent on the city, example based on Warsaw;

[b] Model of the lease with the receipt of the car from the base of the contractor

[c] excluding Katowice and Wieliczka;

[d] with the possibility of derogation, if other car from the fleet occupies the site

[e] with the need for the return to the zone upon completion of the lease.

Source: Own study based on: Opulski P., Carsharing, czyli auto na minuty, Accessible at: https://www.pcworld.pl/news/Carsharing-czyli-auto-na-minuty4100343.html.

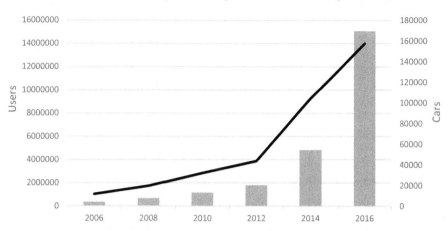

FIGURE 9.5 Global trends on Carsharing services market in 2006–16. *Own study based on: Raportu Innovative Mobility Research Group z kwietnia 2018 r dot. Trendów na rynku usług Carsharingowych, [on-line], [access03.12.2018], Accessible in World Wide Web: https://cloudfront.escholarship.org/dist/prd/content/qt49j961wb/qt49j961wb.pdf? t=pa6fa3&v=lg.*

One should notice that as for the global market (Fig. 9.5) this service still has tendencies of dynamic growth; in the case of Europe (Fig. 9.6) in the last period it is possible to observe the fall in the increase, and even slight imbalance on the market. However, this situation does not apply to the Polish market, since in 2015 the first efforts to implement services of this type were made, and from 2016 the state of the market started to develop.

The following table presents data concerning carsharing systems existing in 28 European cities. If a few operators exist in some city the number of their cars is added up and price for services are averaged and calculated for the first hour of use. However, the results are more credible with a few assumptions made in the examination:

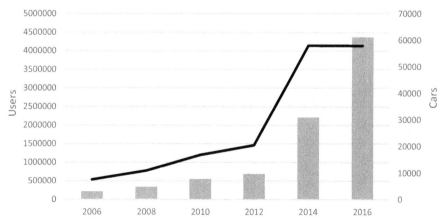

FIGURE 9.6 Trends on European market of Carsharing services in 2006–16. *Own study based on Raportu Innovative Mobility Research Group z kwietnia 2018 r dot. Trendów na rynku usług Carsharingowych, Accessible in World Wide Web: https://cloudfront.escholarship.org/dist/prd/content/qt49j961wb/qt49j961wb.pdf?t=pa6fa3&v=lg.*

- Calculated price is a reflection of the costs of renting the cheapest type of car available in operator's stock with the assumption that it is used by a private person rather than by company
- If the charge includes time and distance, price of covering 50 km will also be added into costs. If the price list predicts costs for time periods, the price of the hour will be counted out on the base of calculation—price for the life cycle through the number of hours.
- The price does not include parking costs, bails, and monthly payments or booking.
- During costs and the number of vehicles calculation services, which by definition include the travel of several people, as Uber or BlaBlaCar and classical car rental firms are not taken into consideration

On the basis of presented data (Table 9.4), it is possible to notice that Paris is a leader amongst European cities, there are 19 vehicles per 10,000 residents, but the price is only 26 PLN/h. Right behind it is Vienna, well known for ecological solutions, in which there are seven cars per 10,000 residents, however, one should pay attention to the fact that in the case of this city costs of the hire are almost 2 to 5 times bigger than in the case of the leader. In the ranking for the lowest charge, Bratislava deserves the title of the pioneer, in which cost of the hire per hour amounts to 6 PLN.

In Poland, carsharing services are actually in early stages, since they appeared relatively recently, while in other countries these services turned up even in years 2000—02. However, the capital city of Poland is not in a bad situation in relation to the ranking because it ranks among the first 10 places and occupies honorable seventh place, with the result of five cars per 10,000 residents and the price within the limits of 66 PLN/h.

The market of carsharing services in Europe constantly gets new supporters. They are also present in Poland, what is due above all to economic reasons, but more and more often also ecological ones. Results from analyses of many examinations show that carsharing services provide for withdrawing from use outdated and very rarely used cars in favor of modern and environmentally friendlier cars with hybrid, electric, or gas drive. Such actions will be certainly a premise to introducing electromobility in the Polish economy and will convince skeptics that it is as comfortable solution as driving internal-combustion cars without the need for buying them.

Sustainable solutions in the carsharing services

Change of times and also the mentality of society and state authorities are starting to see disadvantageous changes in the ecosystem and are trying to do everything in order to slow these changes down or to stop them (Grabara, 2019). It is not surprising that these authorities are creating new changes concerning norms and principles in the area of transport, aspiring in their actions for the creation of sustainable development of the state. This action is based largely on the development plan included in the White Paper of the Transport published by the European Commission in 2011. This document assumes that Europe will become a very competitive area in the transport sector as well as reduce the rate of utilization of nonrenewable raw materials (Igliński, 2011; Oláh et al., 2018). This plan has many specific objectives, for

TABLE 9.4 Summary of Carsharing Services in Europe in 2017.

City	Amount/10,000 residents	Average price/1 h[a]
1. Paris	19	25.90zł
2. Vienna	7	60.70zł
3. Copenhagen	10	61.70zł
4. Brussels	8	57.40zł
5. Berlin	8	70.10zł
6. Bratislava	1	6.20zł
7. Warsaw	5	66.30zł
8. Budapest	1	30.30zł
9. Rome	5	79.00zł
10. Amsterdam	4	79.00zł
11. Dublin	3	34.00zł
12. Vilnius	3	40.80zł
13. Prague	2	46.60zł
14. Madrid	2	61.80zł
15. London	2	46.60zł
16. Bucharest	0.1	46.40zł
17. Lisbon	1	63.50zł
18. Ljubljana	1	76.50zł
19. Stockholm	3	86.30zł
20. Helsinki	3	87.50zł
21. Zagreb	1	90.40zł
22. Luxembourg	1	22.50zł
23. Athens	—	—
24. Riga	—	—
25. Tallinn	—	—
26. Sofia	—	—
27. Valetta	—	—
28. Nicosia	—	—

Own study based on Raportu ShopAlike.pl z lipca 2017 dot. Skali usług carsharingu w Europie, Accessible at: https://www.shopalike.pl/transport-sharing-w-europie.
[a]*In some systems extra fee for registration is needed.*

instance reducing the number of internal combustion cars in cities by 2030 and their total elimination from cities by 2050, or also transferring 30% of cargoes from road transport to other transport branches. However, all actions included in this directive aspire for maximum minimization of CO_2 emissions in Europe.

Therefore state authorities are obliged to introduce new acts, standards, and rules of operation of transport enterprises, and also by implementing systems of payments associated with the adverse influence of transport on the natural environment, aspire in their actions for sustainable development of the state. One of the most often applied ways of implementing changes in the environment is applying privileges both for private persons as well as enterprises, which through the change of certain spheres of living and functioning contribute favorably to changes in the ecosystem.

It is possible to guess this plan not only bothers heads of state but also affects local self-government, exerting great pressure on authorities of major cities, where the system of public transport is characterized by rides on relatively brief routes with very high traffic intensity, which influences the creation of more road congestion and heightened noise level and air pollution. Therefore, local self-government more and more often tries to establish cooperation with large business entities providing transport services within city limits. City halls partially fund the development of individual entities, such as Urban Public Transport, or also enterprises dealing with lending of bicycles and cars. Thanks to such cooperation, authorities can ensure a well-organized system of urban transport, and consequently convince a considerable number of residents to abandon private cars or reducing the frequency of their use. Enterprises undertaking such cooperation can count on additional financing sources of new, more ecological and economical transport fleets. Also, authorities of cities will often assign special parking spaces in the city center to companies acting in a carsharing system, whether also partially funding infrastructure essential for its functioning, for example, battery chargers for electric cars in strategic points of the city.

Carsharing can result in environmental benefits (Susan A. Shaheen et al., 2012) and more sustainable consumer behavior (Pizzol et al., 2017) due to the reduction of vehicle ownership. However, there is still discussion about whether carsharing services actually positively affect the environment (Firnkorn and Muller, 2011; Kopp et al., 2015; Martin and Shaheen, 2011). For example, the CO_2 emission limit may lower carsharing profit; if high demand on new energy-efficient cars can compensate for the loss and is worth being satisfied (Chang et al., 2017) is the question that may need to be answered.

In relation to growing benefits offered by authorities of states and municipal governments, owners of shipping companies also aspire for changes in means of transport to more economical and ecological vehicles with hybrid, gas, or electric drives (Teles et al., 2018; Hernández et al., 2014). They are considerably influencing the reduction of exhaust emissions and the reduction of parking spaces needed in the city center, which earlier constituted one of the biggest problems of the urban infrastructure.

Panek SA is possibly recognized to be an environmentally friendly company, which while designing their fleet decided to choose 300 Toyota Yaris hybrid cars, which are characterized by very low combustion. Additionally, these vehicles are equipped with automatic gearboxes, thanks to which the drive is more economical. Such a solution considerably affects the reduction of CO_2 emissions, which considerably influences the climate causing its warming. This fact affects the reduction of running and the exploitation of cars costs, thanks to which the enterprise can derive much bigger profits from its activity (Kuber, 2018).

However, amongst the carsharing companies, a definite number one is Berlin Multicity Carsharing company, which in 2015 became a pioneer in the field of ecology thanks to the cooperation with Citroën. It could then offer users as many as 350 fully electrified Citroën C-Zero cars. A supplied service by it is also characterized by minimum charges. The registration fee is only EUR 9.90, moreover, users get the first 30 minutes of the ride completely for free and every consecutive minute costs EUR 0.28. The interesting fact is that customers can also take advantage of services in the prepaid system, where they pay determined packages of minutes, e.g., 500 min of riding for EUR 39. It is worthwhile also emphasizing that price-list payments contained costs of parking in paid car parks in the city center, insurance, and exploatation costs (Markisz-Guranowska and Stańko, 2015). It is not so strange then that this service has enjoyed great interest among Berliners. At present, Multicity Cars enterprise has been bought out by other operators in it PSA Group, which has a lot of shares and named this enterprise as Free2Move.

To sum up, an activity of carsharing companies might considerably contribute to the curtailment of obsolete cars with internal combustion worldwide. It is a particularly relevant fact for Poland with reference to recommendations included in the White Paper of the Transport, because one should emphasize that the average car of the Pole is 13 years old and is driven by internal combustion. The report of journalists from "Choice of Drivers" portal shows that this is one of the worst results in Europe, the worst thing Polish cars are 2 years older than the medium European age of vehicles (Średni wiek samochodów w Polsce i Europie, 2018). The fact is that old cars have the highest levels of exhaust emissions, which can be additionally heightened in case of cars in bad technical condition, and this considerably influences the air quality that residents of major cities are breathing. However, the worst thing is that an average Pole cannot afford a new car straight from the car showroom, whether hybrid or electric drive, to help solve the climate change issues.

However, not all is lost, because as the studies in cities like San Francisco and Philadelphia show, one car acting in a carsharing system can replace nine private vehicles (Jaroszyński and Chład, 2015). So, it is possible to acknowledge that clients contribute to a meaningful reduction in exhaust emissions, particularly if they choose electric vehicles, whether hybrid propulsion or not, without the need for the purchase of new cars or incurring running costs of these previous ones.

Conclusion

The carsharing services are related to a wide framework of attempts shaping the modern transportation development strategy, including a new city mobility culture, for example, through the usage of vehicles powered by alternative, renewable energy and mobile applications used by companies serving in this area. The utilization of communication and information technology has radically changed the business mechanism in services. The main study results confirm that the carsharing services in Poland are developing fast, however, they have still not reached the EU average. The increase in the number of vehicles available through carsharing services signifies their growing life cycle on the Polish market. This implies that companies are willing to invest in fleet expansion, and sometimes have to, because that allows them to retain their leading position, and to stay in business. It is also worth noting that electric cars services within carsharing in Poland are still very rare.

The future development of carsharing services is determined by few factors: both the national and European legislation exhibit a tendency to limit the traffic of cars that meet only the obsolete emission standards with excessive emissions, in favor of hybrid, and in particular electric vehicles. Such automobiles are made available by most of the recognizable brands even now, and their prices tend to become more affordable as time goes by. The work- and education-related migration of younger people to larger cities also increases demand for innovative solutions such as carsharing services. An increasing social environment aware- ness can be recognized as an additional determinant for using carsharing. Areas to be researched and analyzed in the future include carsharing price competitiveness in relation to new virtual taxi services, as well as electric vehicle utilization in carsharing services.

References

10 wad i zalet carsharingu, 2018. Jak Sie Jeździ "autem Na Minuty"? [on-line]. https://wysokienapiecie.pl/7693carharing_wynajem_auta_na_minuty_test_vozilla_traficar_panek_4mobility/#dalej.

Brzeziński, K., 2018. Wiedziałeś O Tym, Że Carsharing Ma Prawie 70 Lat? [on-line]. https://www.autofakty.pl/po-godzinach/historia-carsharingu-to-juz-70-lat/.

Carsharing w P., (2018) [on-line]. https://www.forbes.pl/biznes/traficar-i-panek-carsharing-opanowaly-rynek-wynajmu-samochodow-na-godziny/s3sb5l6.

Chang, J., Yu, M., Shen, S., Xu, M., 2017. Location design and relocation of a mixed carsharing fleet with a CO_2 emission constraint. Service Science 9 (3), 205–218.

Czym Jest Car Sharing — Samochód Na Godziny Dla Firm I Klientów Indywidualnych, 2018 [on-line]. https://www.24gliwice.pl/wiadomosci/czym-jest-carsharing-samochod-na-godziny-dla-firm-i-klientow-indywidualnych/.

Dobroczyński, M., 2018. Koszty Eksploatacji Samochodu, Czyli Ile Tracisz Pieniedzy! [on-line]. https://oszczedzanienaetacie.pl/samochod/koszty-eksploatacji-samochodu/.

Firnkorn, J., Müller, M., 2011. What will be the environmental effects of new free-floating carsharing systems? The case of car2go in Ulm. Ecological Economics 70 (8), 1519–1528.

Grabara, J., 2019. Sustainable development — never fulfilled dream. Quality — Access to Success 20, 565–570.

Hernández, J., Didarzadeh, S., Pascual, V., Arrúe, A., 2014. ISHARE - Car Sharing Concept Vehicle 2013 World Electric Vehicle Symposium and Exhibition. EVS 2014, art. no. 6914830.

Igliński, H., 2011. Biała Ksiega 2011- Implikacje Dla Polskiego Systemu Transportu. Logistyka, pp. 33–34. No. 4/2011.

Jamrozik, M., Kisielewski, P., 2018. Innowacje W Systemach Indywidualnego Transportu Miejskiego. Autobusy. No. 6/2018.

Jaroszyński, J.W., Chład, M., 2015. Koncepcja Logistyki Miejskiej W Aspekcie Zrównoważonego Rozwoju. Zeszyty Naukowe Uniwersytetu Ekonomicznego w Katowicach. No. 249.

Józefowicz, M., 2007. Środki przewozowe w transporcie miejskim. In: Wyszomirski, O. (Ed.), Transport Miejski, Ekonomika I Organizacja. Uniwersytet Gdański, Gdańsk.

Katzev, R., 2003. Car sharing: a new approach to urban transportation problems. Analyses of Social Issues and Public Policy 3 (1), 65–86.

Keller, A., Aguilar, A., Hanss, D., 2018. Car Sharers' interest in integrated multimodal mobility platforms: a diffusion of innovations perspective. Sustainability 10 (12) art. no. 4689.

Kopp, J., Gerike, R., Axhausen, K.W., 2015. Do sharing people behave differently? An empirical evaluation of the distinctive mobility patterns of free-floating carsharing members. Transportation 42 (3), 449–469.

Kubera, M., 2018. Geneza i rozwój carsharingu w Polsce. Zeszyty Naukowe Politechniki Częstochowskiej Zarzadzanie 31, 124–125.

Lane, C., 2005. Philly CarShare: first-year social and mobility impacts of car sharing in Philadelphia, Pennsylvania. Transportation Research Record: Journal of the Transportation Research Board 1927, 158–166.

Larisch, R., 2014. Car Sharing, Biblioteka Źródłowa Energetyki Prosumenckiej.

Markisz- Guranowska, A., Stańko, K., 2015. Dobre praktyki wykorzystywania pojazdów elektrycznych jako elementu kształtowania zrównoważonego systemu transportowego na przykładzie wybranych miast. Logistyka 666—667. No. 4/2015, s.

Martin, E.W., Shaheen, S.A., 2011. Greenhouse gas emission impacts of carsharing in North America. IEEE Transactions on Intelligent Transportation Systems 12 (4), 1074—1086 art. no. 5951778.

Modern Living: The Witkars of Amsterdam, 2018 [on-line]. http://content.time.com/time/magazine/article/0,9171,879340,00.html.

Nobis, C., 2006. Carsharing as Key Contribution to Multimodal and Sustainable Mobility Behavior: Carsharing in Germany. Transportation Research Record, (1986), pp. 89—97.

Oláh, J., Nestler, S., Nobel, T., Popp, J., 2018. International characteristics of the macro-logistics system of freight villages. Periodica Polytechnica Transportation Engineering 46 (4), 194—200. https://pp.bme.hu/tr/article/view/11656. https://doi.org/10.3311/PPtr.11656.

Opulski, P., 2018. Carsharing. Marki, Miasta, Rozwiazania [on-line]. https://www.forbes.pl/biznes/traficar-i-panek-carsharing-opanowaly-rynek-wynajmu-samochodow-na-godziny/s3sb5l6.

Pizzol, H.D., de Almeida, S.O., Couto Soares, M., 2017. Collaborative consumption: a proposed scale for measuring the construct applied to a carsharing setting. Sustainability 9 (5).

Popp, J., Kot, S., Lakner, Z., Oláh, J., 2018a. Biofuel use: peculiarities and implications. Journal of Security and Sustainability Issues 7 (3), 477—493. http://jssidoi.org/jssi/uploads/papers/27/Popp_Biofuel_use_peculiarities_and_implications.pdf.

Popp, J., Oláh, J., Farkas Fekete, M., Lakner, Z., Domicián, M., 2018b. The relationship between prices of various metals, oil and scarcity. Energies 11 (9), 1—19, 2392. http://www.mdpi.com/1996-1073/11/9/2392.

Raport Instytutu Keralla Research (2018). dot. *Rynku Carsharingu w Polsce*, [on-line], https://www.keralla.pl/res/files/SYGNALNE/SYGN_12_10_18_455.pdf.

Schmöller, S., Weikl, S., Müller, J., Bogenberger, K., 2015. Empirical analysis of free-floating carsharing usage: the Munich and Berlin case. Transportation Research Part C: Emerging Technologies 56, 34—51.

Shaheen, S., Sperling, D., Wagner, C., 1999. A short history of carsharing in the 90's. Journal of World Transport Policy and Practice 5 (3), 16—37.

Shaheen, S., Sperling, D., Wagner, C., 2001. Carsharing in Europe and North American: Past, Present, and Future. University of California Transportation Center, Berkeley.

Shaheen, S.A., Mallery, M.A., Kingsley, K.J., 2012. Personal vehicle sharing services in North America. Research in Transportation Business and Management 3, 71—81.

Średni Wiek Samochodów W Polsce I Europie, 2018 [on-line]. https://www.wyborkierowcow.pl/wiek-samochodow-w-europie-wiadomosci/.

Story of cities #30, 2018. How This Amsterdam Inventor Gave Bike-Sharing to the World [on-line]. https://www.theguardian.com/cities/2016/apr/26/story-cities-amsterdam-bike-share-scheme.

Teles, F., Gomes Magri, R.T., Cooper Ordoñez, R.E., Anholon, R., Lacerda Costa, S., Santa-Eulalia, L.A., 2018. Sustainability measurement of product-service systems: Brazilian case studies about electric carsharing. The International Journal of Sustainable Development and World Ecology 25 (8), 721—728.

Electromobility for sustainable transport in Poland

Beata Ślusarczyk[1,2]

[1]Faculty of Management, Czestochowa University of Technology, Czestochowa, Poland;
[2]North-West University, Faculty of Economic and Management Sciences, South Africa

Introduction

Electromobility is currently an area of great interest of both EU institutions, national ones but also local authorities or even individual users. And this is not only the consequence of the arrangements established at the legislative level on the share of alternative fuels in transport (Kovacs and Kot, 2016) but also or perhaps, most of all, the concern for the natural environment and excessive exploitation of energy resources in the world.

Energy Transformation towards Sustainability
https://doi.org/10.1016/B978-0-12-817688-7.00010-0

The transport sector of the European Union (EU) is responsible for approximately one-third of the total demand for energy, which is mostly satisfied with petroleum derivative fuels. This situation results in a substantial dependency of the sector on the import of petroleum and makes it a significant factor generating pollution of the natural environment (Kupczyk et al., 2019; Riba et al., 2016). The sector is responsible for approximately 21% of the greenhouse gas (GHG) emissions, with more than half of the emissions produced by passenger cars. The EU Directive (2009/33/EC) on the promotion of clean and energy efficient road transport vehicles has been released to foster a broad market penetration of environmentally friendly vehicles in order to decarbonize the transportation sector and to reduce oil dependency (Gass et al., 2014).

Electricity has the potential to increase the energy efficiency of road vehicles and contribute to the reduction in CO_2 emissions from transport. It is a source of energy necessary for the spread of electric vehicles, which may contribute to the improvement in air quality and reduction in noise level in urban, suburban, and other densely populated areas.

The White Paper of the European Commission "Roadmap to a Single European Transport Area — Towards a Competitive and Resource Efficient Transport System" indicates ways to reduce the dependence of transport on petroleum. The achievement of the goals set will be possible by taking many different policy initiatives, including the development of the strategy concerning sustainable alternative fuels as well as the development of appropriate infrastructure (European Commission, 2011).

Electric-driven vehicles offer a solution to reduce greenhouse gas emissions and local air pollution, but their market penetration is still marginal. The widespread deployment of electric vehicles in the EU depends on a large variety of factors. To be a successful mobility alternative several obstacles and challenges have to be overcome first. Various studies identified the following main obstacles and challenges for a broad market penetration (Bühne et al., 2015):

- Battery technology (costs, energy density, recycling, etc.)
- Availability and preparedness of relevant industrial capacity (business models, new vehicle models, etc.)
- Charging infrastructure (minimum density of charging points, identification of early adopter hotspots, etc.)
- Customers' requirements (awareness, willingness to pay, range anxiety, safety concerns, etc.)
- Electric grid (capacity and connectivity issues, etc.)
- Impacts on energy efficiency and greenhouse gas emissions (electricity mix, load management, etc.)
- Standardization issues (charging plug, billing systems, data protocol, etc.).

Despite the fact that the interlinkages between the aforementioned factors can play a major role for the market penetration of electric cars, the customers' requirements and expectations are the most decisive aspects for the market deployment of alternative propulsion systems.

Historical overview of electric vehicles

It may appear that electric cars are a contemporary invention, however, this type of drive has been used in the automotive industry from the beginning of its existence. The development of cars with diesel and electric propulsions occurred at the end of the 19th century. The first electric car was made in 1834 by an American named Thomas Davenport. The vehicle gained energy from a Volt galvanic battery. A three-wheeler equipped with 45 kg batteries was the first available electric vehicle (Business insider, 2018). The change took place in 1859 when Gaston Plantè invented a lead-acid battery (InfoCar, 2016). Five years later Karl Benz and Gottlieb Daimler became the first inventors of combustion cars.

At the end of the 19th century, the significance of combustion four-wheeled vehicles grew, while the significance of electric ones started to decrease. Discovering new deposits of petroleum had a great impact on the development of the refining industry, which resulted in the decrease in gasoline prices. As a result, more and more petrol stations were being built. The growth of the oil industry also resulted in the development of the automotive industry. More luxurious cars with efficient engines started to be designed and produced. The development of electric cars was characterized by much smaller dynamics. Progression on electric vehicles was started soon after 1945. In 1959, an electric car built by the Henney Kilowatt company based on the transit technology appeared on the American market. The maximum speed of the Kilowatt car was 96 km/h and it could drive for 1 h on one charge (Business insider, 2018). At the end of the 1990s, the automotive industry sensed effects of a fuel crisis. The growth of air pollution with exhaust fumes and promoting the environmental awareness of potential buyers caused those producers to think about the return to the concept of electric cars. In 1990 engineers of General Motors company first introduced the project of the model GM Compact, and after a short time, they started producing EV1. The maximum speed of EV1 was 129 km/h. This model could drive about 257 km at one charging. At the end of 20th century a few other cars, such as the electric as Škoda Favorit ELTRA, Honda EV Plus, and Chevrolet S10 EV entered the market, however, a problem of charging them outside the city existed, since batteries in these models were enough only for a distance of 100—180 km. Electric cars were also a more costly solution than combustion ones. All these factors caused personal electric cars not to have a substantial amount of buyers. In 2008, the Tesla Roadster was introduced, which was the first mass-produced car equipped with lithium-ion batteries. Its maximum speed was 210 km/h and it could cover a distance of 354 km (Polityka Insight, 2018). It initiated the development of electric cars. In recent years electric vehicles have become more and more practical means of road transport. In 2012, Tesla introduced the Tesla S model that became a life event. Maximal speed of this model in the strongest version amounts to 249 km/h and it is able to travel 509 km. Tesla S can also take seven passengers. These features gave the Tesla S an enthusiastic market of buyers despite the high price of at least 70,000 dollars (InfoCar, 2016). From this moment, the number of electric personal cars started to grow.

Technologies used in eco-friendly cars

Drives of electric cars are relatively simple solutions. Instead of internal-combustion engines, electric engines are being applied, which are structurally straighter. These engines are also much more efficient. Their advantage is also a lack of the need to use a gearbox. The energy required to the drive of passenger cars is stored in batteries. The drive of hybrid cars is more sophisticated (Zachariasz and Dybkowski, 2014; Riba et al., 2016; Rahman et al., 2016). Hybrid cars can be divided into HEV, PHEV, MHEV, and BEV.

HEV is the hybrid electric vehicle. It is also a so-called full hybrid. For driving this kind of car power, comes from both an internal-combustion and electric engine. It means that the car is equipped with a fuel tank and a battery, which stores energy taken from the internal-combustion engine. The drive system control unit decides which element of the system should be used at the particular moment and in what step, e.g., electric driving mode can be offered in order to temporarily lower exhaust emissions (Michalak, 2018). Full hybrid also initiated the development of other types of drives.

MHEV (mild hybrid electric vehicle) denotes soft hybrid agreement. This type of drive includes the internal-combustion engine and electric unit. The role of the electric drive is limited to only for performing duties of starter and alternator. Generated electric energy is used for the partial supporting internal-combustion engine since this drive is too weak to accelerate the car.

PHEV (plug-in hybrid electric vehicle) is the next type of electrified vehicle. This kind of car has two engines: combustion and electric. The development of this kind of hybrid is a solution that is closer to the electric vehicle. As the name indicates, plug-in hybrids can be connected to an outdoor source of electric energy, e.g., charging station or wall socket at home (Grauers et al., 2013). In a hybrid car, the electric engine drives the car at low speeds, and combustion engine at high ones, which not only influences fuel saving but also the reduction of carbon dioxide emissions.

BEV (battery electric vehicle) is a type of vehicle that is not equipped with an internal-combustion engine. For the drive purposes, it only uses energy coming from batteries charged from an outside source of electricity. An advantage of these vehicles is that they do not emit harmful substances; however, the main disadvantage, just like in the case of all-electric cars, is that the distance covered by the vehicle is conditional on the capacity of the battery (ORPA, 2018).

The next solution is cars of FCEV type (fuel cell electric vehicles), which are vehicles based on the hydrogen drive technology. They use the technology of fuel cells, which means that during the reaction of hydrogen with oxygen, electric energy is produced, which powers the engine. The result of this process is water, which is dismissed outside the car by an exhaust pipe (ORPA, 2018).

In classification of eco-friendly cars, there is also the ECV (electrically chargeable vehicle). This group includes all kind of vehicles that are being driven by electric energy from batteries. ECV is BEV vehicles, FCEV, PHEV, and EREV (extended-range electric vehicles) (Kowalczuk, 2017), that is the so-called vehicle driven with an electric engine, driven by energy stored in batteries, equipped with an internal-combustion engine for their periodic loading during the ride. They also have a possibility of stationary loading (Hennek, 2018).

To sum up, hybrid cars, thanks to the cooperation of two types of drives, support smaller fuel consumption in particular during urban driving, which can be a more ecological as well as an economical solution. Additionally, PHEV hybrids can also be connected to an outdoor electric energy source, which influences in increasing distance and reducing travel expenses. Straighter and lighter construction of electric vehicles also influences the reduction of the frequency of breakdowns and smaller power consumption. An important advantage of these vehicles is also the possibility of recovering this energy, e.g., during braking, and the lack of the gearbox and the clutch enables a smoother ride.

Electric passenger cars in Europe

In most EU countries there is observed growth in sales of ECV vehicles (Electrically Chargeable Vehicles = battery electric vehicles (BEV) + extended-range electric vehicles (EREV) + fuel cell electric vehicles (FCEV) + plug-in hybrid electric). Table 10.1 includes the data for all the member states as well as for Norway and Switzerland.

When assuming 2013 as the base year, the largest increase in ECV vehicles was recorded in the following countries: Sweden, Belgium, United Kingdom, Portugal, Poland, and Bulgaria. In Poland, every year more and more electric cars are registered, although their share in the group of new car registrations amounted only to: in 2012, 10.29%; 2013, 10.65%; and 2014, 11.29%. In spite of the apparent trend of sales growth, the rate of market development of electric vehicles is still lower than in the European Union (Ministerstwo Energii, 2016). There are also countries in which the popularity of ECV vehicles is slightly lower and their purchase is falling. Among others, this is the case of: Estonia, Denmark, Latvia, and Lithuania. On the other hand, it is worth noting an enormous increase in newly registered ECV vehicles in countries such as Bulgaria, Greece, Romania, and Slovakia in 2017, compared to 2016. The leader among the European countries in the number of registered electric cars is Norway (in the analyzed period, their number amounted to 168,644 vehicles). Norway, where currently 25% of cars are electric ones and which built its wealth on the extraction of oil and gas, is taking next steps toward the intensification of electromobility (Figenbaum et al., 2015). According to *The Independent*, Norwegian policy has reached agreement on the gradual shift from the use of fuels in road transport and to ultimately having only electric cars driven on Norwegian roads. Another example is the Netherlands where they are planning to introduce a ban on sales of fuel vehicles in 2025. Great Britain, France, and Ireland also have such plans.

In the European Union, the number of cars grows every year (Fig. 10.1). Compared to 2013, when 11.9 million passenger cars were registered, there was an increase of approximately 27% in 2017. The percentage share of ECV vehicles in the total number of vehicles is not impressive: in 2013 it was only 0.2%, reaching the value of 1.43% in 2017. However, on the positive side, it should be noted that an increase in the share of ECV vehicles in the total number of vehicles was significantly larger since it reached 600%.

According to statistical data, petrol and diesel are still the most popular fuels in Europe. This results from the fact that these are the most easily available sources and out of drivers' habit to traditional types of drives. On the basis of the data available on the website of

TABLE 10.1   New passenger ECV registrations by market in the EU + EFTA[a] (2013−17).

	2013	2014	2015	2016	2017
Austria	654	1,718	2,787	5,068	7,154
Belgium	500	2,047	3,837	8,958	14,299
Bulgaria	1	2	21	13	106
Czech Republic	37	197	298	200	307
Denmark	650	1,616	5,298	2,063	1,347
Estonia	149	340	34	35	43
Finland	50	445	658	1,431	3,054
France	8779	12,497	22,867	29,189	36,835
Germany	6051	13,118	23,557	25,214	54,617
Greece	0	59	67	32	199
Hungary	10	39	130	343	1192
Ireland	49	256	583	690	948
Italy	864	1,420	2,343	2,827	4,811
Latvia	4	194	35	43	56
Lithuania	n.a.	9	37	64	52
The Netherlands	2619	14,805	44,448	22,801	11,079
Poland	31	141	337	556	1,049
Portugal	166	289	1,166	1,821	4,053
Romania	4	7	24	74	198
Slovakia	6	117	52	59	279
Spain	811	1,405	2,246	3,654	7,476
Sweden	432	4,667	8,663	13,221	19,678
United Kingdom	2719	14,608	28,715	36,917	47,298
European Union	24,586	69,996	148,203	155,273	216,512
Norway	7882	19,771	33,770	44,908	62,313
Switzerland	1156	2,688	6,289	6,403	8,391
EFTA	9038	22,459	40,059	51,311	70,704
EU + EFTA	33,624	92,455	188,262	206,584	287,216

[a]*Only countries for which sourced data is available are listed.*
Source: Authors' elaboration based on ACEA, 2018d. Alternative Fuel Vehicle Registrations. Retrieved from https://www.acea.be/statistics/
tag/category/electric-and-alternative-vehicle-registrations.

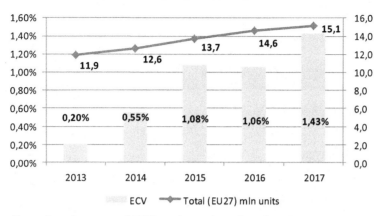

FIGURE 10.1 Share of newly registered ECVs in the total number of newly registered passenger cars. *Source: Authors' elaboration based on ACEA, 2018c. The Automobile Industry Pocket Guide 2018–19. Retrieved from https://www. acea.be/publications/article/acea-pocket-guide.*

European Automobile Manufacturers' Associations, one may observe the drivers' preferences concerning the choice of vehicles powered by specific fuels (Fig. 10.2) in years 2015–17.

The data collected in the graph, due to the limited availability of data, included selected countries. For the countries included in the EU15, the average values indicating the choice between specific fuels showed nearly an equal split between petrol and diesel (in the research period, the share of gasoline-powered vehicles increased by nearly 6 percentage points at the expense of diesel-powered vehicles). The share of ECV vehicles was not too high—at the level of 1.1%–1.5%. On the other hand, among the EU15 countries, two of them are extreme cases, since, in Ireland, diesel-powered vehicles significantly predominate (71% in 2015; 65.2% in 2017). Electric vehicles constitute a marginal number in this country. On the other hand, in the Netherlands, the situation is exactly opposite—drivers choose gasoline-powered vehicles there (as much as 75% in 2017) at the expense of diesel-powered ones (17.5% in 2017). Unfortunately, one cannot observe the growing number of ECV vehicles in this country since, compared to 2016 (when the share of this type of drive amounted to 6% of the total number), a sharp decline was recorded in 2017.

As already mentioned, the undisputed European leader in the field of electromobility is Norway. In this country in 2016, 29% of vehicles had electric propulsion, the same amount had petrol drive. In the following year, already nearly 40% of the total number of vehicles were ECV vehicles (petrol 24.7%, diesel 23.1%). The result of 40% is the highest in Europe and proves that the popularization of electric cars in Norway contributes to smaller consumption of mine fuels in this country.

Data from 2016 shows that in Poland, cars with other types of the drive than gas, petrol, and diesel amount to only 1%. Irrespective of the age of the car Poles preferred diesel (39%) and petrol (45%) at this time. Amongst cars 1–4 years old, 58% of cars are driven by petrol and 37% with diesel. Gas drive among new cars amounts to only 4%. Along with the increase of the age of the car, a tendency of reducing numbers of vehicles driven by diesel exists, and a number of the ones driven on LPG increases. Of cars 11–20 years old, 18% are

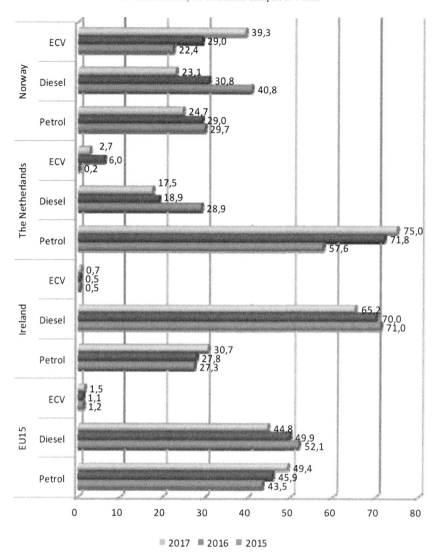

FIGURE 10.2 Types of drives used in selected European countries (%). *Source: Authors' elaboration based on ACEA, 2018e. Passenger Car Fleet by Fuel Type. Retrieved from https://www.acea.be/statistics/tag/category/passenger-car-fleet-by-fuel-type.*

driven with gas, 37% with diesel, and 44% with petrol (PZPM, 2017). It proves that in Poland other types of drives do not enjoy great popularity.

It is also worth supplementing the numerical values presented above with additional commentary. Despite the fact that various organizations, associations, and even car manufacturers talk about departing from internal combustion in passenger cars and the Moody's Corporation estimates that by the end of the next decade over 50% of the registered passenger cars in Europe will have been electrically operated or equipped with a hybrid system (Insight

Automotive, 2016), European drivers have a different opinion. In the research conducted by Mazda Driver Project, 11,000 people participated, of whom 30% expressed the hope that diesel-powered vehicles will continue to exist. In Poland, 65% are supporters of this drive, whereas 60% are supporters in Germany, Spain, and Sweden per each country. In the case of Poland, this opinion is determined by both the price of electric vehicles and very slow development of the infrastructure for charging these vehicles (Bołtryk, 2018a).

To sum up, the balance sheets concerning the popularity of different fuels in countries of the EU show that still there are not many electric and hybrid cars in comparison to cars with diesel propulsions. This data proves that despite the slight participation of electric and hybrid cars, their number gently rises year to year. It can prove that governments' efforts in the form of various subsidies to electric cars and tax breaks cause a positive effect in the form of the systematically increasing number of electric vehicles on European roads.

Electric cars charging infrastructure

The ability to charge the batteries of electric vehicles is a challenge for the development of electromobility. There are various ways of charging, which include: charging with an electric socket or with the battery charger, including fast battery charger. These ways differ in time of charging as well the type of communication with the vehicle. Loading the vehicle at home, or at the office, are commonly done, since electrical wiring of the amperage of 16 A is enough.

Characterizing both normal and high-power vehicle charging stations, it is possible to define them as standalone civil structures with at least one charging point installed. Charging stations must also be equipped with software that will enable the performance of the charging service at parking stands and installations from the charging point to the electrical connection. In the literature, charging points and charging stations are frequently treated as synonyms. They can also be called electrical charging point, a battery charger, or EVSE (electric vehicle supply equipment).

Electrical charging points can be characterized through (Sendek-Matysiak and Szumska, 2018):

- the level describing power input,
- the type of nest and plug,
- mode of the communication protocol between a car and the charger.

Depending on the level of the electric power led by the charging point and the kind of exploited electricity (DC, AC) and of a type of connector, three levels are distinguished: level 1, level 2 for soft chargers; and level 3 for fast chargers. It influences the duration of charging car batteries, which depending on the level, can last from 14 h on the first level, up to 1 h on the third level.

Among the different ways of charging electric personal cars are: mechanical exchange of the set of batteries in the vehicle, solar charging with photovoltaic panels, and wired charging. Connecting the car to a charging device have also defects, e.g., it is necessary to perform this activity every time. Problems with connecting can also appear, particularly during winter, when the temperature is low, but currently, it is a favorable way of loading passenger vehicles (Sendek-Matysiak and Szumska, 2018).

In 2016 the battery amounted almost 48% of the price of the electric car, and in 2018 only by 6% less. According to BNEF analysis,[1] in 7 years electric vehicles will be even cheaper than those with internal combustion. It has been predicted that in 2024 prices of some models of cars with electric and traditional propulsion will be equal. However, a year later prices of electric vehicles will drop. A systematic decline in prices of lithium-ion batteries is the main assumption of the decline in prices of electric cars (Bołtryk, 2018d).

From 2012 to 2016 (Fig. 10.3) the price dropped by 369 dollars for kWh, which proves that speculations of analysts concerning leveling of prices of electric and hybrid cars can be real, since it would level the barrier of high price of the car, however, the still small number of charging stations is the main barrier influencing the purchase of vehicles.

The market for electric vehicle charging services in Europe is still in its initial stage of development, and the activity in this field is conducted by different actors in different business models, often in the form of free service aimed at building the image of a specific company or property (among others, shopping centers, hotels).

Implementing charging infrastructure should be done with a view to meet users' and suppliers' needs. PEV users require access to CSs whenever they need them, accompanied with a high quality of service. Therefore, a lack of charging facilities due to siting them inappropriately or not at all will have a negative impact on drivers' convenience (Alhazmi et al., 2017). Also, the location of these charging points is a very important issue, which has made many authors construct mathematical models of this solution (Alhazmi et al., 2017; Zhu et al., 2016; Rahman et al., 2016; Morrissey et al., 2016; Oláh et al., 2018a, 2018b, 2018c).

At the EU level, the issue of the development of infrastructure for electric vehicles was regulated in the document "The Directive of the European Parliament and the Council 2014/94/UE of 22 October 2014 on the development of infrastructure for alternative fuels" (European Parliament, 2014). In Article 4 (paragraph 1) of this document, the member states

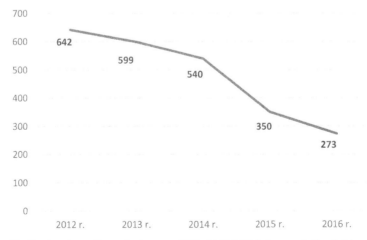

FIGURE 10.3   Decline in prices of batteries in years 2012–16 ($/kWh). *Source: Authors' elaboration based on Spadek cen akumulatorów w latach 2010–2016.*

[1]Bloomberg New Energy Finance.

shall ensure that, by means of their national policy frameworks, by December 31, 2020, there will have been developed an adequate number of publicly accessible charging points. The Directive indicates that there should be one publicly accessible charging point per every 10 registered electric cars (according to the Directive [article 4, paragraph 9] a publicly accessible charging point is one that allows the users of electric vehicles for ad hoc charging without a contract with the electricity supplier or operator).

Fig. 10.4 shows the number of charging stations in the EU28 countries.

There were over 116,000 charging points available across the EU countries in 2018. Almost 30% are located in the Netherlands, with another 22% in Germany, 14% in France, and 12% in the United Kingdom. These four countries together account for 76% of all EV charging points in the European community. By contrast, the same four countries only cover 27% of the EU's total surface area. Luxembourg, which owns 337 chargers for electric cars in the area of 2586 km^2, deserves special attention.

In 2018, Poland had only 552 charging points, and of those a majority are normal battery chargers and only 20 are fast battery chargers (ACEA, 2018a). Unfortunately Polish

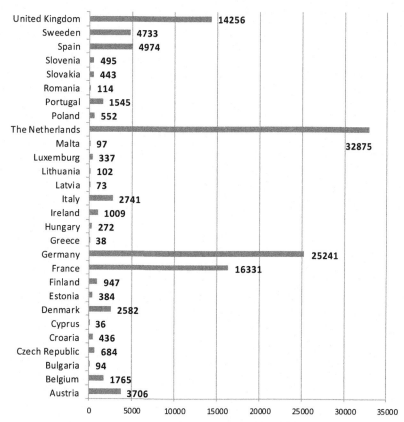

FIGURE 10.4 Number of charging points in EU28 countries in 2018. *Source: Authors' elaboration based on ACEA, 2018a. Interactive Map: Correlation Between Electric Car Sales and the Availability of Charging Points. Retrieved from https:// www.acea.be/statistics/article/interactive-map-correlation-between-electric-car-sales-and-the-availability.*

infrastructure is not as extended as in Western European countries. The problem is also the geographical distribution of charging points. Nowadays, the largest number of them are in Warsaw, Cracow, Poznan, and Gdansk. This low result also affects the rather small sale of electric cars.

For Poland, a turning point in the field of the development of the sector connected with the widely comprehended transport came in 2018 through implementing an act on electromobility and alternative fuels (Dz.U. 2018 poz. 317). According to the Ministry of Energy, spreading electric cars can take place at a relatively rapid pace. According to the government electromobility development plan, in 2025 as many as one million electric vehicles should move along Polish roads, and this is connected also with the development of infrastructure necessary to charge electric cars (Bołtryk, 2018b; Zatoński, 2017). In this act from January 11, 2018 a device enabling to charge both electric and hybrid vehicles was determined as the charging point (Sendek-Matysiak and Szumska, 2018). The term of charging station for eco-friendly was also defined.

According to the provisions of the above Act, since January 2019, *Ewidencja Infrastruktury Paliw Alternatywnych* (The Records of Alternative Fuel Infrastructure) has been operating in Poland, which is the public register kept of the users of electric vehicles and natural gas-powered vehicles. The information included in the register is to facilitate the use of these vehicles and include, among others, the coordinates of public charging stations and the availability of charging points installed in public charging stations.

The development of charging infrastructure is essential for that more eco-friendly cars to move along the streets. Development plans assume 6400 of points to 2020, including 6000 of normal power electric energy charging points and 400 more powerful ones. Also, seven points for fueling up with compressed natural gas are supposed to be established in 32 urban agglomerations. This network of charging points is supposed to enable the development of the electromobility and increase using alternative fuels in the transport in Poland. It is one component of the plan, which assumes the considerable increase in the number of electric cars in Poland.

For example, Greenway company, dealing with establishing and expansion of the network of fast-charging stations, declares that by 2020, 200 new charging stations will be established, among them 10 ultrafast and 135 fast charging stations (Piszczatowska, 2018).

One should, however, remember that the Greenway company is not the only company, but also it declares that it invests in establishing new charging points. National energy groups such as Tauron, PGE, Enea, and Energa also promise to make investments. The PKN Orlen and Lotos oil companies also have declared specific actions in terms of building electric vehicle charging stations at their filling stations (Puls Biznesu, 2018; Mayer, 2018).

These changes are supposed to contribute to wider use of alternative fuels and reduce the dependency of the transport on petroleum, which will also affect the improvement of air quality. The Act implements the European directive on the development of infrastructure of alternative fuels to the Polish law. It assumes that within two years charging stations can develop on market principles, and the state will back them up with funding. And if progress in the development will not be advantageous then self-government bodies and communes will be able to assist this undertaking.

Commune councils will also be able to determine zones of the clean transport only for cars driven with alternative fuels, and the remaining cars with combustion engines will have to pay a fee for entry into the zone of the clear air.

After enforcing these provisions, electric cars are supposed to be exempt from the excise tax, and entrepreneurs will be also able to get higher depreciation allowances for eco-friendly vehicles; after 2025 their drivers will be able to move along buses lanes and to park in paid parking zones for free (Zatoński, 2017).

To sum up, in Poland numerous transformations are taking place, which are supposed to increase the number of electric cars on Polish roads. Thanks to them, Poland not only will meet the expectations of the European Union but also will benefit from improved air quality. This will not only become a more pleasant place for residents of the country but also for tourists.

Combination of privileges for users of eco-friendly cars in model European countries

The European Union, carrying out the plan of lowering the emission level of greenhouse gases by 80%—95% by 2025, is supporting solutions aimed at electrification of the road transport sector. The main purpose is to make Europe independent of internal-combustion engines and fight against polluted air and the production of greenhouse gases. On that account, European countries support actions in favor of the expansion of vehicle charging stations in all of Europe. Governments of individual states are also introducing numerous concessions and facilities in order to encourage drivers to forgo cars powered by traditional diesel propulsion for vehicles using electric energy.

Many states are trying to arouse an interest in hybrid and electric cars particularly through financial incentives, e.g., thanks to ecological extra charges and concessions. Model categories of privileges for of eco-friendly car owners in Europe are shown Table 10.2.

Many countries implemented numerous privileges and benefits for owners of electric and hybrid cars. As can be seen in the above Table 10.2, the most universal conveniences for eco-friendly cars owners are exemptions from different kind of taxes and subsidies to the purchase of the car. Below are presented some incentives to buy and use electric vehicles in selected countries; a broad description of them is available in the following studies: *Raport: Samochody elektryczne i hybrydowe. Kompendium wiedzy kierowcy* (The Report: Electric and hybrid vehicles. Compendium of driver's knowledge); Overview on Tax Incentives for Electric Vehicles in the EU (ACEA, 2018b) and *The Automobile Industry Pocket Guide 2018—19* (ACEA, 2018c).

In France, a system of direct grants was implemented, and the amount fluctuates from 500 to 7000 EUR depending on the carbon dioxide emission. In major cities of Spain, e.g., in Madrid, Barcelona, and Valencia, tax relief was also introduced, e.g., a reduction in the annual road tax (by 75%) for owners of electric and fuel-efficient cars (ACEA, 2018b).

Nationale Plattform Elektromobilität (National Electric Mobility of Platforms), a sort of advisory council of the government, was established in Germany in 2010. This program is being dedicated to electric cars owners, and its main task is to spread the electric technology in cars

TABLE 10.2 List of privileges for eco-friendly cars owners in European countries.

	Tax relief	Subsidies to purchase	Bonuses	Relief in parking charges	Access to charging station
Austria	✔		✔		
Belgium	✔		✔		
Czech	✔				
Denmark	✔			✔	
Estonia		✔			✔
Finland					✔
France		✔			
Germany	✔				✔
Greece	✔				
Iceland	✔			✔	
Ireland	✔	✔			✔
Italy	✔				
Luxemburg			✔		
Monaco		✔		✔	
The Netherlands	✔	✔		✔	✔
Norway	✔			✔	
Portugal	✔	✔			✔
Romania		✔			
Spain	✔	✔			
Sweden	✔	✔			
Switzerland	✔				
Poland				✔	✔
Great Britain	✔	✔			

Source: *Authors' elaboration based on Baranowski, M. (ed.), 2015. Samochody Elektryczne I Hybrydowe. Kompendium Wiedzy. Retrieved from https://eco-driving.info/hybrydy-do-raportu/.*

in the entire country. Electric vehicles are exempt from the annual circulation tax for a period of 10 years from the date of their first registration. From July 2016, the government granted environmental bonuses of 3000–4000 euro for plug-in hybrid and electric vehicles (ACEA, 2018b).

In Austria, an exemption from "fuel" tax was implemented (VAT and excise taxes), taken at the first registration, and annual bonuses up to 800 euro (Przywileje i korzyści, 2015). In

Finland, users of hybrid and electric cars have access to a free network of charging points and pure electric vehicles always pay the minimum level of the CO_2-based registration tax (ACEA, 2018c).

The Netherlands has the greatest number of conveniences for owners and users of hybrid and electric cars. The Dutch government fixed the purpose of the intensive electrification in the automotive industry. The main assumption is so that 200,000 low-emission vehicles will be driving in the Netherlands by 2020, and one million by the end of 2025 (Przywileje i korzyści, 2015). For achieving this task, the government has allocated funds of 500 million euro (Przywileje posiadaczy, 2015). Also, improvements in the form of tax and registration reliefs added up to 5400 euro, for the four first years of using electric or hybrid cars. An interesting convenience is also a possibility of the access to parking spaces free of charge in centers of such cities as Amsterdam or Rotterdam. Rotterdam predicts the possibility of using a car park free of charge in the city center for 1 year, from the moment of registration of an electric vehicle (Przywileje posiadaczy, 2015). Amsterdam has implemented a lot of inducements for using electric cars, e.g., financing from the city budget for the expansion of charging points. When a person living in Amsterdam states that he/she needs charging point in the vicinity of her domicile, he/she informs the proper city authorities, who commit to building an electric car charging point in the place indicated by the resident.

In Stockholm, they are also investing in the development of infrastructure of electric vehicle charging stations. In 2014, the City Council commissioned the establishment of 10 stations of fast loading and 100 AC charging points. Three years later, in 2017, eight fast and 100 ordinary charging points located mainly at petrol stations, by restaurants, or supermarkets were also additionally installed (Fishbone et al., 2017).

Norway is not a European Union member but is strongly connected with it by economic ties. In 2015 in Norway, every fourth sold car had electric propulsion, and an assumption was that in 2018, 50,000 of these cars would drive on the roads. Since 1990 electric cars have also been exempted from customs and excise taxes, and when buying cars users are exempt from 25% value-added tax. Electric car buyers also have a lower road tax (Polityka Insight, 2018). Since the 1900s, the Norwegian government has supported users and buyers of electric vehicles. It means that authorities of the country made every effort in order to achieve a large number of electric cars.

In Great Britain, the government wants to spend 400 million GDP on the development of the electric vehicle charging infrastructure. Additionally, the government believes that the infrastructure development takes place most effectively at the local level, due to public authorities, entrepreneurs, and natural persons, therefore they provide them with appropriate support. There are available separate grants for eco-friendly private cars, taxi cabs, and buses and also subsidies for the purchase of EV chargers for household use (Bołtryk, 2018c).

Currently, in Poland conveniences for electric vehicles have local character. In a few cities improvements have been introduced such as using a charging station or also exemptions from parking fees. In Wrocław owners of eco-friendly cars can use fast-charging stations for free as well as are free from parking fees in two zones, B and C, however in A zone, that is in the city center, electric cars owners must also pay to park. In Gdańsk, Katowice, and Toruń, parking fees for hybrid and electric vehicles were abolished; however, in Szczecin, Cracow, and Tarnów payments are lower than for combustion cars. Additionally, in Gdańsk, it is also possible to use the chain of fast charging stations totally free.

As well, according to the plan of the electromobility development (Ministerstwo Energii, 2018), the following actions are to be taken in Poland in the context of promoting electric vehicles:

1. Introduction of changes in the tax system that are beneficial for the users of electric vehicles, in particular:
   - changes in excise duty
   - changes in VAT
   - better amortization of electric vehicles.
2. Introduction of a fee related to the price and emissivity of a motor vehicle.
3. The use of soft support instruments.
4. Additional subsidies for electric buses.

According to the latest proposals by the Ministry of Energy (February 2019) there are predicted subsidies for the purchase of electric vehicles (up to 36,000 PLN, i.e., about 84,000 euro) and a range of grants for the purchase of electric buses, hydrogen-powered, CNG or LNG -powered, and also trolleybuses (Bołtryk, 2019). If the regulations announced are implemented in the suggested form, the Polish market will take off.

The participation of electric vehicles is significant in countries, in which numerous conveniences for owners of electric cars are offered, although the amount of subsidies and concessions is different throughout Europe. The greatest number of states in the European Union with a low share of electric cars offer only an exemption from the annual road tax from the owned type of the vehicle. On the other hand, as shown by experience, subsidies for the infrastructure and electric vehicles are, in addition to exemption from VAT, the most efficient forms of incentives to buy zero-emission cars.

Automotive concerns development plans connected with electric cars

The largest firms associated with the automotive industry, such as Japanese Toyota, or German Volkswagen and BMW, have announced the production of the latest models of electric cars for upcoming years (see Table 10.3).

Main European manufacturers of passenger cars are, among others, BMW Group, Daimler, Ford, Volkswagen Group, Toyota whether also PSA group.

Toyota has announced plans connected with the electrification of produced cars for the next 12 years. Development activities are supposed to last within all technologies: plug-in type hybrids, HEV not-loaded from the nest,[2] and BEV powered thanks to the battery.[3] The company assumes that in 2030 sales will reach the more than 5.5 million annually and one million will be BEV cars and cars driven by hydrogen cells (Bołtryk, 2017a).

BMW Group announced that in 2025 (Table 10.3) it will release 25 models of electric cars on the market, with 12 of them fully electrified. This company assumes that in the future electric drive should be available in every car of this German concern. Also, sports versions BMW

[2]Hybrid Electric Vehicle.
[3]Battery Electric Vehicle.

TABLE 10.3   Automotive concerns' plans connected with electric cars.

Automotive concern	Predicted time of the project implementation	Plan description
Toyota	Up to 2020	Implementation of the latest model of electric car with few minutes charging time
BMW	Up to 2025	Implementation of 25 types of electric cars into an offer, the transformation of a mini model into electric only
Volkswagen	Up to 2025	Implementation of 80 types of electric cars into offer and production of 3 million units
General Motors	Up to 2023	Presentation of 20 new types of electric cars and the introduction of the purchase of a million electric cars yearly from 2026

Source: *Authors' elaboration based on Bołtryk, M. 2017b. Nadciaga Elektryczne Tsunami. Puls Biznesu "Transport i Logistyka", s.17.*

M and Rolls-Royce are to be electrified. The first electric Mini model will be produced in 2019 (Bołtryk, 2017b).

Daimler also produces an electric brand named EQ. The first model (SUV EGC) is supposed to come to the market in 2019.

The French concern PSA also donated expenses to investments connected with the development of electric car production. The main plan is a cooperation of PSA with Japanese Nidec automotive company, which is supposed to rely on the production of drives for French electric cars. For that purpose, at the end of 2017 companies made a joint venture agreement and agreed on the mutual cooperation for the development of modern, very efficient and competitive lines of electric traditional engines for electric vehicles. The concern invested 220 million euro in this undertaking. It started at the beginning of 2018. In the Paris region in France, Carrières sous Poissy, is a head office of a joint undertaking. It also includes scientifically developmental objects. The main location of production units is Trémery, Metz, region, also in France (PSA Groupe, 2017).

The previous examples prove that automotive concerns are implementing innovative solutions to follow world trends and cope with customers' expectations by setting their production up to include ecological technology. In the coming years, other companies will produce more and more electric models. Concerns wanting to be more competitive will be making the sale of electric four-wheeled vehicles attractive, which in the future can influence the reduction of costs of purchasing these kinds of vehicle and will cause them to appear more and more often on European streets.

Conclusions

In the entire European Union, including Poland, there has been an intensive campaign for electromobility development. The occurrence of electric vehicles (EV) has brought vast opportunities for both the transport and energy sectors. In the transport sector, electric vehicles are considered as the promising alternative technology of vehicles to achieve reduction of GHG emissions. In the case of the energy sector, electric vehicles provide the potential in terms of adjustment of a high level of energy production from renewable sources.

Although it seems that there is no turning back from electromobility, its pace of development depends on numerous factors: acceptance of consumers, technological progress, as well as support systems. The existing holders of electric vehicles indicate a number of inconveniences and high costs incurred for their purchase and use, among others, insufficient infrastructure and a long charging time in PEV vehicles. In order to develop electromobility, a lot of governments have introduced a range of incentives, among which the most popular are: subsidies for the purchase of a car or setting up the charging infrastructure in convenient locations (Dong et al., 2014; He et al., 2015; Nie et al., 2016).

Poland's performance, against the background of Europe, is not impressive, occupying only the 15th position in terms of newly registered electric vehicles. The situation may improve due to the amendment of the Act on electromobility and introduction of facilities for PEV vehicle drivers. This will allow for respecting stricter EU emission standards and, consequently, improve living conditions and environmental protection.

References

ACEA, 2018a. Interactive Map: Correlation between Electric Car Sales and the Availability of Charging Points. Retrieved from. https://www.acea.be/statistics/article/interactive-map-correlation-between-electric-car-sales-and-the-availability.

ACEA, 2018b. Overview on Tax Incentives for Electric Vehicles in the EU. Retrieved from. https://www.acea.be/uploads/publications/EV_incentives_overview_2018.pdf.

ACEA, 2018c. The Automobile Industry Pocket Guide 2018–2019. Retrieved from. https://www.acea.be/publications/article/acea-pocket-guide.

ACEA, 2018d. Alternative Fuel Vehicle Registrations. Retrieved from. https://www.acea.be/statistics/tag/category/electric-and-alternative-vehicle-registrations.

ACEA, 2018e. Passenger Car Fleet by Fuel Type. Retrieved from. https://www.acea.be/statistics/tag/category/passenger-car-fleet-by-fuel-type.

Alhazmi, Y.A., Mostafa, H.A., Salama, M.M., 2017. Optimal allocation for electric vehicle charging stations using trip success ratio. International Journal of Electrical Power & Energy Systems 91, 101–116. https://doi.org/10.1016/j.ijepes.2017.03.009.

Baranowski, M. (Ed.), 2015. Samochody Elektryczne I Hybrydowe. Kompendium Wiedzy. Retrieved from. https://eco-driving.info/hybrydy-do-raportu/.

Biznesu, P., 2018. PKN Orlen Na Rzecz Rozwoju Elektromobilności.

Bołtryk, M., 2017a. Toyota Podpina Sie Do Gniazdka. Puls Biznesu "Transport i Logistyka", p. 15.

Bołtryk, M., 2017b. Nadciaga Elektryczne Tsunami. Puls Biznesu "Transport i Logistyka" s.17.

Bołtryk, M., 2018a. Kierowcy Wierza W Diesla. Puls Biznesu "Transport i Logistyka" s.14.

Bołtryk, M., 2018b. Polska Przestawi Europe Na Prad. Puls Biznesu "Transport i Logistyka" s.14.

Bołtryk, M., 2018c. Brytyjczycy Dali Przykład Polakom. Puls Biznesu s.16.

Bołtryk, M., 2018d. Elektryki (Nie) Stanieja. Puls Biznesu "Transport i Logistyka" s. 15.

Bołtryk, M., 2019. Doczekaliśmy Sie Dopłat Do Elektryków. Puls Biznesu s.3.

Bühne, J.A., Gruschwitz, D., Hölscher, J., Klötzke, M., Kugler, U., Schimeczek, C., 2015. How to promote electromobility for European car drivers? Obstacles to overcome for a broad market penetration. European Transport Research Review 7 (3), 30. https://doi.org/10.1007/s12544-015-0178-0.

Business Insider Polska, 2018. Samochody Elektryczne Maja Już Blisko 200 Lat. Dlaczego Nie Od Razu Odniosły Sukces. Retrieved from. https://businessinsider.com.pl/motoryzacja/pierwszy-samochod-elektryczny-na-swiecie-historia-motoryzacji/2jxm5w0.

Dong, J., Liu, C., Lin, Z., 2014. Charging infrastructure planning for promoting battery electric vehicles: an activity-based approach using multiday travel data. Transportation Research Part C: Emerging Technologies 38, 44–55. https://doi.org/10.1016/j.trc.2013.11.001.

European Commission, 2011. White Paper on Transport. Roadmap to a Single European Transport Area — towards a Competitive and Resource Efficient Transport System. Retrieved from. http://eur-lex.europa.eu/LexUriServ/LexUriServ.do?uri=COM:2011:0144:FIN:pl:PDF.

European Parliament, 2014. The Directive of the European Parliament and the Council 2014/94/UE of 22 October 2014 on the Development of Infrastructure for Alternative Fuels. Official Journal of the European Union No 307.

Figenbaum, E., Assum, T., Kolbenstvedt, M., 2015. Electromobility in Norway: experiences and opportunities. Research in Transportation Economics 50, 29–38. https://doi.org/10.1016/j.retrec.2015.06.004.

Fishbone, A., Shahan, Z., Badik, P., 2017. Infrastruktura Ładowania, Pojazdów Elektrycznych, Wytyczne Dla Miast. Retrieved from. http://greenwaypolska.pl/wp-content/uploads/sites/7/2018/05/GreenWay_Infrastruktura_ladowania_pojazdow_elektrycznych_Wytyczne_dla_miast_www_maj_2018.pdf?fbclid=IwAR1ilPV4Sxp8GaHM PNFQm9TUzXlTKLYT3FxmZ6Pzx9BrVu-2VdKL52IRKkA.

Gass, V., Schmidt, J., Schmid, E., 2014. Analysis of alternative policy instruments to promote electric vehicles in Austria. Renewable Energy 61, 96–101. https://doi.org/10.1016/j.renene.2012.08.012.

Grauers, A., Sarasini, S., Karlström, M., 2013. Why electromobility and what is it?. In: Systems Perspectives on Electromobility. Chalmers Publication Library, pp. 10–21. Retrieved from. http://publications.lib.chalmers.se/records/fulltext/211430/local_211430.pdf.

He, F., Yin, Y., Zhou, J., 2015. Deploying public charging stations for electric vehicles on urban road networks. Transportation Research Part C: Emerging Technologies 60, 227–240. https://doi.org/10.1016/j.trc.2015.08.018.

Hennek, K., 2018. Perspektywy rozwoju i wykorzystania pojazdów elektrycznych. Autobusy–Technika, Eksploatacja, Systemy Transportowe 220 (6), 458–462.

InfoCar, 2016. Historia Samochodu Elektrycznego. Retrieved from. http://info-car.pl/infocar/artykuly/historia-samochodu-elektrycznego.html.

Polityka Insight, 2018. Cicha Rewolucja W Energetyce, Elektromobilność W Polsce. Retrieved from. https://www.politykainsight.pl/_resource/multimedium/20106685.

Insight Automotive, 2016. A Watershed Moment for the Automotive Industry. Retrieved from. https://www.alixpartners.com/media/3759/ap_a_watershed_moment_for_the_automotive_industry_aug_2016.pdf.

Kovács, G., Kot, S., 2016. New logistics and production trends as the effect of global economy changes. Polish Journal of Management Studies 14 (02), 115–126. https://doi.org/10.17512/pjms.2016.14.2.11.

Kowalczuk, D., 2017. BEV, PHEV I HEV — Analiza Sprzedaży Samochodów Niskoemisyjnych. Retrieved from. https://eco-driving.info/bev-phev-i-hev-analiza_sprzedazy-samochodow-niskoemisyjnych/.

Kupczyk, A., Maczyńska, J., Redlarski, G., Tucki, K., Baczyk, A., Rutkowski, D., 2019. Selected aspects of biofuels market and the electromobility development in Poland: current trends and forecasting changes. Applied Sciences 9 (2), 254. https://doi.org/10.3390/app9020254.

Mayer, B., 2018. Elektryczny Prymat Państwa. Puls Biznesu s.3.

Michalak, M., 2018. Samochód Hybrydowy. Co Oznaczaja Skróty mHEV, HEV, PHEV I BEV? Retrieved from. https://www.motofakty.pl/artykul/samochod-hybrydowy-co-oznaczaja-skroty-mhev-hev-phev-i-bev.html.

Ministerstwo Energii, 2016. The National Policy Framework for the Development of Infrastructure for Alternative Fuels. Retrieved from. http://www.me.gov.pl/files/upload/26450/ME_KRAJOWE_RAMY_projekt.pdf.

Ministerstwo Energii, 2018. Plan Rozwoju Elektromobilności. Retrieved from. https://www.gov.pl/web/energia/elektromobilnosc-w-polsce.

Morrissey, P., Weldon, P., O'Mahony, M., 2016. Future standard and fast charging infrastructure planning: an analysis of electric vehicle charging behaviour. Energy Policy 89, 257–270. https://doi.org/10.1016/j.enpol.2015.12.001.

Nie, Y.M., Ghamami, M., Zockaie, A., Xiao, F., 2016. Optimization of incentive polices for plug-in electric vehicles. Transportation Research Part B: Methodological 84, 103–123. https://doi.org/10.1016/j.trb.2015.12.011.

Oláh, J., Nestler, S., Nobel, T., Popp, J., 2018a. Ranking of dry ports in Europe - benchmarking. Periodica Polytechnica Transportation Engineering 46 (2), 95–100. https://doi.org/10.3311/PPtr.11414. Retrieved from. https://pp.bme.hu/tr/article/view/11414.

Oláh, J., Nestler, S., Nobel, T., Harangi-Rákos, M., Popp, J., 2018b. Development of dry ports in Europe. International Journal of Applied Management Science 10 (4), 269–289. https://doi.org/10.1504/IJAMS.2018.10010622.

Retrieved from. http://www.inderscience.com/info/inarticletoc.php?jcode=ijams&year=2018&vol=10&issue=4. http://www.inderscience.com/storage/f311101291864275.pdf.

Oláh, J., Nestler, S., Nobel, T., Popp, J., 2018c. International characteristics of the macro-logistics system of freight villages. Periodica Polytechnica Transportation Engineering 46 (4), 194—200. Retrieved from. https://pp.bme.hu/tr/article/view/11656. https://doi.org/10.3311/PPtr.11656.

ORPA, 2018. Pojazdy Zelektryfikowane, Rodzaje Napedów. Retrieved from. http://www.orpa.pl/samochody-elektryczne-co-musisz-wiedziec/.

Piszczatowska, J., 2018. Nowa Mapa Ładowarek Samochodów Elektrycznych W Polsce. Retrieved from. https://wysokienapiecie.pl/2465-nowa-mapa-ladowarek-samochodow-elektrycznych-w-polsce-ev-auta-baterie/.

Przywileje i korzyści dla właścicieli samochodów hybrydowych. (2015), Retrieved from https://moto.onet.pl/przywileje-i-korzysci-dla-wlascicieli-samochodow-hybrydowych/ffkpgg1.

Przywileje posiadaczy aut niskoemisyjnych. (2015). Retrieved from http://ogarniamprad.weebly.com/strona-g322oacutewna/przywileje-posiadaczy-aut-niskoemisyjnych.

PSA Groupe, 2017. Groupe PSA and Nidec to Set-Up a Leading Joint Venture for Automotive Electric Traction Motor (Retrieved from).

PZPM, 2017. Automotive Industry Report 2017/2018 (Retrieved from).

Rahman, I., Vasant, P.M., Singh, B.S.M., Abdullah-Al-Wadud, M., Adnan, N., 2016. Review of recent trends in optimization techniques for plug-in hybrid, and electric vehicle charging infrastructures. Renewable and Sustainable Energy Reviews 58, 1039—1047. https://doi.org/10.1016/j.rser.2015.12.353.

Riba, J.R., López-Torres, C., Romeral, L., Garcia, A., 2016. Rare-earth-free propulsion motors for electric vehicles: a technology review. Renewable and Sustainable Energy Reviews 57, 367—379. https://doi.org/10.1016/j.rser.2015.12.121.

Sendek-Matysiak, E., Szumska, E., 2018. Infrastruktura ładowania jako jeden z elementów rozwoju elektromobilności w Polsce. Prace Naukowe Politechniki Warszawskiej, Transport 121, 329—340.

Spadek cen akumulatorów w latach 2010—2016. (n.d.). Retrieved from https://elektrowoz.pl/transport/jak-spadaly-ceny-baterii-w-ostatnich-latach-wykres/.

Zachariasz, M., Dybkowski, M., 2014. Analiza układu napedowego pojazdu hybrydowego z silnikiem indukcyjnym. Prace Naukowe Instytutu Maszyn, Napedów i Pomiarów elektrycznych Politechniki Wrocławskiej. Studia i Materiały 70 (34), 225—232.

Zatoński, M., 2017. Elektromobilność Ładuje Akumulatory. Puls Biznesu s.8.

Zhu, Z.H., Gao, Z.Y., Zheng, J.F., Du, H.M., 2016. Charging station location problem of plug-in electric vehicles. Journal of Transport Geography 52, 11—22. https://doi.org/10.1016/j.jtrangeo.2016.02.002.

Further reading

Bjerkan, K.Y., Nørbech, T.E., Nordtømme, M.E., 2016. Incentives for promoting battery electric vehicle (BEV) adoption in Norway. Transportation Research Part D: Transport and Environment 43, 169—180. https://doi.org/10.1016/j.trd.2015.12.002.

Nationale Plattform Elektromobilität, 2018. Electric Mobility— the Background. Retrieved from. http://nationale-plattform-elektromobilitaet.de/en/.

http://www.pzpm.org.pl/Rynek-motoryzacyjny/Roczniki-i-raporty/Raport-branzy-motoryzacyjnej-2017-2018.

Ustawa Z Dnia 11 Stycznia 2018r. O Elektryczności I Paliwach Alternatywnych (Dz.U. 2018 poz. 317).

https://media.groupe-psa.com/en/groupe-psa-and-nidec-set-leading-joint-venture-automotive-electric-traction-motor.

11

Toward solutions for energy efficiency: modeling of district heating costs

Ugis Sarma[1,3], Girts Karnitis[2], Edvins Karnitis[2],
Gatis Bazbauers[1]

[1]Institute of Energy Systems and Environment, Riga Technical University, Riga, Latvia;
[2]Faculty of Computing, University of Latvia, Riga, Latvia; [3]JSC Latvenergo, Riga, Latvia

O U T L I N E

Energy Transformation towards Sustainability
https://doi.org/10.1016/B978-0-12-817688-7.00011-2

219

Nomenclature

Cdep	depreciation costs (1000s €)
Cfu	fuel costs, expressed in energy units (€/MWh)
Cl	heat loss costs (1000s €)
Clab	labor costs (1000s €)
Cpr	total heat production costs (1000's €)
Ctr(lin)	modeled costs, linear modeling (1000s €)
Ctr(nolin)	modeled costs, nonlinear modeling (1000s €)
Ctr(m)	modeled heat transmission costs (1000's €)
Ctr	total transmission costs (1000s €)
Dmax	the largest inner diameter of pipes (mm)
Dmed	inner diameter of most frequently used pipes (mm)
Dmin	the smallest inner diameter of pipes (mm)
Ktf	heat generation costs index
L	network length (km)
n	number of connected users
Pcon	connected load (MW)
Pins	installed heat capacity (MW)
Qpr	produced thermal energy (GWh)
Qus	thermal energy transferred to users (GWh)
R^2	coefficient of determination
Rtf	indicator on proportion of Tfu in Tpr
T	total district heating tariff (€/MWh)
Tfu	tariff component that covers "the best practice" fuel costs
Tpr	heat production tariff (€/MWh)
Tpr(sim)	simulated heat production tariff (€/MWh)
Ttr	transmission tariff (€/MWh)
Π	set of transmission predictors
η	heat generation efficiency

Abbreviations

ANSDM	average normalized squared deviation from the mean value
CAPEX	capital expenditure
CHP	cogeneration
DH	district heating
DH01-DH23	district heating utilities
KPI	key performance indicators
NRA	National Regulatory Authority
OPEX	operating expenses
VIF	variance inflation factors

Introduction

The provision of networked utility services — electricity, gas, heat, and water supply — is naturally divided into two stages, which are characterized by principally different technologies, processes, and competition prerequisites. Some limited competition is possible in the commodity production stage. Actually, all production segments legally are opened for competitors; in practice there are more competitors (e.g., power generation) or only very few activities (e.g., drinking water production) in various branches. At the same time, an expensive infrastructure (network of cables/wires or pipes) with a very long payback period is a strong barrier for competition in transmission/distribution (hereinafter transmission) stage, leaving a monopoly in particular area).

The economic regulation of utility services has been introduced — more general and soft procedures in the production segment and the strong cost control and following tariff setting in transmission segment. Specific input data for regulation, which utilities should declare, actually are insufficiently qualitative (unreliable, mutually incompatible) for the large number of district heating (DH) utilities. Submission of the large number of data is a strong administrative burden even for large utilities. Assessment of costs is a laborious task for the National Regulatory Authorities (NRAs); especially this relates to the water supply and district heating because of their high fragmentation even within a single country, and a large number of the local monopolies (see, e.g., Zuters et al., 2016). The usage of advanced data mining technologies has become more and more popular for investigation, management, and prediction of physical, technological, and economic issues of all utility services, including those relating to regulation (e.g., Mansura et al., 2018; Tutusaus et al., 2018). In recent years simulation and modeling tools have been widely created and offered (e.g., Strzelecka et al., 2014; Berg van den et al., 2016; Molinos-Senante et al., 2017; Schweiger et al., 2018); benchmarking methods, which are based on the aggregation of a large number of key performance indicators (KPIs), are preferred as suitable tools (e.g., Festel and Würmseher, 2014).

DH services naturally are not as comprehensive as electricity or water supply; these are not developed in the European Union's (EU) most influential countries with mild climate, with the exception of some of the largest cities (Werner, 2017). In contrast, the DH is very important in Nordic countries and in Eastern Europe. Latvia ranks second in Europe behind Iceland (Epp, 2016) by population served by DH (65% in 2015); nevertheless, only about two-thirds of Latvian households are using a centralized heating supply.

Studies on the DH pricing models and strategies for implementation of optimal price structure have been carried out, and various models of the final price's structure and price differentiation are described; they are based on users' consumption profile and load demand (e.g., Song et al., 2016; Dyrelund, 2017). Nowadays it is becoming particularly topical to take into account integration of renewable and waste energy sources as well as development of low-temperature DH systems (Schmidt, 2018).

Very few studies are devoted to the DH regulation and tariff-setting principles. A positive exception is rather widely analyzed reform of the DH in Sweden. The reform's vision was to build a common single heat market, in which heat prices in all DH systems will strive to a single level. Unfortunately the conclusion is that the goal was not reached (Magnusson and Palm, 2011; Söderholm and Wårell, 2011; Åberg et al., 2016).

There is a lack of a major normative framework on DH at the EU level. Despite acknowledgment of the DH as a significant contribution to mitigation of the climate change and to reaching the objectives of decarbonization (e.g., Rezaie and Rosen, 2012; Ziemele et al., 2014; Delangle et al., 2017; Song et al., 2017), relatively minor activities had been made to improve and to synchronize the DH regulation at the EU level. Currently the regulatory regime in Nordic countries is more liberal, but in Eastern Europe overregulation can be observed (see, e.g., Zigurs et al., 2015).

Although the interest in simulation and modeling of the DH processes is lower in comparison with electricity and drinking water supply, advanced technologies are used to optimize the DH systems. It is self-explanatory that the direct quantitative calculation of the adequate heat supply costs is practically an impossible approach for regulatory authorities. The alternative option that ensures a well-based compromise between high credibility of the cost evaluation and rapidity of the calculations is usage of the intelligent benchmarking data mining methods and mathematical modeling tools. The DH utilities provide identical service under homogenous geographic, normative, and business environment; because of analogous impact of external factors, the benchmark methodologies are applicable.

A wide range of academic literature relates to the thermohydraulic modeling (e.g., Arce et al., 2018; Vivian et al., 2018; Dahash et al., 2019; Denarie et al., 2019); economic and financial issues (Persson and Werner, 2011; Song et al., 2017; Ahn et al., 2018) as well as the management of DH business (Wang et al., 2018) are less popular. Nevertheless, the tariff-setting process remains very traditional; it is based on careful evaluation of all declared cost items including the very insignificant ones. At the same time, exact and objective tariffs are a critical precondition for efficient and sustainable energy supply (e.g., Lukosevicius and Werring, 2011).

The aim of the current study is development of the methodology for fast and objective evaluation of the actual costs of DH to optimize the regulatory process and to set the adequate tariffs for consumers, as well as to increase the DH systems' efficiency and to reduce the administrative burden on utilities. It would support also a gradual shift from the strict ex ante regulation to flexible ex post monitoring. The concrete developed methodology will be based on the data sets of Latvian DH providers, but the methodological principles will be applicable in another area too.

Previous analysis, performed by the authors, indicated that the heat production and the transmission networks should be analyzed separately because of differences in processes and technologies (Bazbauers and Sarma, 2016).

Simulation of the heat production tariff

In the previous studies (Sarma et al., 2016; Sarma & Bazbauers, 2017a, 2017b) authors have focused on the heat generation segment, which is quite fragmented in Latvia. On the one hand, heat generation is the regulated segment; on the other hand, there are elements of competition — several larger and smaller heat producers operate in a number of the DH systems. The NRA approves a heat production tariff for vertically integrated companies; the legislation also allows an agreed upon price between the independent heat producer and the network operator. Two fuel types dominate the heat production: natural gas and biomass (mainly wood chips, somewhere granules). Also, several technologies are used — heat-only boilers and cogeneration (CHP) plants — the latter being granted state aid under a rather complicated scheme, which sometimes leads to cross-subsidies and also market distortion.

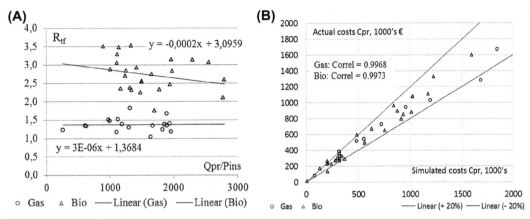

FIGURE 11.1 Dependence of the production tariff structure on Qpr in the concrete plant (A); correlation of actual and simulated heat production costs (B); developed by the authors.

A reciprocal of the specific indicator Rtf, which was offered in the previous stage of the study (Sarma and Bazbauers, 2017a), shows the proportion of the tariff component Tfu in the total heat production tariff Tpr:

$$\text{Tpr} = \text{Tfu} * \text{Rtf} = \frac{\text{Cfu} * \text{Rtf}}{\eta}, \text{where Tfu} = \text{Cfu}/\eta \tag{11.1}$$

The conclusion in Sarma and Bazbauers (2017a) is very informative (see Fig. 11.1A); it is clearly evident that the chosen fuel defines Rtf. In the case of gas use, the mean Rtf = 1.37 is actually independent from the amount of heat Qpr produced in the plant, which can be characterized by the Pins. Small dispersion of Rtf values is caused by specifics of the concrete plant: heat generation efficiency, company's performance, consumption of electricity, etc.

In the case of biomass use, the proportion of other costs slightly decreases as Qpr rises. Nevertheless, in the first simulation we can assume that the proportion is rather constant and use the mean value of Rtf = 2.69; the error of this assumption in the real range of installed capacity utilization does not exceed ±8% as can be seen in Fig. 11.1A (fluctuations of fuel price during the heating season often are higher).

In addition η also is quite constant for adequately maintained and operated plants: $\eta = 0.95$ for gas-fired plants and $\eta = 0.85$ for biomass-fired ones. Thereby:

$$\text{Tpr(sim)} = \text{Cfu} * \text{Ktf} \tag{11.2}$$

where:

$$\text{Ktf} = \text{Rtf}/\eta \tag{11.2a}$$

Ktf = 1.44, using natural gas;
Ktf = 3.16, using biomass.

If the production capacities are selected adequately to the actual demand and heat generation efficiency η is near the optimal one, the heat production tariff using the certain type of fuel will depend on a single external factor — fuel price (Eq. 11.2).

It is obvious that the tariff component Tfu directly depends on Cfu and η (Eq. 11.1). This component is a large part of the heat production tariff Tpr. Its proportion highly depends on fuel type (75%—80% for gas-fired plants, 40%—45% for biomass-fired plants in average, considering the reasonable heat generation efficiency), while the objective differences between costs of heat-only boiler and CHP technologies are practically aligned by the CHP subsidies. On the other hand, the chosen fuel determines production technology, which in turn defines the technical parameters of production, as well as variable costs, CAPEX and OPEX cost items.

Comparison of the heat production costs, which are covered by the actual tariffs Tpr and the simulated ones Tpr(sim) (Fig. 11.1B), shows extremely high correlation between them. It should be noted that for efficient producers (the deviation from benchmark is less than one standard deviation) actual tariffs in real terms are similar in both groups (Fig. 11.1B). This can be explained by the fact that if the gas prices are approximately two times higher than the prices of biomass, then other costs associated with biomass fuel usage are much higher than for the gas-fired technologies. Despite fluctuations of prices in short term, in the medium term the ratio between the gas and biomass prices remains quite stable and consequently the tariffs are similar.

The conclusion is that the fuel type and specific fuel costs can be classified as the KPI for determination of production tariff in the first simulation.

Transmission costs modeling: the paradigm

The transmission segment is very different from the production one in several aspects. The network systems are more homogeneous (assets mainly consist of standardized pipelines) and determined; the design of the networks does not change significantly during long time period. Changes in segment structure are much slower than those in the generation segment; the network renovation projects, which are directed to replacement of the pipelines with those that are better insulated and more adequate for the real heat demand, are planning investments for a long life cycle (30—40 years). It is strictly regulated because of the natural monopoly in a certain area — in the license zone of the DH network system operator.

On the other hand, the DH network system is characterized by many interconnected geometric (lengths and diameters of pipes, branching, number of connecting points of heat producers and customers), thermophysical (heat flows and temperatures), and energetic (amount of delivered heat, heat losses) factors.

The intelligent data mining methods are appropriate for the discovery of existing regularities in the network's data set. The benchmarking data mining algorithms are well suited to investigate cause-and-effect relationships and to compare the performance aspects of companies, to discover the top performing leaders, and to identify how current less successful operators could progress.

The usage of mathematical modeling tools has become a popular and generally accepted method for the prediction of the scenarios in a wide variety of fields. In our case the modeling will determine the exact relationships between the network indicators as independent variables (predictors) and some cost indicator as the dependent variable, which should reflect the technological and economic performance of the network operator. This way we will disclose the impact of indicators on costs and extract the KPI, reducing a large number of potential factors.

There are several multivariate analytical methods that are appropriate in principle for the investigation of cause-and-effect relationships between the input and output variables without investigation of internal aspects of the system (a *black box* modeling). Regression analysis, as the method at the crossroads of data mining and modeling, was chosen as the most preferable statistical modeling tool, which is directly focused on the relationships between the dependent variable and several predictors. The authors already have successfully used regression methodologies for the cost modeling in a related industry — water supply (Karnitis et al., 2017); the results obtained have shown the suitability of the method to solve similar tasks.

Data gathering and exploratory analysis

Predictors

The quality of output data strongly depends on the quality of input data. Therefore, we used only a limited number of clearly and unambiguously defined predictors, which are quantitatively measurable and controllable, and which are obtainable for the NRA without additional administrative burden on utilities.

These predictors were separated from the sets of indicators, declared by 23 DH utilities, which were available for modeling (see Appendix, Table 11.4).

Dependent (target) variable

To perform the modeling, the dependent (target) variable (total transmission costs or several separate cost components) should be determined. An analysis of the structure of declared cost components has been made; it identified a great variety in these components (Table 11.1). There are several objective reasons for it, e.g., utilities are using various business

TABLE 11.1 Proportion of transmission cost components; developed by the authors.

	Percentage of total costs Ctr	
	Min	**Max**
Heat loss costs Cl	20.9%	83.3%
Labor costs Clab	5.5%	58.3%
Depreciation costs Cdep	2.4%	41.3%

models, wide but uneven investments (from both national resources and EU Structural Investment Funds) have been made for extension and efficiency increase of the DH networks; in addition, the national regulations on accounting and bookkeeping are quite general, as well as accuracy and uniformity of data on several cost components remains some challenge for the utilities.

The Ctr can be evaluated as more qualitative (accurate, reliable) data in comparison with the separate cost components. In addition, exact total costs are the main DH efficiency criterion and basic regulatory focus as the determinant of tariff; analysis of cost structures and examination of every cost component leads to overregulation of utilities and is not an NRA function. Consumers also are interested in the total costs only. Therefore, the total transmission costs were chosen as the dependent variable. The modeled Ctr(m) would be defined as a multiparameter function of the set of predictors:

$$Ctr(m) = f(\prod) = f\{i_1, i_2, ..., i_k\} \tag{11.3}$$

where:

k—number of predictors i.

Outliers

One can find several data points in Table 11.4 that deviate significantly from other data points in the columns of indicators Q, L, n, and C. The identification of outliers, i.e., data points that are too far away from the central data cluster, is an essential preparatory activity before benchmarking modeling. Outliers may have a different essence and cannot be used to create general regularities.

Rejection of such data samples from the total data set is necessary to achieve the normal observations on costs of the heat transmission processes in the corresponding network. "Detected outliers are candidates for aberrant data that may otherwise adversely lead to model misspecification, biased parameter estimation and incorrect results. It is therefore important to identify them prior to modeling and analysis" (Ben-gal, 2010).

Using a box plot procedure for normalized (in the range 0–1) values of the Qpr, L, n, and Ctr, we identified three DH utilities (DH01, DH09, DH15) as extreme outliers (Fig. 11.2A). The scale of their business is relatively very high in comparison with other utilities; in total they serve 76.6% of users, maintaining 77% of the total pipe length and operating 83.7% of heat in networks (Fig. 11.2B). Two of them lie just beyond an upper outer fence, and the third one (DH15) lies in extremely abnormal distance from the central data cluster. Nevertheless, all of them should be qualified as bad data sets and excluded; data samples of remaining 20 DH utilities had been used for the modeling.

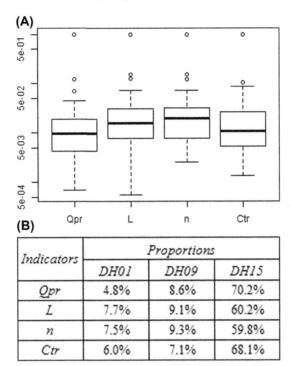

(A)

(B)

Indicators	Proportions		
	DH01	*DH09*	*DH15*
Qpr	4.8%	8.6%	70.2%
L	7.7%	9.1%	60.2%
n	7.5%	9.3%	59.8%
Ctr	6.0%	7.1%	68.1%

FIGURE 11.2 Box plots of indicators (A) and proportion of outliers' business in total business of 23 DH utilities (B); developed by the authors.

Multicollinearity

Having so many independent variables, it is necessary to check their cross-independence level, to make sure how strong are the mutual correlations between the predictors. Usage of strongly correlated predictors would lead to the so-called multicollinearity problem (see, e.g., Allen, 1997). The essence of the problem is the possibility of obtaining unreliable results by usage of interdependent predictors.

To examine the potential problem, the typical tool was used — the variance inflation factors — which characterize the mutual correlation between any pair of predictors. The reliability criterion is:

$$VIF = 1/(1 - R^2) < \delta$$

There are various evaluations of the critical value δ. We followed the strong recommendations of the Allen(1997) and used $\delta = 2.5$ as the value of the threshold.

The VIFs were calculated for full matrix of predictors; the mutual correlations of five variables are above the mentioned threshold (Table 11.2). There is an obvious conclusion: only one of the five strongly correlated predictors can be used for modeling, the others should be excluded from the data set.

TABLE 11.2   Fragment of the VIF matrix; developed by the authors.

	Variance inflation factors			
	Pcon	**Qus**	**Qpr**	**L**
Qus	13.3			
Qpr	15.5	238.1		
L	5.5	11.3	14.3	
n	5.0	6.7	6.7	4.9

Modeling of transmission costs

Multistage linear modeling

For modeling we chose the well-developed, powerful, and at the same time user-friendly R statistics environment, which provides flexible algorithms for modeling. We started modeling by using the simpler linear regression algorithm. The postmodeling analysis of residuals will show the purposefulness of continuation of modeling by usage of more complicated nonlinear regression algorithms to obtain the stronger cause-effect relationship.

For the linear modeling we used the built-in linear regression model, implemented in R as a function lm, which is called by command lm. The target was to detail the general regularity (Eq. 11.3) and to create the benchmarking model as the linear mathematical expression:

$$Ctr(lin) = \alpha + \beta_1 * i_1 + \beta_2 * i_2 + \ldots + \beta_k * i_k \tag{11.4}$$

where:

α— constant

β_k— estimated optimal weight of predictor i_k.

In the first stage we included all nine thermal energy and geometric indicators in the set of predictors with a single goal — to define the most preferable one among the strongly interconnected five predictors and to avoid potential multicollinearity. The general regularity (Eq. 11.4) in this case becomes:

$$Ctr(lin9) = \alpha + \beta_1 * Pcon + \beta_2 * Qus + \beta_3 * Qpr + \beta_4 * Tpr + \beta_5 * L + \beta_6 * n + \beta_7 * Dmax + \beta_8 * Dmin + \beta_9 * Dmed \tag{11.5}$$

The result is very instructive (Table 11.3, column Ctr(lin9)). Although the correlation between any of processed five predictors and Ctr(lin9) is very strong (>0.9), their impact on Ctr(lin9) is different (their mutual correlation is one of the basic reasons). A perusal of *P*-values shows that the network length L is the most significant predictor among the five mentioned mutuallycorrelating ones; there is only 3.8% probability on random impact of

TABLE 11.3 Characteristics of the models and predictors; developed by the authors.

	Characteristics of the predictors						
	Linear models					Nonlinear model	
	Ctr(lin 9)	Ctr(lin 5)		Ctr(lin 2)		Ctr(nolin)	
Indicators	P-Value	P-Value	Model parameters	P-Value	Model parameters	P-Value	Model parameters
Pcon	0.2730						
Qus	0.2730						
Qpr	0.6256						
L	0.0381	5.84e−08		3.05e−10		2.11e−13	
n	0.8499						
Dmax	0.0916	0.00428		0.000309		7.25e−06	
Dmin	0.9038	0.70497					
Dmed	0.3632	0.83559					
Tpr	0.4303	0.79119					
Statistical characteristics of the models							
		1.071e−09		5.89e−13		7.91e−16	
R^2		0.9651		0.9636		0.9747	
ANSDM		0.0736		0.1443		0.0303	

the L on Ctr(lin9). Chances of other indicators to be accidental ones are incomparably higher (27%−85%); the network length L was chosen as the predictor for the next modeling stages.

The second stage is creation of 5-predictor model, using the network length L and remaining four indicators as predictors, in the form:

$$\text{Ctr(lin5)} = \alpha + \beta_1 * \text{Tpr} + \beta_2 * L + + \beta_3 * \text{Dmax} + \beta_4 * \text{Dmin} + \beta_5 * \text{Dmed} \qquad (11.6)$$

The obtained correlation is extremely strong, the P-value for the model is very small (Table 11.3, columns CTR(lin5)). Nevertheless, high P-values of three predictors show that their impacts on costs are insignificant; these predictors only complicate the model unnecessarily.

In addition, the average normalized squared deviation from the mean value (ANSDM) of the actual costs Ctr from those of modeled Ctr(lin5) is quite big; we have used the normalization for evaluation of deviations here and further because of the large range of networks size:

$$\text{ANSDM} = \sum_{1}^{20} \frac{\left(\dfrac{Ctr - Ctr(m)}{Ctr}\right)^2}{20}$$

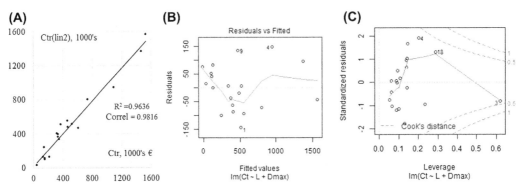

FIGURE 11.3   predictor linear model; scatter plot Ctr(lin2) versus Ctr (A), diagnostic plots (B, C); developed by the authors.

The 5-predictor model approves the paradigm on length of the pipelines and inner pipe-line diameter as the determining (significant) factors for the flow of the heat carrier in the pipes and consequently for network design (Pusat and Erdem, 2014). Costs for laying out pipes (Delangle et al., 2017) as well as the total cost of piping per meter (District, 2013) strongly depend on the pipes' diameter, therefore significance of the largest diameter is well explained. Thus L and Dmax were chosen as the predictors for creation of 2-predictors model (Eq. 11.7) in the third modeling stage:

$$\text{Ctr}(\text{lin2}) = \alpha + \beta_1 * L + \beta_2 * \text{Dmax} \tag{11.7}$$

The final result of linear modeling (Table 11.3, columns Ctr(lin2)) is:

$$\text{Ctr}(\text{lin2}) = -176.7086 + 23.0418 * L + 0.9863 * \text{Dmax} \tag{11.8}$$

The correlation parameters are virtually unchanged in comparison with the Ctr(lin5)). Extremely small P-value confirms the validity of the predictors choice; the abandoned three indicators really have no significant impact on the searched regularity. At the same time the excellent correlation is partly misleading because of too few data points in the 30-fold range of the actual Ctr values (Fig. 11.3A); this is confirmed by the quite large value of ANSDM.

R diagnostic plots also point to incomplete compliance of the linear model Ctr(lin2) with the actual data. The plot 3b shows that residuals are not equally spread around a horizontal line (especially at small fitted values). It is an indication that the linear model doesn't capture the existing nonlinear relationship between predictors and the target variable. The plot 3c identifies the data point that lies very near to the threshold, so-called Cook's distance. This point is not an outlier; nevertheless it can become influential against a general regularity.

Nonlinear modeling

Using the traditional method of the least squares, it was found that the nonlinear trendline in the linear model (Fig. 11.3A) provides a little higher coefficient of determination R^2 in

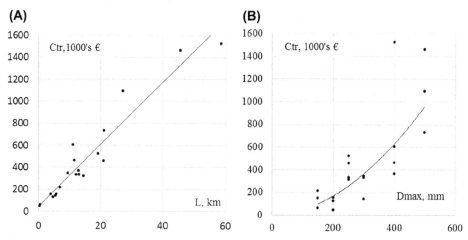

FIGURE 11.4 Scatter plots Ctr versus L (A) and Ctr versus Dmax (B); developed by the authors.

comparison with the standard linear one. It approves a necessity to add some nonlinearity in the model.

To find the correct way, we checked the scatter plots Ctr versus L and Ctr versus Dmax (Fig. 11.4). The impact of each individual predictor on the result of the total mining is, of course, different from the individual correlation (e.g., due to some mutual impact of predictors). Nevertheless, the qualitative differences between both scatter plots provide some comparative indication. The optimum trendline on the scatter plot (Fig. 11.4A) is linear, while on the scatter plot (Fig. 11.4B) one can observe a moderate nonlinearity as well as can indicate that the inclusion of power function is preferable.

A nonlinear regression process was applied to reduce the mentioned inconsistencies and to increase the coincidence of the searched regularity with declared transmission costs, thus improving the quality of the model. As the first step, we used the NLS function, which determines the nonlinear (weighted) least-squares estimates of the parameters of the nonlinear model, to define the nonlinearity that best matches the regularities of the actual costs Ctr:

$$\text{model1} < -\,\text{nls}(\text{Ctr} \sim \beta_1 * L + \beta_2 * (\text{Dmax})^\gamma, \text{start} = \text{list}(\beta_1 = 23, \beta_2 = 1)) \tag{11.9}$$

where the start values of β_1 and β_2 were determined according to the linear model (8).

The optimum values of the coefficients β_1 and β_2 were calculated by NLS function for number of picked γ values using the correlation (Ctr:Ctr(nolin)) as the quality criterion.

As the second step we used linear modeling to find the free term of the final model Ctr(nolin) by processing function (Eq. 11.9) that was found in the first step:

$$\text{Ctr (nolin)} < -\text{lm}(\text{Ct} \sim \text{predict (model 1)})$$

The final result is:

$$\text{Ctr(nolin)} = 25.40652 + 22.994925 * L + 5.77105E - 09 * (\text{Dmax})^4 \tag{11.10}$$

Slightly improved formal correlation and *P*-value (Table 11.3, columns Ctr(nolin)) in comparison with the linear model Ctr(lin2) is not the main achievement. One can see that nonlinear model (Fig. 11.5A) is much more coinciding with the actual costs of utilities. Deviation of actual costs Ctr from the modeled Ctr(nolin) has decreased for 18 utilities, while it has insignificantly increased only for two utilities; ANSDM has decreased significantly as the result. The highlighted confidence interval (at 95% confidence level) is satisfactorily narrow (Fig. 11.5A).

R diagnostic plots show better behavior of the residuals in comparison with the linear model. The residuals are more equally spread around the horizontal zero line (Fig. 11.5C).

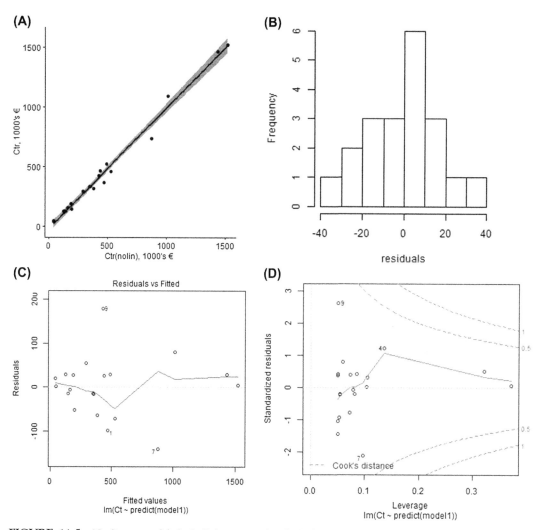

FIGURE 11.5 Nonlinear model Ctr(nolin); scatter plot Ctr(nolin) versus Ctr (A), histogram of residuals (B), diagnostic plots (C, D); developed by the authors.

There are no patterns between the residuals and fitted values, thus the independence assumption might hold as well. No data point is near the Cooks distance (Fig. 11.5D). The histogram on Fig. 11.5B suggests the normality of the residuals around the zero value that is desirable for a good model.

Then the transmission tariff can be determined as:

$$\text{Ttr} = \frac{\text{Ctr}(nolin)}{1000 * \text{Qus}} \tag{11.11}$$

where Ctr(nolin) is defined by (11.10).

The conclusion is: the network length L, the largest inner diameter of pipes Dmax and the amount of thermal energy transferred to users Qus can be classified as KPI for determination of the transmission tariff.

Results and discussion

The achieved accuracy of the results and their coincidence with the parameters of real DH utilities clearly demonstrates correctness of the trend of the current research and its perspective. Several KPIs have been identified, which are decisive for the definition of adequate tariffs of heat production (fuel type and fuel price) and of heat transmission/distribution (length of pipes, the largest inner diameter of pipes, and amount of thermal energy transferred to users).

The analysis of results of heat production simulation (Fig. 11.1B) shows that more than 85% of utilities lie in the corridor ±20%. This proportion is higher for the utilities using gas-fired technologies (94%), and lower for biomass users (82%), thus confirming a thesis on higher share of the fixed costs in the case of biomass use (e.g., much higher labor costs).

The histogram on Fig. 11.5B indicates that there are five utilities with actual Ctr that are more than 10% above the values of Ctr(nolin); it is the combined effect of several causes. The specific heat loss per unit of the virtual pipe surface ($\pi * L * $ Dmax) in each of these networks is significantly higher than the mean level. The heat loss in DH11 network is near one-third of produced heat, while the loss in DH20 network is very high for the very short network (600 m only). In addition, utilities DH10 and DH20 have declared high depreciation costs, i.e., large investments in the networks, which indicate the massive renovation of the networks that should have significantly reduced losses. The labor costs in DH7 and DH18 are incomprehensibly high.

Further improvement of the models certainly is possible and necessary; the impact of other cost items should be assessed in more detail in the next studies. Reduction of the existing information asymmetry is the primary task; the input data problems currently are greater than the inadequacy of the proposed methodology. One can understand that a large number of data sets are required for modeling to select the most significant KPI. Currently data on only 43 of existing 107 heat generation plants and 23 of 56 functioning pipe networks were available for simulation/modeling.

While the total number of transmission systems would be sufficient for further modeling in case of the full coverage due to their comparative uniformity, the number of production plants in each of the segments (natural gas or biomass use, heat-only boiler or CHP) definitely is too small. It was concluded in Sarma and Bazbauers (2017b) that the simulation should be based on a sufficiently large number of possible input values to obtain a more objective picture of potential variations of Tpr. Therefore, a study has been started to build Monte Carlo simulation models for calculation of the parameters forming Tpr values and to integrate them in the data mining algorithm.

Conclusions

The created benchmarking models well reflect the association of fuel parameters (fuel type and fuel price) with the heat production tariff, as well as the association of network parameters (length of pipes and largest inner diameter of pipes) and amount of the thermal energy transferred to users with the tariff for heat transmission and distribution. The performed analysis and the models indicate the KPIs, which determine generation and transmission costs and tariff quite objectively.

The models well illustrate appropriateness of the proposed methodology for development of tools and procedures for the regulatory analysis of operation performance of the DH companies and for evaluation of the DH tariffs. The methodology is a reasonable compromise between accuracy and simplicity, which is its strong advantage for practical applications. Despite the use of a small number of predictors, the correlations between the modeled and the actual costs are extremely strong and average deviation small. Because of constantly developing technologies, growing efficiency and changing external factors, for practical applications the NRA should make annual calibrations of the model according to the data declared by DH utilities.

Based on the proposed methodology, it would be possible to make considerable changes in the regulatory regime. The NRA, as an energy policy implementer, could use this tool for incentive regulation by setting tariff ceilings and permissible deviations of the actual values from the tariffs calculated by the models. Furthermore, following technological and economic developments, the NRA can gradually reduce the permissible deviations, stimulating in this way the DH utilities' approach to the best practice. The goal should be a transition to the model-based tariff setting by using KPIs that are declared in annual reports of the utilities.

Using the described methodology, the NRA could also draw conclusions on economic soundness of the specific DH system in cases with critical deviations from the model value. Such cases are becoming more of a reality due to continuing decrease of heat demand because of implementation of energy efficiency measures in buildings and pipe networks (Persson and Werner, 2011).

DH utilities could use the methodology for the sustainable planning and assessment of investments and activities to increase their efficiency. This will be a very important aspect in the near future when the number of prosumers and their activity will increase (see, e.g., Song et al., 2017).

Overall, by using the proposed methodology the regulation process will become simplified and accelerated; the administrative burden on the DH utilities and the NRA will substantially reduce while NRA efficiency will significantly increase.

Appendix

TABLE 11.4 Performance indicators of district heating utilities.

Utility	Lcon, MW	Qpr, GWh	Qus, GWh	Tpr, €/MWh	L, km	n	Dmax, mm	Dmin, mm	Dmed, mm	Ctr, 1000's €	Cl, 1000's €	Clab, 1000's €	Cdep, 1000's €
DH1	103.00	246.63	215.29	34.36	102.18	990	600	25	100	3008.0	1077.0	376.0	1018.0
DH2	20.26	47.90	40.51	41.50	12.80	192	400	40	80	368.0	306.4	49.9	10.7
DH3	15.09	35.46	30.17	36.81	14.50	158	250	32	100	317.5	194.6	17.5	53.8
DH4	65.46	156.01	133.45	36.85	58.60	349	400	40	200	1524.6	831.4	281.9	157.4
DH5	29.80	70.90	59.18	38.27	27.30	264	500	25	80	1093.4	448.3	201.9	304.7
DH6	6.60	15.17	14.11	49.87	5.39	55	300	20	150	143.3	52.7	12.6	29.7
DH7	15.39	35.02	30.78	27.81	9.50	123	300	25	65	344.5	117.8	116.3	45.8
DH8	28.34	64.50	58.71	37.51	21.20	162	500	50	65	732.3	217.1	155.2	250.1
DH9	429.89	438.85	366.74	37.93	120.68	1238	750	32	200	3596.9	2735.4	268.6	375.7
DH10	4.11	9.63	7.87	54.99	6.91	43	150	65	150	214.6	97.0	31.4	57.9
DH11	34.40	38.07	25.50	32.76	11.00	140	400	32	150	606.1	411.9	79.9	96.0
DH12	15.50	34.77	28.33	34.89	20.97	250	250	32	200	458.7	225.0	97.3	74.4
DH13	4.65	29.35	24.01	30.89	11.45	158	400	15	65	462.6	165.0	81.0	122.8
DH14	3.30	10.08	8.37	31.88	4.70	69	200	32	100	128.6	54.5	23.5	9.8
DH15	3085.00	3576.73	3152.04	31.41	800.00	7941	1200	25	80	34369.0	13341.0	6398.0	5452.0
DH16	14.00	38.20	30.41	36.24	19.29	175	250	25	150	520.2	282.0	96.6	98.6
DH17	9.85	23.29	19.70	32.90	13.00	130	250	25	65	332.3	118.3	49.5	96.5
DH18	15.00	15.01	13.91	34.73	3.96	48	200	40	100	154.0	38.0	73.0	19.0
DH19	1.10	2.46	2.15	57.69	0.44	20	200	65	200	46.0	18.0	5.0	19.0
DH20	1.50	3.49	2.95	60.66	0.60	21	150	65	150	62.0	33.0	7.0	16.0
DH21	76.11	160.23	145.49	44.72	45.59	567	500	32	100	1463.3	659.3	224.3	302.3
DH22	12.00	30.96	25.02	33.97	12.08	143	300	40	150	334.2	201.9	37.6	8.1
DH23	3.80	10.18	9.23	33.64	5.68	35	150	40	150	153.9	32.1	89.7	8.5

Source: Utilities, members of the Latvian Association of Heat Companies.

Acknowledgments

The work has been supported by the State Research Program (VPP) project "Development of heat supply and cooling systems in Latvia" (project No. VPP-EM-EE-2018/1—0002) and by the University of Latvia project AAP2016/B032 "Innovative information technologies." The authors would like to thank members of the Latvian Association of Heat Companies for sharing actual data.

References

Åberg, M., Fälting, L., Forssell, A., 2016. Is Swedish district heating operating on an integrated market? — differences in pricing, price convergence, and marketing strategy between public and private district heating companies. Energy Policy 90, 222—232.

Ahn, J., Chung, D.H., Cho, S., 2018. Energy cost analysis of an intelligent building network adopting heat trading concept in a district heating model. Energy 151, 11—25.

Allen, M.P., 1997. The problem of multicollinearity. In: Understanding Regression Analysis. Plenum Press, New York, pp. 176—180.

Arce, I.H., et al., 2018. Models for fast modelling of district heating and cooling networks. Renewable and Sustainable Energy Reviews 82, 1862—1873.

Ben-gal, I., 2010. Outlier detection. In: Maimon, O., Rokach, L. (Eds.), Data Mining and Knowledge Discovery Handbook. Springer, Heidelberg, pp. 117—130.

Bazbauers, G., Sarma, U., 2016. District heating regulation: parameters for the benchmarking model. Energy Procedia 95, 401—407.

Van den Berg, D., van der Heijden, M.C., Schuur, P.C., 2016. Allocating service parts in two-echelon networks at a utility company. International Journal of Production Economics 181, 58—67.

Dahash, A., Mieck, S., Ochs, F., Krautz, H.J., 2019. A comparative study of two simulation tools for the technical feasibility in terms of modelling district heating systems: an optimization case study. Simulation Modelling Practice and Theory 91, 48—68.

Delangle, A., Lambert, R.S.C., Shah, N., Acha, S., Markides, C.N., 2017. Modelling and optimising the marginal expansion of an existing district heating network. Energy 140, 209—223.

Denarie, A., Aprile, M., Motta, M., 2019. Heat transmission over long pipes: new model for fast and accurate district heating simulations. Energy 166, 267—276.

District Heating, January 2013. Retrieved from: https://iea-etsap.org/E-TechDS/PDF/E16_DistrHeat_EA_Final_Jan2013_GSOK.pdf.

Dyrelund, A., 2017. District heating tariffs — a way to communicate. HOT|COOL 3, 23—25.

Epp, B., July 1, 2016. Top District Heating Countries — Euroheat&Power 2015 Survey Analysis. Retrieved from: https://www.euroheat.org/news/district-energy-in-the-news/top-district-heating-countries-euroheat-power-2015-survey-analysis/.

Festel, G., Würmseher, M., 2014. Benchmarking of energy and utility infrastructures in industrial parks. Journal of Cleaner Production 70, 15—26.

Karnitis, E., Karnitis, G., Zuters, J., Bobinaite, V., 2017. Modelling of water supply costs. Procedia Computer Science 104, 3—11.

Lukosevicius, V., Werring, L., 2011. Regulatory Implications of District Heating. https://www.yumpu.com/en/document/view/35150691/regulatory-implications-of-district-heating.

Magnusson, D., Palm, J., 2011. Between natural monopoly and third party access — Swedish district heating market in transition. In: Karlsen, R.W., Pettyfer, M.A. (Eds.), Monopolies: Theory, Effectiveness and Regulation. Nova Science Publishers Inc, New York, pp. 1—33.

Mansura, H., Rafiqul, I., Shawkat, A., 2018. A component based unified architecture for utility service in cloud. Future Generation Computer Systems 87, 725—742.

Molinos-Senantea, M., Gómez, T., Caballero, R., Sala-Garrido, R., 2017. Assessing the quality of service to customers provided by water utilities: a synthetic index approach. Ecological Indicators 78, 214—220.

Persson, U., Werner, S., 2011. Heat distribution and the future competitiveness of district heating. Applied Energy 88, 568—576.

Pusat, S., Erdem, H.E., 2014. Techno-economic model for district heating systems. Energy and Buildings 72, 177—185.

Rezaie, B., Rosen, M.A., 2012. District heating and cooling: review of technology and potential enhancements. Applied Energy 93, 2–10.

Sarma, U., Bazbauers, G., 2017a. District heating tariff component analysis for tariff benchmarking model. Energy Procedia 113, 104–110.

Sarma, U., Bazbauers, G., 2017b. Algorithm for calculation of district heating tariff benchmark. Energy Procedia 128, 445–452.

Sarma, U., Zigurs, A., Bazbauers, G., 2016. Correlation analysis for district heating tariff benchmarking model. In: 13th International Conference on the European Energy Market (EEM), Portugal, Porto, 6-9 June. https://ieeexplore.ieee.org/document/7521208.

Schmidt, D., 2018. Low temperature district heating for future energy systems. Energy Procedia 149, 595–604.

Schweiger, G., et al., 2018. District energy systems: modelling paradigms and general-purpose tools. Energy 164, 1326–1340.

Söderholm, P., Wårell, L., 2011. Market opening and third party access in district heating networks. Energy Policy 39 (2), 742–752.

Song, J., Li, H., Wallin, F., 2017a. Cost comparison between district heating and alternatives during the price model restructuring process. Energy Procedia 105, 3922–3927.

Song, J., Wallin, F., Li, H., Karlsson, B., 2016. Price models of district heating in Sweden. Energy Procedia 88, 100–105.

Song, J., Wallin, F., Li, H., 2017b. District heating cost fluctuation caused by price model shift. Applied Energy 194, 715–724.

Strzelecka, A., Skworcow, P., Ulanicki, B., 2014. Modelling, simulation and optimisation of utility–service provision for households: case studies. Procedia Engineering 70, 1602–1609.

Tutusaus, M., Schwartz, K., Smit, S., 2018. The ambiguity of innovation drivers: the adoption of information and communication technologies by public water utilities. Journal of Cleaner Production 171, S79–S85.

Vivian, J., Uribarri de, P.M.A., Eicker, U., Zarrella, A., 2018. The effect of discretization on the accuracy of two district heating network models based on finite-difference methods. Energy Procedia 149, 625–634.

Wang, H., Duanmu, L., Lahdelma, R., Li, X., 2018. A fuzzy-grey multicriteria decision making model for district heating system. Applied Thermal Engineering 128, 1051–1061.

Werner, S., 2017. International review of district heating and cooling. Energy 137, 617–631.

Ziemele, J., Vigants, G., Vitolins, V., Blumberga, D., Veidenbergs, I., 2014. District heating systems performance analyses. Heat Energy Tariff. Environmental and Climate Technologies 13, 32–43.

Zigurs, A., Sarma, U., Ivanova, P., 2015. Implementation of the energy efficiency directive and the impact on district heating regulation. In: 12th International Conference on the European Energy Market (EEM), Portugal, Lisbon, 29-22 May. https://ieeexplore.ieee.org/document/7216630.

Zuters, J., Valainis, J., Karnitis, G., Karnitis, E., 2016. Modelling of adequate costs of utilities services. In: Dregvaite, G., Damasevicius, R. (Eds.), Communications in Computer and Information Science, vol. 639. Springer, Basel, pp. 3–17.

12

Small business energy challenges and opportunities

Ruth C. Hughes[1], Marleen A. Troy[2]

[1]Sidhu School of Business and Leadership, Wilkes University, Wilkes-Barre, PA, United States; [2]Department of Environmental Engineering and Earth Sciences, Wilkes University, Wilkes-Barre, PA, United States

OUTLINE

Energy Transformation towards Sustainability
https://doi.org/10.1016/B978-0-12-817688-7.00012-4

Introduction — overview of small business

Small businesses need energy to operate, market, and implement nearly every aspect of their business plans. Small businesses also need to continually minimize their expenses in order to remain competitive, particularly in a global economy. Thus, energy efficiency in small business should be a core part of any business plan, but unfortunately that assumption is not universally included in all of those small business operations plans. In addition, small business energy use collectively has an enormous impact on the environment. This chapter will outline the rationale for encouraging all small businesses to implement operational energy efficiency measures and review the tools currently available to achieve the goals of energy efficiency.

The Small Business Administration (SBA) in the United States administers many programs for small businesses in the United States and sets the size standards for defining "small business" for most programs benefitting the small business community, including energy efficiency programs. The SBA sets size standards based on either annual sales or number of full-time equivalent employees. For example, for most manufacturing and industrial concerns, regulations set a maximum number of employees between 1000 and 1500 to be categorized as a "small business," with the exact number depending on the specific type of industry. In contrast, most wholesale, retail, and service businesses have a maximum set between 100 and 200 employees (Size Standards, 2011).

According to SBA statistics, the United States had 30.2 million small businesses comprising 99% of all businesses in the country (2018 Small Business Profile, SBA). With such a large impact on the economy, one would expect a similarly large environmental impact. One study found that globally small businesses cause 70% of pollution (Johnson and Schaltegger, 2016). Thus, any comprehensive efforts by this group of businesses to reduce environmental impact should be encouraged.

Small businesses span all industries with most businesses falling in the general category of service businesses (42.93%), including professional services and health care services. Retail and wholesale businesses that sell products rather than services comprise 11.66% of businesses while construction businesses make up 10.25% of business nationwide. Small manufacturing businesses are about 2% of total small businesses (2018 Small Business Profile, SBA). Each of these small businesses has energy needs and all can work to achieve energy efficiency.

Research on small businesses incentives and obstacles to implement sustainable practices has indicated trends that would also apply to energy efficiency in particular. Most small businesses have no employees, leaving business owners to perform multiple tasks, particularly regarding management of the business. In turn, this multitask responsibility makes it difficult for those owners to easily innovate when that innovation takes time and attention away from the daily management of the business (Hughes and Troy, 2017). Furthermore, small businesses cite various other reasons that they have not implemented environmental sustainability and, by inclusion, energy efficiency such as risk of inadequate return on investment, initial cost of project, lack of clear results from implementation, and lack of impact due to size of business (Hughes and Troy, 2017; Federation of Small Business UK, 2007; Neamtu, 2011).

However, many small businesses are implementing environmental management and energy efficient measures and have indicated various motivators, including competitor and community pressure (Hughes and Troy, 2017; Redmond et al., 2008). Some other motivators applicable to energy efficiency include cost savings and carbon footprint reduction, as well as the personal views of the owner (Federation of Small Business UK, 2007; Jabbour and De Olivera, 2012).

Operational energy issues common to all small businesses

Electricity use

In the United States, small businesses collectively spend more than $60 billion per year on energy (US EPA, 2007b). Internationally, this trend has also been observed (Fleiter et al., 2012; Henriques and Catarino, 2016). No matter what their type of operation or service, small businesses will have in common basic electricity use for day-to-day office operations, as well as any associated manufacturing or production activities. Common sources of energy requirements for most small businesses typically will include energy use for lighting, heating and air conditioning, hot water, computers, printers, copiers, coffee/tea makers, and refrigerators and microwave ovens. Although energy usage by these items may seem trivial when considering the overall energy demands of an organization, studies have shown (Fanara et al., 2006) that over time they can be a large energy sink and can contribute to an organization's operational costs. Reducing overall energy usage and becoming more energy efficient is also an important component of the goal of reducing greenhouse gas emissions (Fleiter et al., 2012).

Business premises

The condition and location of the facility housing the small business can also present an energy use challenge. Very often, the small business does not own their facility. Because they are not the primary owner, their control over what can be modified or upgraded may depend on the leasing arrangement. Lighting, the condition of the windows, doors, roof, and insulation, the type and condition of the heating, ventilation and air conditioning system, and the overall age and condition of the building as well as its plumbing and utility infrastructure may be out of their control. If the landlord agrees to allow for any efficiency changes, the costs for any upgrades may need to be borne by the small business owner.

Wearing many hats — time management and skill set issues

Most small business owners and their employees have many responsibilities in addition to running their businesses. They are primarily concerned about keeping their customers/clients happy as well as staying viable. Because of this time demand, taking on the additional challenge of monitoring their operational energy usage can be a burden and may not be feasible for them. An additional obstacle frequently reported is that owners/employees will not have the knowledge base or skill set to monitor their energy use. Small businesses are often unaware of energy efficiency choices and strategies available to them (International Energy Agency, 2015). Their work schedules can preclude the opportunity to learn about these options and opportunities. Lack of motivation has also been reported as an obstacle (Henriques and Catarino, 2016). Adequate training and experience by both owners and employees with energy use monitoring and energy efficiency implementation techniques has also been identified as a significant challenge (Drake, 2014).

Tools/resources available to assist small businesses

Fortunately, there are many resources available to small businesses to help them with their energy efficiency improvement goals (US EPA, 2007a). Many of the resources available to small businesses are free or low cost. Resources include technical assistance as well as financing to help fund the start-up costs of implementing energy efficiency. The following describes some of the assistance available to businesses in the United States:

Technical assistance

Energy Star

Energy Star (www.energystar.gov/) is a United States Environmental Protection Agency program to increase awareness about energy efficiency as well as provide the tools and resources to meet energy efficiency goals. This program provides resources to increase awareness about energy efficiency; how to identify, prioritize, and manage projects; how to obtain funds; and how to select contractors (as needed) (US EPA, 2007a).

No matter what the type of small business, the first step in the improvement journey is benchmarking. Benchmarking is the process of assessing the organization's current status and quantifying where they currently are now regarding utility usage, inventorying the quantities and types of equipment being utilized, and identifying operational conditions and parameters. This information is important to identify where the organization is now regarding current energy usage as well as providing a mechanism for prioritizing energy efficiency projects. This benchmarking process involves gathering information about the organization's energy and water consumption and any associated cost data. A year's worth of utility bills is needed to start the process. Energy Star provides a free online resource to assist small businesses with the benchmarking process. Energy Star Portfolio Manager is a free software program that facilitates an organization's tracking and measuring of their energy and water use over time (US EPA, 2018). Portfolio Manager facilitates the calculation of energy intensity and cost, water use, and carbon emissions.

Energy Star also provides information regarding resources available to small businesses including technical assistance, facilitating a small business network, and providing recognition for achievements via an annual awards program (US EPA, 2007b).

Resources available from local utilities

Many electricity and natural gas providers provide resources that are available to assist the small business with both tracking their energy use metrics as well as providing opportunities and resources to assist with becoming more energy efficient. Examples can be found on the websites of two Pennsylvania utilities:

PPL https://www.pplelectric.com/
UGI https://www.ugi.com/

Utilities may offer resources to facilitate the performance of an energy audit. An energy audit is a thorough review of the actual performance of a business's systems and equipment against their designed performance or against best available technologies (US EPA, 2018). These audits are performed by energy professionals and/or engineers to assist the business in finding opportunities for energy savings. The benefits of energy audits are that they can provide the small business with important information about their current energy usage patterns and costs and also provide options for where changes can be made to promote operational improvements (Redmond and Walker, 2016).

Other resources

In addition to utility companies, private-sector companies, energy service companies, and ENERGY STAR service and product providers are options for the performance of energy audits (US EPA, 2018). Many federal and state agencies also will apply resources and assistance, particularly for financing.

There are also opportunities for small businesses to partner with local universities (Hughes and Troy, 2016) to get assistance with benchmarking and energy saving opportunities. The case study detailed later in this chapter provides an overview of this small business resource.

Financing

Small businesses cite cost as a primary obstacle to implementation of energy efficiency and other sustainable business practices (Hughes and Troy, 2017). Initial costs for energy efficient upgrades vary based on the business and include building envelope improvements, heating and air conditioning upgrades, and lighting upgrades, among others. In cases where there is significant cost, businesses need financing in order to install and maintain these systems.

The following are the most common types of financing programs available for small businesses to implement energy efficiency measures. These programs have a variety of sources, but many are government subsidized programs intended to encourage small business energy efficiency. One database containing comprehensive information on the government programs is the Database of State Incentives for Renewables and Efficiency located at www.dsireusa.org/.

Traditional debt financing

As with any project, small business may access traditional debt financing, typically from a financial institution, to pay the up-front costs of implementing energy efficient upgrades. However, these loans will charge interest at the market rate, depending on business's credit rating. Further, this type of traditional financing often requires that businesses meet other requirements, such as credit history and operating history, in order to qualify for the financing.

Government subsidized low-interest loans

Many state and local authorities offer low-interest loans to encourage small business energy efficiency. These loans typically have interest charged at a rate far below the market rate and so will allow for a return on investment that is faster than traditional financing. A typical example of a state loan program would be the Small Business Pollution Prevention Assistance Account Loan Program offered by the Pennsylvania Department of Environmental Protection (http://www.portal.state.pa.us/portal/server.pt./community/financial_

assistance/10495/ppaa_loan/553247). This program focuses on investing in technology to improve business energy efficiency and will finance 75% of a project cost up to $100,000 with interest rates charged at a fixed rate of 2% per year for a maximum term loan of 10 years.

Grant programs

While not as common as low interest loans, several states and utilities do offer grant programs that offer financing that a business does not have to repay as long as they fulfill all grant directives. These programs tend to be very specific and often require that the recipient business match their investment, requiring that the business have the funding in place at the time of the project. The Small Business Advantage Grant program offered by the Pennsylvania Department of Environmental Protection will provide successful business applicants with 50% of qualified project costs up to $9500, https://www.dep.pa.gov/Citizens/Grants LoansRebates/SmallBusinessOmbudsmanOffice/Pages/Small%20Business%20Advantage%20 Grant.aspx.

Thus, a business could have a $19,000 project and pay for only 50% of the costs of that project using this program, significantly reducing the initial financial burden.

Bonds

Another option offered by some states to encourage investment in energy efficient projects is a real estate–based bond financing program. These low interest bonds use business real estate as collateral and combine that with a government guarantee to allow for public financing in small business energy efficiency. While not commonly used for relatively small projects, this tool should not be overlooked by small businesses seeking resources (Office of Energy Efficiency and Renewable Energy).

Tax credits and deductions

Because all small businesses in the United States pay federal income tax, having tax credits, deductions, and other programs to reduce that income tax should encourage investment in energy efficient measures. Tax deductions allow the business to deduct from taxable net income qualified energy efficiency expenditures. Tax credits provide the added benefit of allowing the business to subtract a portion of the investment from the income tax due.

The Business Energy Investment Tax Credit allows a business to claim between 10% and 30% of a qualifying project cost as a credit against taxes due, depending on the type of energy project (https://www.irs.gov/pub/irs-pdf/i3468.pdf).

Another tax incentive at the federal level is accelerated depreciation allowed for certain capital investments, including those for energy efficiency upgrades with the Modified Accelerated Cost Recovery System (MACRS). However, it should be noted that MACRS is not a program that specifically targets energy investments but encompasses all small business capital investments (Internal Revenue Service Publication 946 (2016)). However, it should be noted that another program, the Energy Efficient Commercial Buildings Tax Deduction, is a subsection of the MACRS program and allows for a deduction based on square footage for certain energy efficient commercial buildings placed in service during the taxable year (29 U.S. Code Section 179D).

In addition, many states offer similar tax deductions and credits for those businesses required to pay state income tax.

Utility-based financing

Other more localized programs offered by energy utilities to encourage energy efficiency include rebate programs, feed in tariffs and on bill financing. Utility rebate programs allow businesses to apply for a rebate against their energy bills paid if they have implemented certain energy efficient measures, such as purchase of qualifying energy efficient equipment. Feed in tariffs are usually based in state regulation and require that a business implementing renewable energy capture be able to sell the excess energy produced by the on-site system but not used by the business back to the utility, thus allowing an income stream for the business. Implementing energy efficiency in conjunction with that renewable system has the potential to increase the amount of surplus energy for sale (Ottinger and Bowie, 2016). On bill financing, programs have participating utilities front the cost of the efficiency project for the business, which would then repay the utility each month. Some research shows that the savings from the project may offset the repayment amount under this program (Ottinger and Bowie, 2016).

Crowdfunding

Finally, another relatively new source of financing for small businesses for any project, including energy efficiency upgrades, is crowdfunding (Ottinger and Bowie, 2016). Crowdfunding allows a business to solicit funding from small investors using Internet sites typically organized specifically for the purpose of funding. If the site is not registered with the appropriate U.S. government agencies, the funding must be donation based. However, if the Internet site hosting the solicitation is registered with the U.S. Securities and Exchange Commission, the business may get debt or equity financing for the project using this method.

Operational considerations — case study

A sole-owned and operated small gift shop located in Wilkes-Barre, Pennsylvania, contacted the Wilkes University Small Business Center (WUSBDC) for consulting advice to help them look at opportunities to reduce operational costs. The WUSBDC operates as a part of the university to provide free, confidential consulting services to small businesses in northeastern Pennsylvania. The center uses students, faculty, and professional staff to deliver these services.

In conjunction with the WUSBDC, a team of three senior Wilkes University students (1 - marketing/management, 1- finance and 1- environmental engineering) were assigned this business as their semester project for an interdisciplinary small business consulting class (Hughes and Troy, 2016). The students met with the owner to identify their concerns and goals. In addition to improvements with the shops' marketing and Internet and social media presence, an additional goal identified was to look at opportunities to reduce the high energy costs associated with operations in the building where the business was housed.

Background

When the students performed this study, the shop had been in operation for 5 years. It was open 5 days a week, Tuesday–Friday 10 a.m.–4 p.m. and Saturday 11 a.m.–4 p.m. The students also assisted the gift shop with upgrading their Internet presence and sales opportunities. The shop prides itself on their unique selection of gifts and friendly customer service. At the time of the study there were only two employees and annual sales were approximately $108,000. The shop was leasing this retail location. This location is part of a strip mall, with the gift shop occupying a space approximately 1000 square feet. Electricity is used for lighting, air-conditioning and equipment operation, while natural gas is used for heating. The store is comprised of a glass door and window store front, open retail space with lighting and shelves, and an enclosed small office space and restroom in the rear. As part of their project, the students obtained a year's worth of electricity bills and natural gas bills to analyze as well as conducted an on-site energy use assessment.

The students analyzed 12 months of electricity use and observed that peak electricity usage was in January (725 kWh), followed by August (640 kWh) and September (510 kWh). Twelve months of natural gas usage were also evaluated. Natural gas was only used for heating in this location and was only utilized for 6 months out of the year: January, February, March, April, November, and December. The average high for November was 49°F, and average low was 33°F. Once temperatures started reaching the upper 60s°F in late April, heating was no longer utilized.

In addition to analyzing the utility bills, the students also performed an on-site energy assessment. The assessment entailed performing an inventory of all energy-using items in the shop: lighting number and types; and equipment such as computers, printers, printers, mini-refrigerator, and phone. The students also had access to a thermal-imaging camera, which they utilized to evaluate the integrity of the windows and doors by infrared scanning. Thermal imaging cameras detect radiant heat, also known as infrared radiation. Viewing various building components through an infrared scanner allows differences in the temperature of building components inside of building to be observed (Tenter, 2009).

As a result of the assessment the students recommended low-cost options to the shop to improve their energy efficiency. The inventory indicated that upgrading the type of light bulbs used for lighting the main retail area to newer Energy Star—approved options could result in savings of approximately $80/year.

The images obtained with the thermal imaging camera indicated that a significant amount of heat was being lost during the heating season via the windows at the storefront, as well as from the two swinging doors at the entrance of the store. The students recommended the consideration of weather stripping for the entrance doors and caulking of the perimeter of the storefront windows.

A student entrepreneurial club on campus followed up with this project and applied for and received a small business partnership grant of $1500 to fund some of the suggestions. The majority of the grant was used to build a website for the gift shop, with the remaining funds utilized to purchase new light bulbs and supplies to seal and caulk the windows. The students switched out the light bulbs to Energy Star—approved options as well as caulked and sealed the windows. They estimated a savings of $300/month as a result of more efficient and environmental business practices.

Conclusion

Small businesses are the backbone of the economy and their success is crucial to the future. Although small in individual stature, collectively they have a great impact on the environment. Despite small business owners having to wear many hats to perform their day-today operations, there are many resources available to them to facilitate improvements to their energy efficiency and to reduce their carbon footprints. Many of these resources are free or low cost. Implementation of operational energy efficiency practices and programs can have significant benefits to the small business and can be valuable components of their success, reputation, and longevity.

References

Drake, T., September 17, 2014. A New Approach to Small Business Energy Efficiency. Retrieved from: https://www.betterenergy.org/blog/a-new-approach-to-small-business-energy-efficiency/.

Fanara, A., Clark, R., Duff, R., Polad, M., 2006. How Small Devices Are Having a Big Impact on U.S Utility Bills. Environmental Protection Agency and ICF International, Washington, DC. Retrieved from: https://pdfs.semanticscholar.org/23eb/ce06e3d1d6754e91ebed62bde0bb30cb8c24.pdf.

Federation of Small Business (UK), 2007. Social and Environmental Responsibility and the Small Business Owner.

Fleiter, T., Schleich, J., Ravivanpong, P., 2012. Adoption of energy-efficiency measures in SMEs — an empirical analysis based on energy audit data from Germany. Energy Policy 51, 863—875.

Henriques, J., Catarino, J., 2016. Motivating towards energy efficiency in small and medium enterprises. Journal of Cleaner Production 139, 42—50.

Hughes, R.C., Troy, M.A., 2016. Teaching sustainability to undergraduates: evolution of a course structure and development of a model for partnering with local small businesses. Sustainability 9 (4), 200—204.

Hughes, R.C., Troy, M., 2017. Motivations of one small business community to implement environmental sustainability as a business practice. Journal of Environmental Sustainability 5 (Issue 1). Article 2. Available at: http://scholarworks.rit.edu/jes/vol5/iss1/2.

Internal Revenue Service Publication 946, 2016. How to Depreciate Property- Section 179. Deduction at. https://www.irs.gov/publications/p946.

International Energy Agency, 2015. Accelerating Energy Efficiency in Small and Medium-Sized Enterprises. Retrieved from. www.iea.org/publications/freepublications/publication/SME_2015.pdf.

Jabbour, C., DeOlivera, J., 2012. Barriers to environmental management in clusters of small businesses in Brazil and Japan: from a lack of knowledge to a decline in traditional knowledge. The International Journal of Sustainable Development and World Ecology 19 (3), 247—257.

Johnson, M., Schaltegger, S., 2016. Two decades of sustainability management tools for SME's: how far have we come? Journal of Small Business Management 54 (2), 481—505.

Neamtu, B., 2011. Public-private partnerships for stimulating the eco-efficiency and environmental responsibilities of SME's. Transylvanian Review of Administrative Sciences 34, 137—154. E/2011.

Ottinger, R., Bowie, J., 2016. Innovative financing for renewable energy. In: Manzano, J., Chalifour, N., Kotze, L. (Eds.), Energy, Governance and Sustainability. Edward Elgar Publishing, Cheltenham, UK, pp. 125—145.

Redmond, J., Walker, B., 2016. The value of energy audits for SMEs: an Australian example. Energy Efficiency 9, 1053—1063.

Size Standards Used to Define Small Business Concerns, 2011, 13 C.F.R. § 121.201.

Tenter, D., 2009. Saturn Energy Auditor Field Guide. Saturn Resource Management, Inc. Ver. 070109. www.srmi.biz.

United States Environmental Protection Agency, 2007a. Putting Energy into Profits: Energy Star® Guide for Small Business. Retrieved from. https://www.energystar.gov/ia/business/small_business/sb_guidebook/smallbizguide.pdf.

United States Environmental Protection Agency, 2007b. Energy Star® — Small Businesses: An Overview of Energy Use and Energy Efficiency Opportunities. Retrieved from: https://www.energystar.gov/sites/default/files/buildings/tools/SPP%20Sales%20Flyer%20for%20Small%20Business.pdf.

US EPA Energy Star® — Energy Star® Action Workbook for Small Business — Appendices, September 2018. Retrieved from : https://www.energystar.gov/sites/default/files/tools/ENERGYSTAR_Small_Business_AWB_Appendices_508_September_2018.pdf.

Further reading

Andrews, R.N.L., Johnson, E., 2016. Energy use, behavioral change, and business organizations: reviewing recent findings and proposing a future research agenda. Energy Research & Social Science 11, 195–208.

Backman, F., 2017. Barriers to energy efficiency in Swedish non-energy-intensive micro- and small-sized enterprises — a case study of a local energy program. Energies 10, 100. https://doi.org/10.3390/en10010100.

Bowman, T., October 7, 2010. How One Small Business Cut its Energy Use and Costs. Yale Environment 360. Retrieved from: https://e360.yale.edu/features/how_one_small_business_cut_its_energy_use_and costs.

Coles, T., Dinan, C., Warren, N., 2016. Energy practices among small-and medium sized tourism enterprises: a case of misdirected effort? Journal of Cleaner Production 111, 399–408.

Energy Efficient Commercial Buildings Deduction, 29 U.S.C. §179D, 2015.

Huebner, G., Shipworth, D., Hamilton, I., Chalabi, Z., Oreszczyn, 2016. Understanding electricity consumption: a comparative contribution of building factors, socio-demographics, appliances, behaviors and attitudes. Applied Energy 177, 692–702.

Jansson, J., Nilsson, J., Modig, F., Hed Vall, G., 2017. Commitment to sustainability in small and medium-sized enterprises: the influence of strategic orientations and management values. Business Strategy and the Environment 26, 69–83.

Jossi, F., January 3, 2017. Sustainable: reducing the energy waste of 'plug load.'. Retrieved from: https://finance-commerce.com/2017/01/sustainable-reducing-the-energy-waste-of-plug-load/.

Minnesota Department of Commerce, Division of Energy Resources, April 30, 2018. Small Commercial Characterization. Prepared by Seventhwave. Contract 104450. Retrieved from: http://mn.gov/commerce-stat/pdfs/card-small-commercial-characterization.pdf.

North Carolina Clean Energy Technology Center, 2018. Database of State Incentives for Renewables and Efficiency ®. http://www.dsireusa.org/.

Office of Energy Efficiency and Renewable Energy, 2018. Bonding Tools. Retrieved at https://www.energy.gov/eere/slsc/bonding-tools.

Palmer, K., Walls, M., 2017. Using information to close energy efficiency gap: a review of benchmarking and disclosure ordinances. Energy Efficiency 10, 673–691. https://doi.org/10.1007/s12053-016-9480-5.

Redmond, J., Walker, E., Wang, C., 2008. Issues for small businesses with waste management. Journal of Environmental Management 88, 275–285.

United States Small Business Administration Office of Advocacy, 2018. Small Business Profile retrieved from: https://www.sba.gov/sites/default/files/advocacy/2018-Small-Business-Profiles-US.pdf.

Walker, B., Redmond, J., 2014. Changing the environmental behavior of small business owners: the business case. Australian Journal of Environmental Education 30 (2), 254–268.

Williams, S., Schaefer, 2013. Small and medium-sized enterprises and sustainability: managers' values and engagement with environmental and climate change issues. Business Strategy and the Environment 22, 173–186.

CHAPTER

13

Promotion of renewable energy in Morocco

Agnė Šimelytė

Department of Economics Engineering, Faculty of Business Management, Vilnius Gediminas Technical University, Vilnius, Lithuania

OUTLINE

Introduction to global problems of energy consumption

The energy sector represents a fundamental economic sector, as its efficiency strongly affects competitiveness of the entire national economy, particularly as regards industry (Jankauskas et al., 2014). Environmentalists warn society and business companies that natural resources may run out much faster than is currently expected. Even more, burning fossil resources increases the level of CO_2, which is recognized as exerting a negative impact on the environment and causes the greenhouse effect (Sarkis and Tamarkin, 2008). Thus, this has

Copyright © 2020 Elsevier Inc. All rights reserved.

resulted in asearch for alternative energy sources. However, presently fossil fuel-based technologies predominate across most countries, especially in countries with well-developed industry. As the world embarks on the transition to a truly sustainable energy future, the world's renewable resources and technologies increasingly offer the promise of cleaner, healthier and economically and technically feasible power solutions and sustainable energy access for all (IRENA, 2015). Historically, the choice of energy has been based on economics and domestic conditions. Society has been driven to choose inexpensive energy (Marano and Rizzoni, 2008). However, nowadays, the technical superiorities of energy systems may fail to describe for instance renewable energy systems or its technology properly. Still, the primary decision-making criteria of investing into renewable energy systems are economic. These criteria are focused on renewable energy systems and thesupport ancillary infrastructure technical superiorities, such as efficiency and cost, which arereasonable in the context of generous financial support schemes (Azzopardi, 2014).

As the global population and global economy have been increasing, so hasthe global energy consumption. In the period of 2004—2008, the global population increased by 5%, consequently total energy generation and annual emission of CO_2 increased by approximately 10% per year (International Energy Agency, 2016; Frankl et al., 2010). Despite the global financial crisis, G20 states reported a decrease in energy consumption of 1.1% only in 2009, which grew by 5% in 2010 (Enerdata, 2011), while CO_2 emissions increased by 5.8% due to the energy generation. According to the estimates of the International Energy Agency (IEA) (2015), if the current trend continues, in 2030 the global energy demand might increase by approximately 60% while the emission of carbon dioxide might expand by 62%. The International Energy Agency (2015) has warned that if urgent measures arenot taken for reducing greenhouse gases, the earth's temperature might rise by 3.5°C by the end of the century. In order to decrease the volume of greenhouse gases, some researchers (Boharb et al., 2016) suggested improving efficiency of energy consumption in industry and buildings. However, it is a quite challenge as the industrial sector accounts for 30%—70% of the total global energy consumption and it is certainly responsible for a great part of the global greenhouse emissions. For example, in Germany (Fig. 13.1) the consumption of fossil energy sources was 45.6% in 2018, while the country expects to reduce this number to 17% by2050. Still the number is high as the total consumption of fossil and nuclear fuels would be 31%. In 2018, renewable energy accounted for 35% of all produced energy power in Germany. Hence, the consumption of renewable energy sources was only 14% of the total.

Thus, the perspectives and energy development trends in developed countries have suggested that renewable energy technologies are rapidly growing. This tendency is significant as the consumption of fossil resources has led to economic and environmental problems (Klevas et al., 2013). On the international level, during the last 2 decades great interest has been devoted to preserving the environment and managing energy resources. These highly discussed issues have been commonly founded atthe top of the international policy agenda due to their tremendous significance for the next generations. Several important decisions and conventions have been issued to promote greenhouse gas (GHG) emissions reduction and minimize fossil fuel dependency (Hamdaoui et al., 2018). Thus, several countries around the globe have set double targets to promote and improve energy efficiency by implementing renewable energy strategy. However, it might be too costly to tackle the challenge of climate change if the world as a whole delays in taking actions (Kousksou et al., 2015b). Thus, in

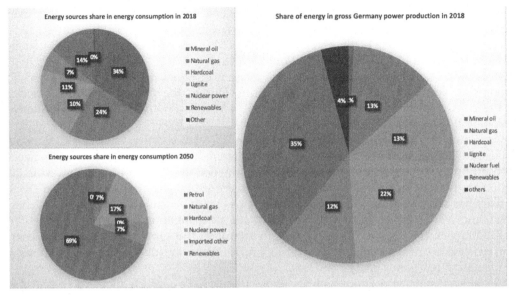

FIGURE 13.1 Comparison between share of gross German energy power production in 2018 and energy source share in energy consumption in 2018 and 2050. *Sources: Data from Samus, T., Lang, B., Rohn, H., 2013. Assessing the natural resource use and the resource efficiency potential of the Desertec concept. Solar Energy 87, 176–183. https://doi.org/10. 1016/j.solener.2012.10.011; Appun, K., Bieler, F., Haas, Y., Wettengel, J., 2019. Germany's energy consumption and power mix in charts. Clean Energy Wire CLEW. Available on the internet at: https://www.cleanenergywire.org/factsheets/germanys-energy-consumption-and-power-mix-charts.*

order to reduce greenhouse gas emissions at global, national, and regional levels, several regulatory instruments have been introduced, such as the Kyoto Protocol, an international agreement linked to the United Nations Framework Convention on Climate Change (UNFCCC). Under the Kyoto Protocol, in order to help countries achieve their greenhouse gas reduction targets, the Clean Development Mechanism has been developed. In the line with this mechanism, highly developed countries might fulfill part of their commitment tothe protocol by carrying out the projects in developing countries. The main objective of Sustainable Energy for All (SE4ALL), aninitiative launched by the United Nations, is to double the proportion of renewable energies in the global energy mix by 2030. Although at the global level these regulatory instruments have been employed, fossil fuels are the most used as energy sources in a largenumber of countries. The consumption of renewable energy as the primary energy had slightly increasedin advanced and emerging economies by the 2016. However, the most industrial countries have not been consuming a high volume of renewable energy (RE). For example, China even has reduced the contribution of renewable energy to total primary energy supply (TPES) from 19.5% in 2000 to 9.01% in 2016. The average contribution of renewables to TPES in the Organization for Economic Cooperation and Development (OECD) countries is 9.88%, whereas the world average is 13.67%. The average contribution of RE to TPES in the European Union (EU) is similar to the world average and was 13.55% in 2016. South Korea stands out fromboth OECD countries and industrial countries as it contributes the least amount of renewables to TPES. During the last 16 years South Korea has increased RE to TPES from 0.4% to 1.51%. The second country among the most industrial countries in the world thatcontributes the least renewable energy to TPES is Russia (2.54% in 2016). Even Japan, one of the most advanced economies with outstanding technological development, contributes RE to TPES of only 5.24%.

Meanwhile, Iceland consumes the highest amount of renewable energy sources to TPES among the Nordic and all the OECD countries. During the period of 2000–17, Iceland increased its contribution of RE to TEPS from 77.4% to 88.45% (Fig. 13.2), by productively exploiting its potentialgeothermal energy.

These numbers have been steadily increasing in all Nordiccountries since 2000. Norway is the second largest consumer of renewable energy sources among Nordiccountries; however, the contribution of RE to TPES fluctuated during the period of 2000–17 from lowest in 2004 (38.15%) to highest in 2017 (52.84%). Meanwhile, since 2013 the contribution of RE to TEPS in Norway has been constantly growing. Denmark has made remarkable changes in its energy sector, and from contributing RE to TEPS 9.6% in 2000, this number jumped to 34.91% in 2017. Denmark has set along-term target to stimulate the growth of consuming energy from renewable sources in the future. Fig. 13.3 presents the worldwide situation with respect to consuming renewable energy sources.

Global leaders, in accordance with the above trends, have conflicted ideas aboutthe objectives of sustainable development. Therefore, as early as 2002, global leaders came to an agreement at the World Summit on Sustainable Development to have the share of the renewable energy increased substantially in the global context. The EU placed particular emphasis on renewable energy and increase of energy efficiency. In 2008, renewable energy sourcesaccounted for as little as 10.3% in the total EU energy balance, which within 8 years reached only 13.55%. In recent years increasing energy consumption in the EU countries has caused serious development constraints. In the next 20 years energy consumption is going to continue to grow. It has been estimated that global energy demand will increase around 50 percent.

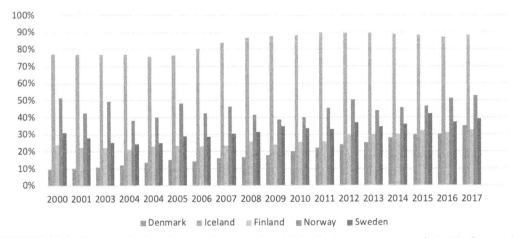

FIGURE 13.2   The contribution of renewable energy in total % of primary energy supply in Nordiccountries during the period 2000–17. *Source: Compiled by the author according to retrieved data from EUROSTAT database.*

Share of renewables in total energy production (%) (2016)

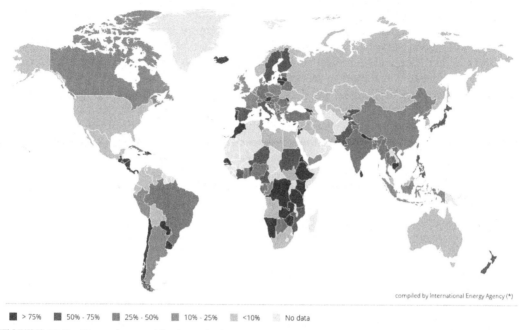

■ > 75%   ■ 50% - 75%   ■ 25% - 50%   ■ 10% - 25%   ■ <10%   □ No data

FIGURE 13.3   Share of renewables in total energy production (%). *Source: International Energy Agency, 2016. Renewables Information. Available on the Internet at: http://www.iea.org/bookshop/668-Renewables_Information_2015. Atlas. http://energyatlas.iea.org/#!/tellmap/-1076250891/1. All rights reserved.*

Another important determinant in choosing renewable energy sources instead of fossil fuels, is energy security, which is extremely significant for countries that do not have natural resources. Moore (2017) pointed out the geopolitics of energy as the uneven geographical distribution of energy reserves, such as oil and natural gas, with financial implications and undue power and influence for countries possessing these resources. Overall, the realist school of thought in international relations claimed that a"global energy landscape" exists thatdivides the world into blocs competing for resources. Following this concept, the emphasis has been put on dependencies on primary energy fuels, such as hydrocarbons, coal, and biomass, and the potential for sabotage of these primary fuels. Meanwhile, risk of sabotage has been also reflected in the policy discourse on energy security as independence from reliance on other states for primary fuels (Sovacool, 2011). Energy security is based upon four components and threats—availability, reliability, affordability, and sustainability—adopted from four criteria that include availability, affordability, energy efficiency, and social and environmental stewardship. For that reason, the promotion of renewable energy has become one of the most important national goals to ensure sustainable development of the energy sector and acountry's energy independence. The consumption of renewable energy sources is equally important for reducing environmental pollution or climate change mitigation, and for the country's economic development, which promotes creation of new jobs, and encourages innovation and research development. However, the competitiveness of renewable energy decreased as the price of oil has dropped off. On the other hand, weighted average costs of electricity from biomass for power, geothermal energy, hydropower, and onshore wind are all now in the range, or even span a lower range, than estimated fossil fuel-fired electricity generation costs. Especially, onshore wind energy is a very promising renewable source as the recent price has dropped to 0.03 US dollar per kWh. It is expected that by 2020 and beyond, renewable energy sources might become even more competitive energy sources for commercial use than fossil-based ones (IRENA, 2018).

Thus, the European Parliament has decided to cut off the CO_2 pollution until 2020 by 20% across whole the EU. At the same time, the European Parliament seeks to increase the efficiency of energy use by the same rate and to obtain up to 20% of energy from the renewable energy sources. Meanwhile, in 2014 the share of renewables in electricity production in the EU amounted to 30% (Enerdata, 2015). Global investment into renewable energy has demonstrated a steep growth over the last few years. Europe has dedicated most funding to the development of combating climate change and development of competitive and safe energy, which has become a key funding source for the renewable energy projects. Over the last few years, the date of loans extended by the European Investment Bank for the development of the renewable energy sources has increased by several times and totaled €6.2 billion ($6.68 billion) in 2010, while the share of the renewable energy in the total portfolio of energy investment tripled in the period of 2006–10, i.e., an increase from 10% to 30%.

The loans are largely dedicated to the development of wind and solar energy generation. Given the intention of the EU member states to accomplish the tasks set, the coming decade would clearly require more funding for the development of the renewable energy sources, while the efficient use of the same funding might bring a tangible impact on the development of the region and the well-being of the public. The member states must first of all estimate the benefit created by the same investment and channel the same investment to those technologies that generate most benefits.

Assessment of the added value in the sector of renewable energy is not a simple task, as it concerns not the energy generation alone but instead covers a much wider field, including protection of the environment, sustainable development, creation of jobs, and impact on economic development. Direct added value represents a single component of the value, however, given that this component is the easiest to measure, it usually predominates over any other values.

Moroccan economy and energy consumption

In the countries of the Middle East and North Africa (MENA) region, especially in Gulf Cooperation Council countries, excluding Qatar and Oman, the highest level of gas production comes from associated gas. Only relatively recently some MENA countries have started exploring the potential of alternative energy sources. Since the 2000s, the demand for energy in the MENA region has been growing at much higher rate than was predicted. According to Hamdaoui et al. (2018), in 2011 the gas demand in the MENA countries grew faster than in other regions. It has been increasing almost 9% yeartoyear, reaching just under 490 Bcm. For example, in Egypt gas consumption has been growing at an annual average rate of 7%–10%. The main reason for the rising demand in the MENA countries has been domestic low gas prices, which are kept artificially low by MENA governments. Thus, with time the supply became lower than demand, resulting in a costly deficit and these countries started to develop nonassociated gas reserves to meet their growing domestic needs market. Even more, the rapidly growing demand for power (annual average rate of 8%−10%) has led to the increased demand for gas. Furthermore, consumption of both renewable and nonrenewable energy has impactedeconomies.

Scientific literature provides four hypotheses regarding energy consumption and economic growth: growth, conservation, feedback, and neutrality (Table 13.1). The growth hypothesis suggests that energy plays a vital role in economic growth directly, so an increase (decrease) in the use of energy leads to an increase (decrease) in the growth of an economy (Šimelytė and Dudzevičiūtė, 2017). In this case unidirectional causality exists between energy use and income. The conservation hypothesis confirms that the consumption of energy plays a vital role in economic development in both direct and indirectways. According to the feedback hypothesis, a bidirectional relationship exists between output and energy use. This relationship leads to the conclusion that energy conservation has a negative impact on economic growth and vice versa. The last neutrality hypothesis suggests that the reduction of energy consumption does not influence economic growth.

Kahia et al. (2017) proved existence of a bidirectional causal relationship between economic growth and both renewable and nonrenewable energy in the longterm while examining economies of the MEAN countries during the period of 1980−2012. According to Kahia et al. (2017) findings, in the MENA countries bidirectional causal association existed between economic growth and renewable energy in the short and long term while bidirectional causal relationship between economic growth and nonrenewable energy remains only in the longterm.

Allouhi et al. (2017) noticed that due to the strong dependence on imported energy sources, mostly on fossil fuels, Morocco has been placed in a very energetically insecure and precarious financial situation. Thus, it has become more vulnerable to the effects of global climate change. Morocco mainly imports electricity from Algeria and Spain. The current energy system in Morocco depends on hydrocarbons (Table 13.2). The demand for energy in Morocco constantly is expanding and it is expected to triple by 2030. Morocco is the only North African country without nearly any fossil reserves (Hanger et al., 2016); however, Morocco has an abundance of renewable resources, especially wind and solar power, and is a leader among the MENA countries in deploying clean energy technologies. Moreover, there is the most potential for further renewable energy development in Morocco (Table 13.3), in line with the country's policy goals and resource endowments.

Meanwhile, according to the IEA, the volume of Morocco's greenhouse gas emissions has reached 42.1 $MtCO_2$ in 2008 and continuously is increasing in the energy consumption with an average annual rate of 5.7%. Moreover, IEA predicted that greenhouse gas emissions might double by 2020 in Morocco. Inthe meantime, Morocco has installed electric power stations, of which 70% of electric capacity is based on fossils fuels. Coal and heavy oil are main fuels in the electricity mix, which represent 68% of current capacity (Alhamwi et al., 2015).

TABLE 13.1 Summary of previous research on renewable energy and economy.

Authors	Countries	Period of research	Confirmed hypothesis
Kraft and Kraft (1978)	United States	1947−74	Conservation hypothesis
Apergis and Payne (2010)	OECD	1985−2005	Feedback hypothesis confirmed in short and longrun
Yildirim et al. (2012)	United States	1949−2010	Conservation hypothesis only for biomass-waste-derived energy Neutrality hypothesis for other renewable energy sources
Al-Mulali et al. (2013)	103 countries	1980−2009	79% of the countries confirmed feedback hypothesis; 19%−neutrality hypothesis; 2%−conservation hypothesis
Aslan (2016)	United States	1961−2011	Growth hypothesis
Kahia et al. (2017)	MENA Net Oil Exporting Countries (NOECs)	1980−2012	Growth hypothesis confirmed in shortrun. feedback hypothesis confirmed in longrun.
Šimelytė and Dudzevičiūtė (2017)	28 EU countries	1990−2012	Neutrality hypothesis in 4 countries; conservation hypothesis in 6 cases; growth hypothesis in 12 cases; feedback hypothesis in 6 cases
Adams et al. (2018)	30 Sub-Saharan African (SSA) countries	1980−2012	Growth hypothesis
Tugcu and Topcu (2018)	G7 countries	1980−2014	Neutrality hypothesis in United States, Japan, and UK Growth hypothesis in Germany, Canada, France, and Italy
Ozcan and Ozturk (2019)	17 emerging countries	1990−2016	Neutrality hypothesis in 16 cases Growth hypothesis confirmed in Poland case
Maji and Sulaiman (2019)	West African countries	1995−2014	Growth hypothesis
Gorus and Aydin (2019)	oil-rich MENA countries	1975−2014	Conservation hypothesis holds in the short and the intermediaterun Growth hypothesis is valid in the longrun
Luqman et al. (2019)	Pakistan	1990−2016	Growth hypothesis

Source: compiled by the author.

Empirical analysis: data and statistics

Energy is considered as the basic input used in the production process and a precondition of sustainable economic development. Consumption of energy affects economic growth by promoting the domestic producers to use better resources. For example, the industrial sector is

TABLE 13.2 North African hydrocarbon resources.

	Algeria	Egypt	Libya	Morocco	Tunisia
Oil proven reserves million bbl (est. January 1, 2011)	12,200	4400	46,420	0.68	425
Oil production (million bbl/day—2010 est.)	2.078	0.6626	1.789	0.003938	0.08372
Oil exports (million bbl/day—2009 est.)	1.694	0.163	1.385	0.025	0.091
Oil imports bbl/day (2009 est.)	18,180	177,200	575	221,000	78,460
Natural gas proven reserves trillion m^3 (January 1, 2011 est.)	4502	2.186	1.548	0.0014	0.065
Natural gas production billion m^3 (2010 est.)	85.14	62.69	15.9	0.060	3.6
Natural gas exports billion m^3 (2010 est.)	55.28	18.32	9.89	0	0
Natural gas imports billion m^3 (2010 est.)	0	0	0	0.5	1.25

Source: Hawila, D., Mondal, M.A.H., Kennedy, S., Mezher, T., 2014. Renewable energy readiness assessment for North African countries. Renewable and Sustainable Energy Reviews 33, 123—140. https://doi.org/10.1016/j.rser.2014.01.066.

TABLE 13.3 Renewable energy potential in the MENA countries.

Resource	Algeria	Egypt	Libya	Morocco	Tunisia
Direct solar radiation (kWh/m^2/year)	1700—2600	2000—3000	2590—2960	1800—3500	1500—1900
Wind (m/s)	Wind speed in >4 and >6 in Adrar	6—11	6—11	4.7—10.4	7—10
Hydro	Low potential	Big potential for large hydro but not adequate for small and medium projects	No discovered substantial potential	Technical potential 2500 MW. Good potential for small and minihydro.	Poor potential for large hydro. Three identified sites with good potential for small and medium projects.
Geothermal	200 sources in the Northeast and Northwest. Confirmed capacity >700 MW from Albian sources.	Potential for development of geothermal resources along Red Sea and Gulf of Suez coast	No discovered substantial potential	Significant potential in the Northeast	Limited to direct utilization, but not electricity generation.
Biomass	3.7 million toe coming from forests and 133 millions of toe per year coming from agriculture and urban waste.	16 million tons per year of agricultural residue and urban waste of 47 million tons per year	Biomass potential is estimated at 2 TWh/year. Unsuitable for large scale power generations	Urban domestic wastes estimated 3 milliontons per year	Organic waste annually produced in Tunisia is more than 30 million tons

Source: Hawila, D., Mondal, M.A.H., Kennedy, S., Mezher, T., 2014. Renewable energy readiness assessment for North African countries. Renewable and Sustainable Energy Reviews 33, 123—140. https://doi.org/10.1016/j.rser.2014.01.066.

responsible for 20.7% of energy consumption in Morocco. From 1990 to 2015 energy consumption in the transport sector has increased from 23% to 34.8%. Meanwhile, the agricultural sector consumed only 7.4% in 2015, which increased by 2.7% from 1990. In 2015 energy consumption in services made up 8.17% out of total energy consumption in Morocco, which is 3.03% less than in 1990. The share of energy consumption in transport has increased significantly during the period of 1990–2015 (from 23% to 34.8%). The other business sectors have been responsible for 28.89% of all energy consumption in Morocco. Currently the country imports more than 96% of the energy it consumes. For instance, petroleum represents roughly 61% of the overall national energy consumption in Morocco (Ministere de l'Energie, 2012). The massive growth rate of electricity demand is 6%–8% per year and increasing population has put pressure on the energy supply in Morocco. Kousksou et al. (2015b) noticed that about 81% of the country's electricity production is based on fossil fuels like coal, oil, and natural gas. Thus, it might be assumed that Morocco's energy profile is dominated by importing fossil fuels. Even more, currently, Morocco imports about 96% of its supplies of energy resources, which means that exploiting imported fossil fuels for other purposes such as industry production increases CO_2 emissions. Production-based CO_2 emissions from 1990 to 2015 increased 1.55 times from 19.6 to 54.9 million tons. Table 13.4 presents descriptive statistics of main indicators of renewable energy and energy consumption in Morocco.

Speaking of Morocco, a positive strong relationship between real GDP per capita and production-based CO_2 emissions has been noticed after calculating the correlation coefficient (0.98) (Table 13.5). Meanwhile, between production-based CO_2 emissions and value added in industry are weakly dependent (-0.212).

The result indicated that increasing value added created in industry would be fairly affected by production-based CO_2 emissions in the future. Even more, the increase in production-based CO_2 emissions would reduce value added in industry. Furthermore, real GDP per capita has been growing every year from 1990 to 2016. Since then Moroccan real GDP per capita has increased nearly twice. At the same time, the population in Morocco constantly becamelarger and larger. Presently, Moroccan population has reached 36.8 million inhabitants; 64% out of all Moroccansare young people; 12 million are active people. As the agency Invest in Morocco stated (2019), Morocco has launched numerous sectorial plans to ensure sustainable economic growth. Thus, during the period of 1990–2016, the supply of total primary energy rose 2.5 times from 7.62 million tons of oil equivalent to 19.32 million tons. Even more, the supply of total primary energy and real GDP per capita is highly dependent as the estimated correlation coefficient is equals to 0.98. Similar result (correlation coefficient 0.99) has been observed analyzing links between supply of total primary energy and production-based CO_2 emissions. Thus, development of economics and consumption of fossil fuels are highly linked. Moreover, it can be stated that productionbased on CO_2 emissions strongly impacts economic growth in Morocco. This means that expanding business sectors thatproduce more greenhouse gas would promote economic development in Morocco. Energy intensity is strongly linked to real GDP per capita (0.986), productionbased on CO_2 emissions (0.986), and energy consumption in agriculture (0.821). Meanwhile, energy consumption in industry and real GDP per capita is moderate dependent. During the period of 1990–2015, energy consumption in agriculture, in percentage of total energy consumption increased 1.57 times. The estimated correlation coefficient disclosed (0.84) strong dependence between energy consumption in agriculture and real GDP per capita. In the meantime, the

TABLE 13.4 Descriptive statistics of main indicators of renewable energy and energy consumption in Morocco.

	Min	Max	Mean	Standard deviation	Variance	Skewness		Kurtosis	
							Standard		Standard
	Statistic	Statistic	Statistic	Statistic	Statistic	Statistic	error	Statistic	error
Renewable energy supply	8.63	17.51	11.6844	2.45392	6.022	0.734	0.536	0.364	1.038
Renewable electricity	5.05	17.43	9.6822	3.80628	14.488	0.517	0.536	−0.973	1.038
Energy consumption in agriculture	4.72	7.44	6.3539	0.62159	0.386	−0.799	0.536	1.718	1.038
Energy consumption in services	8.17	12.31	9.5500	1.18315	1.400	0.947	0.536	0.221	1.038
Energy consumption in industry	20.58	34.73	23.6911	3.27838	10.748	2.425	0.536	7.517	1.038
Energy consumption in transport	23.01	34.78	31.5706	2.91614	8.504	−1.579	0.536	3.196	1.038
Energy consumption in other sectors	26.32	31.32	28.8339	1.36540	1.864	0.186	0.536	−0.815	1.038
Production-based CO_2 emissions	19.60	54.90	40.3889	10.13827	102.785	−0.306	0.536	−0.641	1.038
Energy productivity	10917.59	13024.00	11962.9889	582.33713	339116.537	0.090	0.536	−0.667	1.038
Energy intensity	0.31	0.56	0.4739	0.08190	0.007	−0.645	0.536	−0.820	1.038
Value added in industry	27.30	31.49	29.0056	0.99477	0.990	0.621	0.536	1.060	1.038
Value added in services	51.29	60.10	56.6789	1.71092	2.927	−1.472	0.524	5.551	1.014
Real GDP per capita	3866.34	7257.42	5756.7820	1105.04051	1221114.535	−0.262	0.524	−1.104	1.014

Source: Author's estimation.

relationship between consumed energy in agriculture (% in total energy consumption) and production-based CO_2 emissions is strong as the calculated correlation coefficient is 0.824. Thus, it might be stated that energy consumed in agriculture is based on fossil fuel.

The peak of RE supply in TPES was reached in 2004 and it made up 17% out of total primary energy supply. Meanwhile in 2015, this number compared with 1990 shrank 1.66 times and supply of renewable energy made up only 8.63% oftotal primary energy supply. Thus, estimated correlation coefficient between real GDP per capita and RE supply in TPES is moderate but negative (−0.776). Meanwhile, a relatively positive link exists between real GDP per capita and renewable electricity in total electricity generation; the estimated correlation coefficient is equal to 0.53.

TABLE 13.5   Correlation coefficients.

	Renewable energy supply	Real GDP per capita	Production-based CO_2 emissions	Energy consumption in agriculture	Energy consumption in services	Energy consumption in industry	Energy intensity	Value added in industry	Value added in services
Renewable energy supply	1.000								
Real GDP per capita	−0.766**	1.000							
Production-based CO_2 emissions	−0.777**	0.994**	1.000						
Energy consumption in agriculture	−0.842**	0.839**	0.824**	1.000					
Energy consumption in services	0.926**	−0.647**	−0.641**	−0.752**	1.000				
Energy consumption in industry	0.171	−0.538*	−0.501*	−0.468	0.102	1.000			
Energy intensity	−0.769**	0.986**	0.986**	0.821**	−0.657**	−0.464	1.000		
Value added in industry	0.141	−0.228	−0.212	−0.315	0.181	0.123	−0.244	1.000	
Value added in services	−0.256	0.386	0.467	0.300	−0.156	−0.115	0.476*	−0.637**	1.000

Source: Author's estimation.

For example, research of Adom et al. (2012) showed that carbon dioxide emission acts as a limiting factor to economic growth in Morocco. Two other researchers, Narayan and Popp (2012), while analyzing 93 countries found that Morocco is among of 44 countries having significant long-run causality relationships of GDP to total primary energy consumption. On the other hand, Apergis and Tang (2013) discovered that, in the 85 countries studied, Morocco was among nine countries thatreject the strong support to the energy-local economic development growth hypothesis. The study of Bruns and Gross (2013) provided the results that total energy-GDP causality tests frequently coincide with the results of energy type (electricity, petroleum products, or renewables). These researchers proved that Morocco is among the 82% of countries for which at least two energy-type GDP tests match with the total energy-GDP test. Bennouna and El Hebil (2016) forecasted Moroccan energy consumption until 2030. These two researchers developed two models and compared their results. The first model was based on the energy intensity (IE) and another one was linked with the country's urbanization rate (URB). The energy intensity model was applied to segment energy consumption in four posts while URB model only applied to two post. Bennouna and El Hebil (2016) stated that retrospective correlations of both models were excellent but future extrapolations provided a bit different result. Thus, it was found that in Morocco annual average growth of electricity should still be between 4.9% and 7.1% during 2020−30.

In conclusion, it might be stated that links between real GDP per capita and energy consumption of both fossil-based and renewable ones exist on different levels. In order to indicate causality among these factors, more detailed research should be carried out.

Renewable energy potentials in Morocco

The present situation in polluting environments and shrinking fossil resources makes it obvious that humans should search for alternative energy resources. Thus, it might be pointed out that the future must relyon nuclear and renewable energy. For example, Saidi and Mbarek (2016) revealed that the expansion of production technologies based on nuclear energy and renewable energy would significantly reduce future emissions of greenhouse gases emissions. Wind energy as one the best prospects for renewable energy sources was seen by Italian scientists. Savino et al. (2017) claimed that although big wind power plants have reached a relative maturity, there has been a lack of research on profitability of medium wind turbines and their environmental perspective. In the meantime, Bortolini et al. (2014) maintained that small- and medium-sized wind power plants have required more investments compared to large ones. Thus, the cost of electricity made by small- and medium-sized wind power plants might have risen. As Sebri (2015) noted,the sharp and continuous increase in energy prices, global warming, and running out primary energy sources requires that renewable energy willbe appropriately managed and used to sustain economic development. Nevertheless, most of the studies proved that consumption of energy stimulates economic growth, however, at the same time even renewable energy causes environmental degradation. In addition, renewable energy sometimes is a vital strategic decision for countries, especially those that have limited fossil energy resources and are dependent on other energy importing countries. The worst situation is when a country relies on one particular country for their energy imports (de Arce et al., 2012).

Samus et al. (2013) revealed that in order to consider global warming, growth of commodity prices and the consequences of the overexploitation and overuse of resources, a transition period to renewable energy is required. Thus, this case demands an efficient and renewable supply of operating reserves.

Morocco has great potentialforexploiting renewable energy sources, which would solve two problems of the country: energy security and climate change problems. Table 13.6 presents current renewable energy potentials in power generations. Morocco enjoys a rich potential in renewable energies, mainly solar and wind. On a global scale, Morocco occupies the ninth position for solar and the 31st for wind (Kousksou et al., 2015a).

Meanwhile, Morocco has several ongoing renewable energy projects. Moroccan Solar Plan is one of the world's largest solar energy projects, and it is estimated that it will cost €7.8 billion ($8.4 billion). Moroccan energy efficiency projects, solar and wind programs have

TABLE 13.6 Renewable energy potentials in power generations (Kousksou et al., 2015a).

Energy resource	Current level of generation (MW)	Potential level of generation
Wind	500 MW with 1500 MW under construction	25,000 MW
Hydro	1770 MW with 1625 MW under construction	–
Solar	20 MW with 500 MW under construction	20,000 MW
Biomass	5 MW	950 MW
Geothermal	–	5 MW

been supported and funded by the African Development Bank (AfDB), the World Bank (WB), and the European Investment Bank (EIB).

Five renewable energy resources are going to be exploited. The highest potentiality is observed in wave-wind energy and solar energy. According to Power Engineering International (PEI) (2018), in 2017 Morocco reached 1000 MW capacity of wind power, overtaking Egypt, and has become the second largest wind power producer after South Africa, which installed 1582 MW.

Kousksou et al. (2015a) especially emphasized the potential of biomass energy in Morocco because of its enormous generation of agricultural, animal, and municipal waste. In addition, Naimi et al. (2017) noted that Morocco has a potential of considerable biomass with a forest estate of around 9 million hectares, and a herd of around 7 million units of large cattle. Meanwhile, in Morocco geothermal direct use is mainly limited to balneology, swimming pools, and potable water bottling. Morocco is presently offering many investment opportunities for the development of thermal power generation projects.

In Morocco's case, increasing contribution of renewable energy into the total primary energy supply would solve both environmental problems and energysecurity issues. As a result, expansion of energy security could also aid in understanding the interlinkages among energetic security, sustainable development, and climate change (Moore, 2017). For example, in the Middle East, the Gulf Cooperation Council Interconnection Authority connects Saudi Arabia, Kuwait, Qatar, Oman, and United Arab Emirates, with plans for expanded integration.

Solar energy

Solar energy is considered as one of the most valuable renewable energy alternatives. Solar technologies convert solar radiations into electrical energy. Solar power stations usually are based on two different technologies: photovoltaic technology (PV) and concentrating solar power (CSP). Currently in the market two types of PV technology are available (Bahadori and Nwaoha, 2013):

— crystalline silicon-based PV cells
— thin film technologies made out of a range of different semiconductor materials, including amorphous silicon cadmiumtelluride and copper indium gallium diselenide.

Usually two types of PV systems are installed: grid connected or centralized systems and off grid or decentralized systems. Decentralized systems are more common for small power stations, which operate as local centralized systems. Samus et al. (2013) noticed that gas power plants allow to quickly react when there are sudden changes in energy demand. Even more, gas power plants are beneficial when there aresudden cuts in energy supply from fluctuating sources, such as wind energy, as it is possible to adapt their power production within seconds. Samus et al. (2013) stated that amongst others, concentrating solar power in combination with thermal storage can fill this gap. In addition, Bouhal et al. (2018) agreed with Samus et al. (2013) and maintained that increasing competitiveness of CSP with fossil fuels might perform a determinant role in the future. One principal focus of interest, and a big contributor to the actual popularity of CSP, concerns the desert of North Africa.

Morocco has great potential for the production of solar energy, applying both photovoltaic and solar thermal technologies (Nfaoui and El-Hami, 2018). Morocco has already started building a solar power plant based on both technologies. Besides solar plants, Morocco is implementingprojects exploiting other renewable resources such as wind, wind-wave energy, and hydro. These renewable energy-based plants would supply energy for the general population at an affordable and competitive price, achieve sustainable development through the promotion of renewable energy, increase productivity and competitiveness, strengthen regional integration through opening up to Euro-Mediterranean energy markets, and harmonize energy legislation (Kousksou et al., 2015a). The CSP plant of AinBeni Mathar is already supplying electricity to the grid (Nfaoui and El-Hami, 2018). Morocco is characterized by havingintensive solar radiation. In the south, intensive solar radiation is more than 5500 Wh/m^2 under annual sunshine of approximately 3500 h (Fig. 13.4). Meanwhile, in the north intensive solar radiation is less than 4.5 kWh/m^2 under annual sunshine durations varying from 2700h.

Based on the data of the Moroccan Agency for Energy Efficiency (AMEE) and the National Center of Meteorology, Bouhal et al. (2018) divided Moroccan cities according to climate zone (Table 13.7). These new zones have divided the climate of Morocco into six parts with the same solar irradiation, altitude, and other significant indicators. Table 13.6 regroups the different cities of each climatic zone.

Based on climatic zones and the present situation in Morocco, Bouhal et al. (2018) did detailed research on parabolic trough collector (PTC) technology for concentrated solar power. The study was carried out in six chosen areas in Morocco. Meanwhile the parabolic trough collector, which is a subtechnology of CSP systems, is the lowest-cost large-scale and most proven solar power alternative. The study of Bouhal et al. (2018) proved that the most effective application of PTC as a subtechnology of CSP system would be achieved in the sixth zone, i.e.,Quarzazate (Fig. 13.5). The researchers explained their results as the success of CSP in the sixth zone by the huge solar irradiation and "ideal climatic conditions."

Zone 4 is also promising for CSP. The results of other zones are similar. The efficiency in Agadir (zone 1) has been slightly fluctuating during the summer time. The situation in Tangier (zone 2) was quite the same as in zone 3 (Fez). In the meantime, Nfaoui and El-Hami (2018) distinguished several features that influence annual production of solar energy. Especially, they emphasize the following:

— incidental solar radiation at the installation site
— tilt and orientation of panels
— presence where there is no shading
— technical performance of system components and inverters.˙

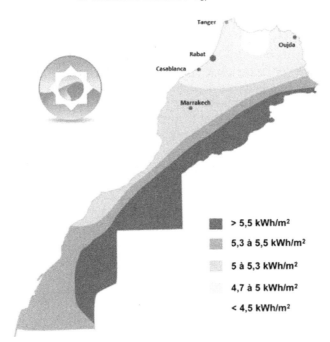

FIGURE 13.4   Moroccan solar potential. *Source: Kousksou, T., Allouhi, A, Belattar, M.Jamil, A., El Rhafiki, T., Arid, A., Zeraouli, Y., 2015b. Renewable energy potential and national policy directions for sustainable development in Morocco. Renewable and Sustainable Energy Reviews 47, 46−57. https://doi.org/10.1016/j.rser.2015.02.056.*

TABLE 13.7   Moroccan cities in accordance with climatic zoning.

Climatic zone	Cities
Zone 1	Agadir, Tiznit, Sidi Ifni, Laaryoune, Dakhla, Guelmim, TanTan, Kenitra, Rabat, Sale, Casablanca, El Jadida, Safi, Essouira
Zone 2	Tangier, Tetouane, Larache, Al Houceima, Nador
Zone 3	Fez, Meknes, Sidi Slimane, Chefchaouen, Taza, Oujda, Berrechid, Settat, Fkih Ben Salh, Khouribga, Beni Mellal
Zone 4	Ifrane, El Hajeb, Azzrou, Khenifra, Immouzzer Du Kandar, Midelt
Zone 5	Marrakech, Benguerir, KalaaSeraghna
Zone 6	Errachida, Taroudant, Ouarzazate, Smara, Bouarfa

Source: Bouhal, Y., Agrouaza, T.Kousksoua, A., Allouhib, T., El Rhafikic, A., Bakkasc, J., 2018. Technical feasibility of a sustainable Concentrated Solar Power in Morocco through an energy analysis. Renewable and Sustainable Energy Reviews 81, 1087−1095. https://doi.org/10. 1016/j.rser.2017.08.056

Rodríguez et al. (2016) analyzed another potential storage solution that might be implemented into a CSP plant for electricity generation operating at temperatures between 170°C and 300°C. These researchers carried out a study on the potential gains that could

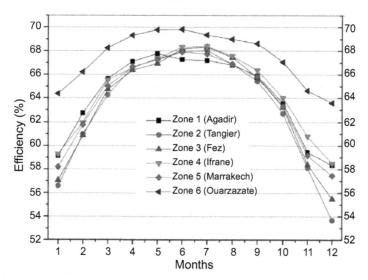

FIGURE 13.5 Monthly solar energy efficiency for various climatic zones in Morocco. *Source: Bouhal, Y., Agrouaza, T.Kousksoua, A., Allouhib, T., El Rhafikic, A., Bakkasc, J., 2018. Technical feasibility of a sustainable Concentrated Solar Power in Morocco through an energy analysis. Renewable and Sustainable Energy Reviews 81, 1087–1095. https://doi.org/10.1016/j. rser.2017.08.056.*

be obtained from different thermal energy storage technologies if these were implemented in a 1 MW organic Rankine cycle (ORC) power plant thatis currently under construction in Morocco. ORC differs from traditional CSP or PV solar plants. It has principally been used in the past couple of decades to recover medium-grade heat from sources such as geothermal, biomass, and the exhaust of a realm of different industrial processes (waste heat recovery) (Rodríguez et al., 2016). The results of this study led to the conclusions that ORC is mostly suitable for small- and medium-sized solar plants where the sizes of solar field (i.e., solar multiple) and thermal energy storage systems must be optimized simultaneously in order to attain the maximum efficiency gain. Furthermore, economic analysis confirmed that the thermocline technology enables a significant reduction in the cost of the thermal energy storage (33% on average). This lower specific cost resulted from the combining of effective thermal performance and very low capital cost and translates into the potential to reduce the cost of electricity of the solar thermal performance.

Biomass energy

Morocco also has high potential for biomass energy by exploiting waste of agricultural products and biological household products. Biomass can be dry or wet and it might be produced in thermochemical transformation, which involves combustion, pyrolysis, direct liquefaction, and gasification. Biomass might be derived from municipal solid waste thatmight be used in a power plant. The average annual production of urban waste is 270 kg per inhabitant. Naimi et al. (2017) noticed huge potential in the Rabat region since the amount of waste was roughly calculated about 714,350 Kg in 2014, which is equivalent to 0.96 Kg per inhabitant a day.

In order to define the concept of waste management, Morocco introduced law 28-00, issued on December 7, 2006. This law classified different types of waste, specified their management, including hazardous waste. This law also defined waste recovery and energy recovery form waste, which is important in using waste as material for producing biomass that later might be used in a power plant. The law 28-00 specifies two ways for how to recover energy from waste. The first one is anaerobic digestion (methanization) andthe other is incineration. The methane might be used for the same applications as natural gas or transformed into gasoline and fuel oil. Incineration as the process might be used for various types of waste. During waste incineration process, in special ovens, the heat reaches high temperatures. Thus, this heat might be used for heating power stations and electric systems. However, waste incineration pollutes the environment and releases high volumes of CO_2. Meanwhile Kousksou et al. (2015a) noticed that the predominant use of biomass in Morocco is as traditional fuels for cooking and heating. Recently, a lot of attention was paid tobiomass conversion into bioethanol as a promising option as arenewable energy source. Morocco has high potential in use of agricultural biomass as a 9 million ha area is covered with agricultural activities, which covers 7 million large livestock units, with more than 500,000 farms. Another source of biomass energy might be household waste (biological). According to, Agence Marocaine pour le Développement des Energies Renouvelables et de l'Efficacité Energétique, *the theoretical potential is* more than 700 million m^3 of biogas per year, which would be 4.2 MW h annually. In addition, the theoretical potential of biomass from waste water is 230 million $m^3$of biogas per year, or 1376 MW h annually. Wood also might be considered as a source of energy. In this case, Moroccan deforestation rate is only 8%, which makes up nearly 9 million hectares.

Kamzon et al. (2016) advised to use sugar cane to produce biomass as acleaner energy source. In order to select the most efficient processes for a large-scale production, these researchers carried out astudy that covered the technical, economic, and environmental criteria. Kamzon et al. (2016) proved that it is plausible to establish a cost-effective biorefinery in Moroccan sugar plants. Further, the sugarcane bagasse and beet pulp might be used to produce bioethanol from lignocellulosic biomass. However, the conversion of lignocellulosic material into bioethanol is quite an expensive process and the yield is low. Thus, there is a challenge to increase the competitiveness of biomass energy and yield from it.

Sibley and Sibley (2015) provided another idea how to use biomass energy while analyzing energy consumption of hammams in Casablanca. They estimated that economy on biomass consumption was at 166 tons per year per hammam, i.e., a cost of 106,240 Dhs/year, whereasa ton of wood for the furnace costs 640 Dhs. The data showed that an average size hammam consumes 30 m^3 of water per day and a total of 1500 kg of wood, of which 300 kg is for underfloor and space heating and 1200 kg for water heating. Thus, Sibley and Sibley (2015) suggested that renewable energy might cover some part of energy demanded for hammams.

Pantaleo et al. (2018) assessed a novel hybrid solar-biomass power-generation system. The system is based on an externally fired gasturbine fueled by biomass (wood chips) and a bottoming organic Rankine cycle plant. "The main novelty is related to the heat recovery from the exhaust gases of the externally fired gas-turbine via thermal energy storage, and integration of heat from a parabolic-trough collectors' field with molten salts as a heat-transfer fluid" (Pantaleo et al., 2018). However, installing such type of technology into power station would require lot of investments as the cost is very high in concentrating solar power section.

Several presented examples (Sibley and Sibley, 2015; Pantaleo et al., 2018) showed that burning wood is one of potential biomass energy sources. Wood even is used in a novel hybrid solar-biomass generation system. Although Morocco might benefit from nearly 9 million hectares of forest, only 30,000 ha are exploited; however, burning wood releases greenhouse gas, which pollutes air and additionally reduces forestry area. In Morocco 90% of firewood is used in rural areas for cooking, baking bread, or heating water. Group for the Environment, Renewable Energy and Solidarity (GERES) is an institution that specializes in sustainable energy and environmental protection. GERES with partners, during the period of 2008–11, implemented a strategy for wood biomass energy in Morocco. The main objective of the strategy was to develop a biomass energy management framework in order to preserve natural resources.

The strategy was supposed to solve several level problems regarding the consumption of wood as material for biomass energy. The first problem covered health and social problems. The second one was oriented toward environment degradation. It was expected that at the end of the project the consumption of wood would decrease by 2600 tons per year and expected CO_2 reduction was 3600 tons per year. GERES implemented a strategy for biomass in Chefchaouen rural region. GERES divided the strategy for wood biomass energy into three actions. The first action was dedicated to appliance efficiency. This meant that new types of multifunctional ovens and stoves were supposed to be produced and introduced into two pilot households (field validation) and predissemination in 20 other homes. In line with this action, monitoring and quality control has been performed. The second action was dedicated to biomass sustainable management in the targeted area. This action contributed to identify risk areas, implement a biomass management project, develop partnerships with local authorities, and support the constitution and the organization of a forest cooperative. In line with the second action of the strategy for wood biomass energy, monitoring tools have been introduced as well as controlling actions every month. The third category of action was large-scale extension. The goals of this action were to create communication tools, to train the persons in charge for large-scale projects. This action assisted to organize and coordinate biomass-energy network on the national level, which is based on previously acquired experience. Under this action, local financial partners have been identified for further large-scale projects. The implementation of strategy was funded by ADEME, AFD (French Development Agency), MFP-GEF/UNDP (Global Environment Facility Micro-Finance Program), MEMEE (Moroccan Ministry of Energy, Mines, Water and Environment), Prince Albert II of Monaco Foundation, Poweo Foundation, Macif Foundation, Veolia Environment Foundation, and Provence-Alpes-Côte d'Azur Regional Council. In the future, GERES and the partners intend to extend implementation of strategy for wood biomass energy in the northern of Morocco and then all other parts of the country.

The total solid bioenergy potential in Morocco is estimated at 12,568 GWh/year, and an additional 13,055 GWh/year can be supplied from biogas and biofuel (ONEE, 2015).

Wind energy and wave-wind energy

Morocco might take an advantage of another renewable energy resource by exploiting wind energy, especially wave-wind energy, whose potential is estimated at 25,000 MW (Fig. 13.6). Morocco has the potential to develop wave-wind energy power plants along a

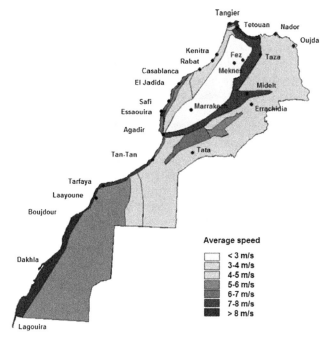

FIGURE 13.6   Moroccan wind energy map. *Source: Kousksou, T., Allouhi, A., Belattar, M., Jamil, A., El Rhafiki, T., Zeraouli, Y., 2015a. Morocco's strategy for energy security and low-carbon growth. Energy 84, 98–105. https://doi.org/10. 1016/j.energy.2015.02.048.*

3500 km coastal area. For example, annually at 10 m above sea level, wind for the Tangier zone in the north is 10 m/s, and for Dakhla zone in the Laâyoune-Sakia El Hamra region, it varies between 7 and 8.5 m/s. Wave energy powers between 2 kW/m and 6 kW/m with an estimated mean wave power of 6.3 kW/m at a point in the Moroccan Mediterranean (Sierra et al., 2016).

Currently, Dakhla is the most suitable site for installing wind power plants. Meanwhile Laayoune has been identified as the second most suitable site. It was discovered that both Dakhla and Laayoune are suitable for connecting in a wind power conversion grid system. Meanwhile, Assila and Essouira have been characterized to be appropriate for stand-alone wind power application (Allouhi et al., 2017).

Morocco's total potential of wind energy is 7936 TWh/year, which covers both onshore and offshore. At the moment Morocco's total installed wind power capacity is 850 MW. Worldwide modern special turbines have been used to convert wind power into electricity. This power might vary from 750 kW to 2.5 MW. Wind energy projects have been planned or developed at the following locations: Tanger II 150 MW; Midelt 100 MW; JbelLahdid (Essaouira) 200 MW; Tiskrad (Laayoune) 300 MW; and Boujdour 100 MW. Some of the mentioned projects are still in process. Alhamwi et al. (2015) and Allouhi et al. (2017) supported an idea to promote wind energy as a renewable, friendly-to-environment, and low-cost energy source. Although Alhamwi et al. (2015) found wind energy advantage as low cost, however, the investment is high. The major costs covering wind energy generation are linked with capital including wind turbines, foundations, transportation, road construction, and grid connection. Some variable costs occur as well, including operation and maintenance, land acquisition, insurance and taxes, management, and administration. Meanwhile, Pérez-Collazo et al. (2012), in their study, stated that synergy of offshore winds and waves might increase the yield. Even more, combination of these two sources might reduce operation and maintenance costs. Energy generated from two different sources and different technologies at a single array site also might increase global yields per array and contribute to more sustainable natural resources. Thus, it might be stated that combining energy harnessing from winds and waves, the efficiency would increase as well. However, compared to wind energy, wave energy is at an early stage of development. Moreover, included wave energy in a wind-only system has reduced balancing costs up to 35%. For example, after combining wave and wind energy, 13 million euros ($14 million) have been saved in Denmark. In this way, harnessing energy from a combined wind and wave system at the same location avoids sudden energy supply reductions.

Another advantage of wave energy is that waves are much more predictable than winds. Pérez-Collazo et al. (2012) found that waves were 23% more predictable than winds, the power output of WECs was 35% more predictable than for wind turbines. However, according to Sierra et al.'s (2016) study, Moroccan Atlantic shore has shownhigh variability in wave energy. The most energetic waves are originally arriving from the Azores, from north, northwest. Even more, in the Morocco western coast clear seasonal patterns have been determined. The highest wind-wave energy was noticed in winter with an average of 43% during January, followed by December and February. The least wind-wave energy was in summer, withonly 10% variability. August and June were determined as the mildest months. Morocco has a potential of annual wave energy up to 262 MWh/m.

In the context of sustainability, combining harnessing from two renewable sources would have less negative impact on the environment. According to Allouhi et al. (2017), another advantage of extracting energy from wind is that it does not need water, which makes generation of wind energy much more attractive then thermal power stations. Thus, it is believed that wind energy might become as spread as hydro energy.

As all with all technologies,application of these energy sources might have some disadvantages as well. Especially, some risks and challenges might occur when two different technologies forharnessing renewable energy are combined. Pérez-Collazo et al. (2012) considered that whentwo renewable energy industries are in different development stages there may be the following risks and challenges:

— Technology readiness of wave energy. However, in a couple of years this risk might be reduced to minimum as wave energy is expending and becoming more known to investors.
— Uncertainty of mooring lines. At this moment there is a shortage of mooring lines thatwould be adopted for wave-wind energy technology.
— Failure due to lack of experience. The third risk is highly linked with the first one. As there is a shortage of knowledge and there is not enough experience, there is the possibility that a failure may occur.
— Impact risk.
— Project insurance.

The last two risks and challenges to reduce such risk also have interrelationships with a lack of technology readiness, knowledge, and experience to implement such projects.

In the meantime, in accordance with the Moroccan National Energy Strategy, a strong legal framework has been developed.

Hydropower energy

Hydropower energy has been generating electricity worldwide for 139 years. Hydropower is verypopular due to its flexible technology. It can provide energy for a single home, and at its largest it can supply industry and the public needs with renewable electricity on a national scale.

Four hydropower technologies exist:

— *Run-of-river hydropower* provides a continuous supply with some flexibility of operation for daily fluctuations in demand.
— *Storage hydropower is* typically a large system that uses a dam to store water in a reservoir. It can offer enough storage capacity to operate independently of the hydrological inflow for many weeks or even months.
— *Pumped-storage hydropower* provides peak-load supply, harnessing water thatis cycled between a lower and upper reservoir by pumps thatuse surplus energy from the system at times of low demand.
— *Offshore hydropower* is a less established but growing group of technologies that use tidal currents or the power of waves to generate electricity from seawater.

According to the International Hydropower Association (IHA) (2019), technologies can often overlap. For example, storage projects can often involve an element of pumping to supplement the water that flows into the reservoir naturally, and run-of-river projects may provide some storage capability.

The same as on the global level, in Morocco hydropower has a long history. Primary orientation of water management started with the construction of dams in 1945. Hydropower has become especiallyimportant since the 1960s. The contribution of installed hydroelectric power is 1800 MW. Depending on rainfall, hydroelectric power accounts for 5.1% to 13.7% of net electricity consumption.

At this time, hydropower installed capacity is 1.77 GW with production of 216 kilo tons per year, equaling 3689 GWh (2017).

It is expected that energy demand willtriple by 2030, and the Moroccan government is planning to accelerate the pace of energy system reform and implementation of a National Energy Strategy by enabling private and public operators to develop more than 10,100 MW of additional capacity. It is expected that 1330 MW of hydropower would be generated and that 550 MW would be developed by the private sector and about 100.6 MW in small hydro would be developed by public operators. In this way Morocco would reach 2000 MW of installed hydro capacity by 2020. Currently, 10 various capacity hydropower stations are under construction.

French renewable energy producerVoltalia has obtained permits for the development of two hydropower plants with a capacity of 9.8 and 7.2 MW, respectively, in Morocco. The

French company VINCI Construction was awarded a $346 million contract to build a pumped storage hydroelectric plant near Agadir.

The government-owned Office National de l'Electricitéetde l'Eau Potable (ONEE), which is the main player in the power sector, is implementing an environmental and social management system (SMES) at the Al Wahada and Afourer hydropower plants. As well, ONEE, Moroccan Ministry of Energy are cooperating with the German government in a study to assess the viability of seawater-pumped storage. In 2017 ONEE announced construction of two new pumped storage stations with a total capacity of 600 MW. The first, the El Menzel II station, will be located in the upper Sebou; and the second, the Ifahsa station, will be built on the right bank of Oued Laou. These projects will each have an installed capacity of 300 MW. The Ifahsa project is expected to be finished by 2025. Another one in the progress is Abdelmoumen pumped-storage project, which is expected to become operational in 2020 with the capacity of 350 MW and to provide electricity to the country. The project of the Khénifra hydropower plant has started, and it is expected that this hydropower station willprovide 128 MW. Additionally, five small hydropower stations with the capacity from 8 to 30 MW are under construction.

Geothermal energy

The first time the potential of geothermal energy in Morocco was evaluated was in 1969. At this time, researchers made a negative conclusion regarding exploiting geothermal energy. Since1980, the interest in geothermal energy has been increasing, but only at the academic level.

Morocco's thermal water application is mainly limited to balneology, swimming pools, and potable water bottling, despite the potential market. Most of thermal waters are applied in agriculture. Barkaoui et al. (2015) stated that harnessing geothermal energy would help Morocco become more self-sufficient, while producing enough food for its domestic market, as well as for export. Additionally, Barkaoui et al. (2015) found out that Morocco has the potential for harnessing geothermal energy in the northeast. Even more, a few areas such as Berkane and Oujda have been identified as having an average geothermal gradient of more than $120°C/km$ at depths greater than 300 m. Thus, in line with the National Energy Strategy, Morocco's government might create additional programs and promote geothermal energy. Estimated geothermal energy capacity is 5 MW in the country, however, Morocco has no single project underway from this renewable energy source.

Moroccan policy and legal regulation on renewable energy promotion

Various different policies are employed to promote research, development, demonstration, and commercialization of renewable energy. Developmentof renewable energy technologies in a country depends on its renewable energy readiness (RE-Readiness) that indicates the gaps and strengths of their development (UNEP, 2012). In order to increase the share of electricity generated from renewable resources, different countries implement a wide range of strategies and regulatory instruments (Table 13.6) (Simelyte et al., 2016). Alhamwi et al.

(2015) stated that the majority of energy strategies that governments implement are based on the results of projections. Governmental energy strategies are based on the results of projections. Furthermore, in order to develop energy strategy and regulatory instruments, the traditional tools for decision-making simulation are used such as cost-benefit analysis and the analytical hierarchy process.

In general, the policies promoting renewable energy might include (Table 13.8):

— *regulatory policies* (feed-intariffs, quotas or portfolio standards, priority grid access, building mandates, and biofuel blending requirements)
— *fiscal incentives* (tax policies and direct government payments such as rebates and grants)
— *public finance mechanisms* (includes mechanisms such as loans and guarantees)
— *climate-led policies* (include carbon-pricing mechanisms, cap and trade, emission targets, and others).

Morocco, as a part of regulatory instruments linked to environment protection, has introduced two types of environmentally related taxes. The first one is an energy consumption tax and the other one is a road transport–related tax. Since the year 2000, the percentage of environmentally taxes in GDP has been fluctuating from the highest at 2.46% in 2000 to 1.98% in 2004. At the same time as being the highest percentage in total tax revenue, environmentally related taxes were10.56% in 2001; the lowest percentage collected from environmentally related taxes was 6.59% in 2008.

Energy-related taxes have becomethe largest part in total environmentally related taxes. In total environmentally related taxes, the highest percentage of energy-related taxes was 90.18% in 2001 and the lowest was 86.06% in 2014. Road-related taxes make much less an impact on the volume of total collected environmentally related taxes. The highest percentage of road-related taxes was collected in 2013 and made up 10.79% in total environmentally related taxes. The lowest volume to road-related taxes was collected in 2011 and made up 8.31% of total environmentally related taxes. Since the Moroccan government started implementing aNational Strategy for Sustainable Development and Environment (PNAP), an

TABLE 13.8 Fundamental types of regulatory instruments.

		Price-driven	Quantity-driven
Regulatory	Investment focused	Investment subsidies Tax credits Loan interests/Soft loans	Tendering system for investment grant
	Generation based	Fixed (feed) in tariffs Fixed premium system	Tendering system for long-term contracts Tradable green certificate system
Voluntary	Investment focused	Shareholders programs Contribution programs	
	Generation based	Green tariffs	

Source: Haas, R., Resch, G., Panzer, C., Busch, S., Ragwitz, M., Held, A., 2011. Efficiency and effectiveness of promotion systems for electricity generation from renewable energy sources — lessons from EU countries. Energy 36, 2186–2193. https://doi.org/10.1016/j.energy.2010.06.028.

advanced tariff structure to incentivize conservation has been introduced. All residences and local collectives that have consumption 20% below established targets would get a new 20% tariff discount. It is expected that employing this measure would save 300 MW of energy (Climate investment funds, 2009).

The first signs of promoting renewable energy have been the development and implementation of a Global Rural Electrification Program (PERG). In the 1990s only 18% of rural areas were covered by electricity grid. PERG has received 155million euros ($167 million) from the African Development Bank Group. Currentlyalmost 100% of the whole country's territories are electrified.

Additionally, in 1995 Morocco started to develop itsPNAP. Meanwhile, the National Energy Strategy (NES) was developed only in 2008. Moroccan NES covers introduction of a series of legislative, regulatory, and institutional reforms, and improving tax benefits to encourage investment in this sector (Table 13.9). The aim of NES is to meet 20% of renewable energy forthe country's domestic energy needs and to increase energy efficiency 12% by 2020 and 15% by 2030. Moroccan government has expectations that 42% of its total energy mix would be generated from solar, wind, and hydroelectric sources by 2020. At this moment, Morocco implements six energy efficiency programs and solar and wind energy programs.

In 2006, Decree 1-06-15 was issued, which opened the possibility to employ the call for-tenders and awarding of energy projects. This law applied to municipalities that may wish to contract with wind farms or other sources of electricity from renewable energy (Leidreiter and Boselli, 2015). In the same year, on December 7, Law no. 28-00 was published, which allowed to manage waste. Although this law was not issued for implementing the NES, later this law served in the biomass energy projects, especially those in which the generation of biomass energy was based on solid waste.

Further, on April 15, 2008, King Mohammed VI launched the development of NES and PNAP.

The PNAP is based on four strategic pillars (Climate investment funds, 2009):

— Security of supply with diversification of fuel types and origins.
— Access to energy for all segments of society at competitive prices.
— The promotion of renewable energy and energy efficiency.
— Regional energy integration among the Euro-Mediterranean markets.

In 2008, Law 16-08 increased the ceiling for self-generation by industrial sites from 10 to 50 MW. The law was conceived principally to support wind power, but also applied equally to other technologies. This law adopted the 1963 Decree, which allowed establishment of the Office National de l'Electricité et de l'Eau Potable (ONEE) and attributed to ONEE a monopoly of production above 10 MW. The government-owned ONEE is the main player in the power sector. ONEE is implementing an environmental and social management system at the Al Wahada and Afourer hydropower plants.

In 2009, renewable energies represented 4% of the Moroccan energy mix (without biomass) and produced 10% of the total electricity demand. The same year, the Moroccan government introduced Law 13-09 on renewable energy with the target to increase these shares from 4% to 10% and from 10% to 20% of electricity production by 2012. Law 13-09 partially opened the electricity market to competition for the production and commercialization of electrical energy from renewable energy sources for customers. The law introduced an authorization

TABLE 13.9 The laws relating to the national energy strategy.

Law	Objective
Decree 1-06-15 issued in 2006	Law: — applied to municipalities that may wish to contract with wind farms or other sources of electricity from renewable energy — placed under obligation of public institutions to employ competitive calls for tender in the award of projects.
Law no. 28-00 published on December 7, 2006, related to managing waste.	The law defines and specifies principles of waste management: — defines the different types of waste — specifies their management — specifies the level of waste support — introduces the concept of hazardous waste and management of this type of waste by subjecting it to a system of prior authorization at all stages of management: collection, transport, storage, and disposal.
Law 16-08 issued in 2008. The law adopted the 1963 Decree.	The law was introduced to support wind power, but also applied equally to other technologies. According to this law,THE Office National de l'Electriciteet de l'Eau Potable (ONEE) was established, and a monopoly of production above 10 MW was given to ONEE.
Law no. 13-09 promulgated on February 11, 2010, relating to renewable energy	This law allows: — the opening of generation to competition — access to the electricity grid — export of green electricity — the construction of a direct line for export.
Law no. 57-09 creating the Moroccan Agency for Solar Energy (MASEN) setting out a specific framework for solar projects (January 14, 2010)	MASEN is the most important actor for the Moroccan solar energy sector. The joint stock company, with a board of trustees and asupervisory board, is independent from MEMEE.
Law no. 47-09 relating to energy efficiency, dated September 29, 2011	This law aims to increase the efficiency of energy resource consumption, to reduce energy costs on the national economy, and to contribute to sustainable development.
Law no. 16-09 creating the National Agency for Promotion of Renewable Energy and Energy Conservation (ADEREE), dated January 13, 2010	This law defines the reorganization and renaming of the existing Center for the Development of Renewable Energy (CDER). The ADEREE is primary active in the Corporate Energy Efficiency Program.
Draft law on Public-Private Partnership (PPP)	This draft is strongly inspired by the French Ordinance of June 17, 2004, on PPPs, but also follows the approach used by the UK Private Finance Initiative experience.
Decree No. 21 781-13-2 published on the September 28, 2013	The decree was dedicated to energy efficiency programs. The time GMT +1 was applied in Morocco since 2008 during the summer to improve the power reserve margin during peak hours. The decree has set the period of GMT +1 schedule, which runs from April to October each year with the exception of the month of Ramadan.

Sources: Data from Kousksou, T., Allouhi, A, Belattar, M.Jamil, A., El Rhafiki, T., Arid, A., Zeraouli, Y., 2015b. Renewable energy potential and national policy directions for sustainable development in Morocco. Renewable and Sustainable Energy Reviews 47, 46–57. https://doi. org/10.1016/j.rser.2015.02.056; Agence Marocaine pour le Développement des Energies Renouvelable set de l'Efficacité Energétique. Available on the internet: http://www.aderee.ma/; Moroccan Ministry of Energy, Mines, Water and Environment, 2016. Moroccan Solar Energy Program. Available on the internet: http://www.mem.gov.ma/SitePages/GrandChantiersEn/DEREESolarEnergy.aspx.

scheme for producing, exploiting, extending of the capacity, or modification of installations that produceenergy from renewable energy sources, wheninstalled power is greater than or equal to 2 MW (Ghezloun et al., 2014). However, Law 13-09 did not put a limit on the installed capacity per project or per type of energy, and provided a legal framework for clean energy export. In the same year, NES was adopted and set energy efficiency as a national priority. The aim of Law 13-09 is to promote generating energy from renewable resources to the local market and to export by public entities or private companies. Furthermore, "this law allows an operator to produce electricity from renewable sources on behalf of a consumer or a group connected to the national of minimum (MV) voltage, high voltage (HV) or extra high voltage (EHC)" (Invest in Morocco, 2019). Since Law 13-09 has allowed the commercialization of producing energy from renewable resources, the operators benefit from the right of access to the national electrical grid.

Law 16-09, voted in 2009, was introduced for the establishment of the Agence Nationale pour le Développement (ADEREE). The agency is currently finalizing the regulatory framework on energy efficiency in passive building. After working on technical aspects of this type of building, the agency is currently working on its regulatory framework together with the Moroccan Energy Ministry.

Law 47-09 on energy efficiency was published on November 17, 2011. The aim of the law is to increase energy efficiency while using renewable energy sources, avoiding waste, reducing the burden of energy costs on the Moroccan economy to sustainable development in integrating energy efficiency measures, and supporting industrial companies in reducing their energy consumption. Law 47-09 allows to implement four main objectives, which are dedicated to implementing a code of conduct on energy efficiency, promoting solar heaters, disseminating energy-saving lamps, and promoting and adopting public lighting equipment. In order to implement Law 47-09, the partnership agreement on energy efficiency in the industry sector was signed by the Moroccan Ministry of Energy and Ministry of Industry. The agreement is focused on assessing the current energy consumption in the industry sector, realization of energy audits, optimizing energy consumption of audited companies, training, capacity building and monitoring, establishing of funding mechanisms, certification, and standardization (AMEE, 2019).

In order to implement the Moroccan Solar Plan, the Moroccan Agency for Solar Energy (MASEN) was established by Law 57-09. MASENis the group responsible for managing renewable energy in Morocco. The aim of this agency is to lead programs of integrated projects which when implemented would generate an additional 3000 MW of clean electricity capacity by 2020 and a further 6000 MW by 2030. Société d'Investissements Energétiques (SIE) references investors in the energy strategy of Morocco; it was founded in February 2010 in accordance with the guidelines of the national energy strategy aimed at the diversification of resources, promotion of renewable energy, and energy efficiency (Moroccan Ministry of Energy, Mines, Water and Environment, 2016).

In 2012, Law 48-15 was introduced. This law has ensured the proper functioning of the free market for electricity generated from renewable sources and regulation of the access of self-producers to the national electricity transmission grid. In line with this law, the possibility to develop electricity generation projects from renewable energy sources for private operators has opened. Furthermore, private operators may sell electricity to a consumer of their choice with a guaranteed right of access to the national power grids within the limit of available technical capacity of networks.

Although, NES was adopted in 2009, the implementation of energy efficiency programs started in 2008 (Arce et al., 2012). Energy efficiency programs implemented in parallel with the development of renewable energy will form a major part of Morocco's energy strategy and hydroelectric sources by 2020.

Programs and projects

Renewable energy programs and projects in Morocco are arousing increasing interest from international and local investors, allowing Morocco to attract more investment in, this sector, such as Siemens's investment of nearly €100 million ($107 million) for the manufacture of wind turbine blades in Morocco (Table 13.10).

Morocco also aims, through the exploitation of its natural renewable energy resources, to contribute to the preservation of the environment and the assurance of the durability of its economic and social development. In addition to the guarantee of a clean earth for future generations, Morocco should comply with its international commitments made at COP22 in 2016 in Marrakesh, within the framework of the Paris Climate Change Agreement, to reduce greenhouse gas emissions, mainly carbon dioxide.

The implementation of NES is finance by the Fond de Développement del' Energie (FDE). The FDE was established for resources mobilizing and monitoring the transparency of fundraising. The Kingdom of Saudi Arabia donated 500 million US dollars, United Arab Emirates donated 300 million US dollars. The contribution of Hassan II Fund for Economic and Social Development was 200 millionUS dollars. The year 2010 was highlighted by the establishment of the Energy Investment Corporation (EIS) with a capital of one billion dirhams endorsed by the state (71%) and the Hassan II Fund for Economic and Social Development (29%) (Invest in Morocco, 2019). African Development Bank has approved the largest project with a loan of €359 million ($387 million) and €125 million ($134.5 million) funding from the Clean Technology Fund for Morocco's Integrated Wind/Hydro and Rural Electrification Program.

In order to develop the largest concentrated solar power plant in Africa (Quarzazate solar complex), in 2015 African Development Bank provided a loan of €200 million ($215.6 million), and the same amount was provided by the Climate Investment Fund's Clean Technology Fund. As well, by 2030 Morocco intends to invest 30 million US dollarsinto renewable energy.

The DESERTEC concept is one of the major projects developing in Morocco. The aim of the project is to build concentrating solar-thermal power (CSP) plants and export renewable energy from MENA to European countries. DESERTEC involves the development of a transnational super grid that integrates all types of renewable energies:

- CSP in desert regions
- wind power in coastal areas
- hydro power in mountainous regions
- photovoltaics in sunny areas
- biomass and geothermal power where geographic conditions are favorable (Norton Rose Fulbright, 2012).

TABLE 13.10 Renewable energy projects.

Resource	Planned projects	Location	Situation	Potential level of production	Investment cost
Wind	Taza	Taza	Under construction	Capacity: 150 MW Annual production: 430 GWh	3 billion DH ($310.2 million)
	JbedLahid	Essaouira	Invitation for tenders	Capacity: 200 MW	850 MW integrated
	Tiskarad	Tarfaya	Invitation for tenders	Capacity: 300 MW Annual production: 1000 GWh	Program: 28.5 billion DH ($2.9 billion)
	Boujdour	Boujdour	Invitation for tenders	Capacity: 100 MW Annual production: 325 GWh	
	Midelt	Midelt	Invitation for tenders	Capacity: 150 MW	
	Tanger II	Tanger	Invitation for tenders	Capacity: 100 MW Annual production: 450 GWh	
	Tarfaya	Tarfaya	Under construction	Capacity: 300 MW	450 million euros ($485.1 million)
	KoudiaLbaida I	Tetouan	Planned project	Capacity: 100 MW	
	KoudiaLbaida II	Tetouan	Planned project	Capacity: 200 MW	
	Jbel Khaladi	Tangier	Under construction	Capacity: 120 MW	1.9 billion DH ($196.44 million)
	Akhfenir II	Tantan	Planned project	Capacity: 100 MW	
Solar	Querzazate	Querzazate	Under construction	Capacity: 580 MW Annual production: 1150 GWh	70 billion DH ($7.23 billion)
	Ain Beni Mathar	Beni Mathar	Under construction	Capacity: 572 MW Annual production: 3538 GWh	4.6 billion DH ($475.6 million)
	Foum El Qued	Laayoune technology park	Finished	Capacity: 500 MW	16.8 billion DH ($1.737 billion) (including technology park)
	Boujdour	Laayoune	Planned project	Capacity: 100 MW Annual production: 230 GWh	
	Sabkhat Tah	Tarfaya	Planned project	Capacity: 500 MW Annual production: 1040 GWh	

(Continued)

TABLE 13.10 Renewable energy projects.—cont'd

Resource	Planned projects	Location	Situation	Potential level of production	Investment cost
Hydro power	M'dez Wl Menzel	Sefrou	Under construction	Capacity: 300 MW	2.87 billion DH ($296.74 million)
	Step Abdelmoumen	Agadir	Under construction	Capacity: 350 MW	2.4 billion DH ($248.1 million)
	Aferer	Agadir	Under construction	Capacity: 465 MW	
	Khénifra		Under construction	Capacity: 128 MW	
	Ifahsa	Oued Laou	Under construction	Capacity: 300 MW	
	Bar Ouender	Taounate	Under construction	Capacity: 30 MW	
	Boutferda	Azilal	Under construction	Capacity: 18 MW	
	Tillouguitaval	Benin Mellal	Under construction	Capacity: 26 MW	
	Tillouguitamont	Benin Mellal	Under construction	Capacity: 8 MW	
	Tamejout	Benin Mellal	Under construction	Capacity: 30 MW	
Biomass	Landfill Fez	Fez		Capacity: 5 MW	
Geothermal	Nothing to report				

Sources: Kousksou, T., Allouhi, A., Belattar, M., Jamil, A., El Rhafiki, T., Zeraouli, Y., 2015a. Morocco's strategy for energy security and low-carbon growth. Energy 84, 98–105. https://doi.org/10.1016/j.energy.2015.02.048; International Hydropower Association, 2018; AMEE, 2019.

It is expected that the DESERTEC plan would also help solve environment pollution problems. Moore (2017) emphasized theimportance of reducing air and water pollution. He claimed that if nearly all of fossil fuel would be removed from generating energy, in this way, greenhouse gas emissions would be captured 0.25 Gt by 2050 in a high electricity demand scenario and 0.15 Gt in a low demand scenario. Samus et al. (2013) described plan DESERTEC as "harvesting" renewable energies. These researchers especially emphasized that this plan allowed converting solar irradiation into electricity by means of CSP for Moroccan and European use. After successful DESERTEC implementation, electricity will be transmitted into the European grid, i.e., the Network of Transmission System Operators for Electricity grid.

In 2009, under the plan of DESERTEC, MASEN launched the development of the 500 MW Quarzazate plant. CSP power station would allow to reduce 240,000 t of CO_2 equivalent emissions per year.

Ain Beni Mathar project is a part of DESERTEC; it plans to build a 470 MW hybrid solar-gas plant. The African Development Bank is financing two-thirds of the cost of the plant, approximately €187.85 million ($201.5 million). In2010,the Moroccan government implementedthe Integrated Wind Energy Project, with an estimated investment of€3.2 billion ($3.45 billion); it aims to increase the share of wind power in the national energy balance to 14% by 2020 (Simelyte et al., 2016).

At present Morocco uses only 1% of its potential in biomass energy. Several energy harnessing projects from biomass are in progress: Biogaz Agricole, BenSergao—Agadir. Projects and some treatment stations are in Fez and Marrakesh (Kousksou et al., 2015b). For generating energy from bioethanol, which is produced from sugarcane, COSUMAR has been established. COSUMAR is the only national operator, which is made up of five companies specialized in the extraction, refining, and packaging of sugar in all its forms. This operator produces sugar from both sugar beet and sugarcane, with an estimated production of around 400 million tons per year. "In cane sugar factory, bagasse is used as an energy source by burning. Molasses byproduct is sold by COSUMAR to another company that produces ethanol liquid, carbon dioxide, and biomethane for its own energy consumption" (Kamzon et al., 2016).

Efficiency plan

Despite implementing various programs and projects to increase generation of renewable energy sources, at the same time Morocco is implementing energy efficiency plans. Energy plansare highly linked to the expected results of the DESERTEC plan.

— The energy efficiency program GMT +1 started in 2008. The aim of the program is to increase power reserve margins during peak hours. The schedule of the summer period is from April to October. The result of implementing this program was clear in 2014 as the potential gains reached 92 MW and the total energy savings of 29.5 GWh. In this way, CO_2 emissions of around 27,658 tons were avoided.
— After installing 6 million low-consumption lamps, the national program of low consumption lamps reduced 3.3% of energy consumption.

- The other program aims to upgrade energy efficiency in 15,000 mosques. This program has two objectives:
  - awareness of energy efficiency techniques to citizens;
  - reducing the energy consumption of mosques by the introduction of efficient lighting, solar water heaters, thermal management in the mosque, and the use of photovoltaic solutions for electricity production.
- Another program is energy efficiency in street lighting. In order to achieve the aim of the program, the equipment to achieve energy savings was installed, which included LED lamps, stabilizers, etc. As aresult of the first project in public lightning, the public lighting installations in the cities were renovated, the network and widespread public lighting to the entire urban territory was extended. Even more, lower energy costs reduced communal budgets.
- Regionalization program *JihaTinou*. For developing this program, the municipalities of Agadir, Oujda, and Chefchaouen were chosen as pilots. *JihaTinou* is an attempt to get citizens directly involved, to turn the Kingdom's energy strategy into a process led by and for the benefit of the local economy and population.
- CEEB programaims to introduce mandatory minimum energy efficiency performance requirements in buildings, through the introduction of an energy efficiency building code in the key sectors (housing, education, health, hospitality industry).
- PEEI program aims to implement energy efficiency measures in the industrial sector, one of the biggest energy consumers in Morocco. It is expected to optimize energy consumption in the industrial sector to an estimated total savings of 2,000,000 toe. Additionally, it is expected to reduce CO_2 emissions at 7,594,335 tons of CO_2 equivalent.
- Shemsi program. To reach NES target, ADEREE works on the decentralized solar thermal energy production to replace the current centralized power production. The first incentive project,"PROMASOL," helped install 160,000 m^2ofsolar water heaters within 8 years in Morocco. PROMASOL includes a solar water heater program. The aim of this program is to increase supply, to reduce equipment costs, and improve the quality of solar-powered equipment. PROMOSOL is a United Nations—funded initiative.

Morocco invests in and promotes renewable energy sources for several reasons. The first one is to eliminate energyinsecurity from Algeria and to exploit its own potential for renewable resources. The second reason for investing in renewable sources is the possibility to reduce consumption and demand forenergy, or in other words, to save energy. The third reason is the result of the second. Due to saving energy and consuming less fossil fuel, air and water pollution are reduced. Thus, implementation of the National Energy Strategy helps Morocco solve all three problems.

Obstacles and challenges of implementing renewable energy promotion plan

Although the Moroccan government has started to implement policiesand regulations on renewable energy promotion, still some barriers for developing renewable energy in Morocco exist. Hanger et al. (2016) identified one primary barrier as a lack of financing and risk of the project implementation for both private and public investment. Especially this is a problem for the small-scale projects. Moreover, foreign companies do not tend to invest in this area, as it looks too risky. Both foreign and local private companies prefer investing in high-profitability projects and require returns in a short time period rather than in the longterm, which is more common for projects of renewable energy sources. Thus, most of the financial support comes from international funds and the government. The World Energy Council (2015) found the short termism to be one of the major economic and financially limiting factors.

Due to the fact that fossil fuels are more subsidized than renewable energy sources, it becomes even more difficult to encourage local and foreign companies to invest in renewable energy. In this way, renewable energy sources lose their competitiveness. Competition in cost-driven markets increases when market penetration is large. However, the Moroccan energy market is occupied by large monopolies, which expect higher profits and usefossil fuel that ismuch cheaper than installing renewable energy technologies.

The Moroccan government introduced various national programs regarding promoting renewable energy. Particular attention is paid to Agadir, Marrakesh, and Quarzazate. However, the World Energy Council (2015) noticed some ambiguities in strategyand found that it is not clear how decisions are made. Furthermore, the World Energy Council (2015) emphasized that Morocco faces a shortage of businesses, entrepreneurs,trained workers, and specialized industries that would be able to promote and expand the market of renewable energy. Alhamwi et al. (2015) maintained that Morocco lacks experienced decision-maker specialists. Another problem Alhamwi et al. (2015) noticed is that obvious inefficient coordination exists between two independent agencies thatare responsible for implementing energy policy in Morocco. Thus, one of the obstacles might be that the Office of Hydrocarbons and Mining (ONHYM) and the ONEE double functions of each other. Moore (2017) stated that Morocco has limited capacity and coverage of the power grid, which might be one of the barriers to successfully implementing the National Energy Strategy. Thus, structural changes are required of the infrastructure of the energy system. At the same time, it needs investments and funding.

According to Mahia et al., (2014) research, the sociopolitical instability in the country and/ or region is one of top five policy-related barriers. Historically, the statutory framework and institutional infrastructure were one of the weakest links in developing renewable energy sources and the enhancement of energy efficiency (Table 13.11).

The other researchers (Medina et al., 2015) didresearch on investment barriers in the renewable energy sector. Three groups of factors were considered:

− The micro- and macroeconomic characteristics of a country, which might influence decision-making.
− Geopolitical situation, which might also change the decision to invest.
− Specific characteristics of the sector thatis under consideration to invest in.

Even more, the reduction of lobbyism and corruption are two other challenges Morocco faces. Lobbyism serves in favor of large monopolies that use fossil fuels. Thus, there is a lack of transparency and clearness over the management of the energy sector. The transparency in implementing policies regarding renewable resources would increase if the society would receive more information about new national programs, availability of funding for the RE projects. Due to the lack of information, local investors do not recognize benefits and challenges of using renewable energy. Even more, key stakeholders do not know how

TABLE 13.11   Barriers thatmay influence decision-making to invest in the renewable energy sector.

Business barriers

Uncertainty of regional or country-level market development and prospects

Not enough specialization in renewable energy sector

Insufficient developed infrastructure

Low level of logistic networks in the country

Low logistic skills of local companies

Low capabilities of engineering companies

Low management skills of local companies

High informality level

Higher wages for international experts

Low level of automatization/modernization of local industries

Inadequate size of local industries

Lack of regional/local funding

Lack of international (private or multilateral) financial resources for new financing

Higher capital costs (risk premium) for initiatives in the area

---
Political barriers

Low level of regional economic and political integration

Social political instability in the country or region

Not enough long-term security for planning

Lack of industrial R&D support

Lack of coordination between energy policy and industrial policy

Unclear/undefined energy policy

Weak connection in the business-political network

Administration and legal barriers in a country

Distorting presence of public actors in value chain

No fiscal and legislative framework for renewable energy sources or unstable framework

Low level of European institutionalcommitment/support in order to promote regional initiatives

Low level of multilateral institutional commitment/support in order to promote regional initiatives

High level of taxation or difficulty in organizing taxation

Corruption

Market barriers
Low level of regional demand for renewable energy
Volatility or instability in renewable energy market
Low level of interconnection/networking among renewable energy market
High level of competition (price driven) among renewable energy
High level of competition with other foreign stakeholders already present in the regional market
High level of competition with other emerging countries

Source: Mahia, R., Arce, R., Medina, E., 2014. Assessing the future of a CSP industry in Morocco. Energy Policy 69, 586–597. https://doi.org/10.1016/j.enpol.2014.02.024.

RE technology works. The World Energy Council (2015) distinguished between thecultural and behavioral barriers. The organization noticedthe public apathy to support renewable energy. In addition, the Moroccan society is inflexible to accept many changes. This might be explained as the consequence of illiteracy and lack of knowledge and understanding of added value of RE technology. Wolsink (2007) and Wüstenhagen et al. (2007) stated that acceptance of RE in society is determined by various factors such as expected costs and benefits; social, economic and environmental risks; trust and perceived fairness; distance to the proposed power plant; and the regulatory context. Meanwhile, Hanger et al. (2016) found that the community of Quarzazate seemed to be overwhelmingly in favor of the project: 91% being completely in favor or partially in favor.

Thus, in the Morocco case, information plays a major part in introducing renewable energy technologies. Furthermore, technical problems might occur as well. For example, the renewable energy has only been incorporated in the electricity sector, as a cross-sectorial system between heating/cooling and transport does not exist. There are unsolved issues with integrating renewable power onto the transmission grid system. Such issues may be successfully addressed by investment to reinforce the transmission grid. More problems might arise for large-scale projects, whereas there are not enough highly qualified staff that could operate renewable energy facilities.

Concluding remarks

Energy is consumed in the various sectors of modern society under different forms. Even more, it is proved that energy consumption impacts economic growth. On the global level,economies are growing, as well as global energy consumption. Demand for energy is growing every day, which is resulting in higher volumes of CO_2. To protect the environment, countries started to explore alternative energy sources. Despite that the share of renewable energy in total energy consumption is supposed to increaseworldwide, the most industrial countries, such as China, Japan, the United States, and the United Kingdom, are not atthe top ofthe countries that arepromoting renewable energy and implementing strategiesfor developingrenewable energy.

Morocco is intensively implementinga National Energy Strategy thataims to reduce energy consumption by exploiting the potential of renewable energy. Even more, this strategy will solve the energyinsecurity problem of Morocco, as Moroccodoes not have any natural resources. However, it has the highest potential to exploit renewable resources out of all of the continent of North Africa.

Morocco has high potential forgenerating electricity from four renewable sources: solar energy, wave-wind energy, hydropower, and biomass energy.

To implement its National Energy Strategy, Morocco liberalized its energy market and established several organizations for administrating implementation of NES. At the same time, these organizations take care of funding, monitoring, and transparency. As well as promoting generation of energy from renewable sources, Morocco is implementing energy efficiency plans.

However, Morocco faces some obstacles in implementing NES, programs, projects. Some problems might occur because Morocco has no long-run experience in running such projects. Thus, in the near future Morocco might face a lack of specialists who are competent in these fields. This problem might occur as Morocco runs more than one large-scale project at one time.

The agencies that administrate programs and projects thatare in process sometimes repeat the work of each other. In this way, bureaucracy increases.

References

Adams, S., Klobodu, E.K.M., Apio, A., 2018. Renewable and non-renewable energy, regime type and economic growth. Renewable Energy 125, 755–767. https://doi.org/10.1016/j.renene.2018.02.135.

Adom, P.K., Bekoe, W., Akoena, S.K.K., 2012. Modelling aggregate domestic electricity demand in Ghana: an autoregressive distributed lag bounds cointegration approach. Energy Policy 42 (March), 530–537. https://doi.org/10.1016/j.enpol.2011.12.019.

Agence Marocaine pour le Développement des Energies Renouvelable set de l'Efficacité Energétique. Available on the internet: http://www.aderee.ma/.

Alhamwi, A., Kleinhans, D., Weitemeyer, S., Vogt, T., 2015. Moroccan National Energy Strategy reviewed from a meteorological perspective. Energy Strategy Reviews 9, 39–47. https://doi.org/10.1016/j.esr.2015.02.002.

Apergis, N., Tang, C.F., 2013. Is the energy-led growth hypothesis valid? New evidence from a sam-ple of 85 countries. Energy Economics 38, 24–31. https://doi.org/10.1016/j.eneco.2013.02.007.

Allouhi, A., Zamzoum, O., Islam, M.R., Saidur, R., Kousksou, T., Jamil, A., Derouich, A., 2017. Evaluation of wind energy potential in Morocco's coastal regions. Renewable and Sustainable Energy Reviews 72, 311–3241. https://doi.org/10.1016/j.rser.2017.01.047.

Al-Mulali, U., Fereidouni, H.G., Lee, J.Y., Che Sab, C.N.B., 2013. Examining the bi-directional long run relationship between renewable energy consumption and GDP growth. Renewable and Sustainable Energy Reviews 22, 209–222. https://doi.org/10.1016/j.rser.2013.02.005.

de Arce, R., Mahia, R., Medina, E., Escribano, G., 2012. A simulation of the economic impact of renewable energy development in Morocco. Energy Policy 46, 335–345. https://doi.org/10.1016/j.enpol.2012.03.068.

Apergis, N., Payne, J.E., 2010. Renewable energy consumption and economic growth: evidence from a panel of OECD countries. Energy Policy 38 (2), 656–660. https://doi.org/10.1016/j.enpol.2009.09.002.

Appun, K., Bieler, F., Haas, Y., Wettengel, J., 2019. Germany's energy consumption and power mix in charts. Clean Energy Wire CLEW. Available on the internet: https://www.cleanenergywire.org/factsheets/germanys-energy-consumption-and-power-mix-charts.

Aslan, A., 2016. The causal relationship between biomass energy use and economic growth in the United States. Renewable and Sustainable Energy Reviews 57, 362–366. https://doi.org/10.1016/j.rser.2015.12.109.

Azzopardi, B., 2014. Green energy and technology: choosing among alternatives. In: Jahang Hussein, Apel Mahmud (Eds.), Large Scale Renewable Power Generation: Advances in Technologies for Generation, Transmission and Storage, Green Energy and Technology. Springer.

Bahadori, A., Nwaoha, C., 2013. A review on solar energy utilization in Australia. Renewable and Sustainable Energy Reviews 18, 1–5. https://doi.org/10.1016/j.rser.2012.10.003.

Barkaoui, A., Zarhloule, Y., Rimi, A., Correia, A., Voutetakis, S., Seferlis, P., 2015. Geothermal country update report of Morocco (2010-2015). In: Proceedings World Geothermal Congress 2015, Melbourne, Australia, 19-25 April 2015.

Bennouna, A., El Hebil, C., 2016. Energy needs for Morocco 2030, as obtained from GDP-energy and GDP-energy intensity correlations. Energy Policy 88, 45–55. https://doi.org/10.1016/j.enpol.2015.10.003.

Boharb, A., Allouhi, A., Saidur, R., Kousksou, T., Jamil, A., Mourad, Y., Benbassou, A., 2016. Auditing and analysis of energy consumption of an industrial site in Morocco. Energy 101, 332–342. https://doi.org/10.1016/j.energy.2016.02.035.

Bortolini, M., Gamberi, M., Graziani, A., Manzini, R., Pilati, F., 2014. Performance and viability analysis of small wind turbines in the European Union. Renewable Energy 62 (February), 629–639. https://doi.org/10.1016/j.renene.2013.08.004.

Bouhal, Y., Agrouaza, T., Kousksoua, A., Allouhib, T., El Rhafikic, A.,J., Bakkas, C., 2018. Technical feasibility of a sustainable concentrated solar power in Morocco through an energy analysis. Renewable and Sustainable Energy Reviews 81, 1087–1095. https://doi.org/10.1016/j.rser.2017.08.056.

Bruns, S.B., Gross, C., 2013. What if energy time series are not independent? Implications for energy-GDP causality analysis. Energy Economics 40, 753–759. https://doi.org/10.1016/j.eneco.2013.08.020.

Climate Investment Funds, October 27, 2009. Available on the internet: https://www.climateinvestmentfunds.org/sites/cif_enc/files/meetingdocuments/ctf_3b_morocco_investment_plan_100708_0.pdf.

Enerdata, 2011. Sluggish Growth of World Energy Demand in 2011. Available on the internet: http://www.enerdata.net/enerdatauk/press-and-publication/publications/2011-g-20-energy-demand-decrease-vs-china-increase.php.

Enerdata, 2015. Global Energy Statistical Yearbook 2015. Available on the Internet: https://yearbook.enerdata.net/#renewable-in-electricity-production-share-by-region.html.

Frankl, P., Nowak, S., Gutscher, M., Gnos, S., Rinke, T., 2010. Technology Roadmap: Solar Photovoltaic Energy. International Energy Agency (IEA), France.

Ghezloun, A., Saidane, A., Oucher, N., 2014. Energy policy in the context of sustainable development: case of Morocco and Algeria. Energy Procedia 50, 536–543. https://doi.org/10.1016/j.egypro.2014.06.065.

Gorus, M.S., Aydin, M., 2019. The relationship between energy consumption, economic growth, and CO_2 emission in MENA countries: causality analysis in the frequency domain. Energy 168, 815–822. https://doi.org/10.1016/j.energy.2018.11.139.

Haas, R., Resch, G., Panzer, C., Busch, S., Ragwitz, M., Held, A., 2011. Efficiency and effectiveness of promotion systems for electricity generation from renewable energy sources — lessons from EU countries. Energy 36, 2186–2193. https://doi.org/10.1016/j.energy.2010.06.028.

Hamdaoui, S., Mahdaoui, M., Allouhi, A., El Alaiji, R., Kousksou, T., El Bouardi, A., 2018. Energy demand and environmental impact of various construction scenarios of an office building in Morocco. Journal of Cleaner Production 188, 113–124. https://doi.org/10.1016/j.jclepro.2018.03.298.

Hanger, S., Komendantova, N., Schinkec, B., Zejli, D., Ihlal, A., Patta, A., 2016. Community acceptance of large-scale solar energy installations in developing countries: evidence from Morocco. Energy Research & Social Science 14, 80–89. https://doi.org/10.1016/j.erss.2016.01.010.

Hawila, D., Mondal, M.A.H., Kennedy, S., Mezher, T., 2014. Renewable energy readiness assessment for North African countries. Renewable and Sustainable Energy Reviews 33, 123–140. https://doi.org/10.1016/j.rser.2014.01.066.

International Energy Agency (IEA), 2015. Energy and Climate Change: World Outlook Special Report. https://www.iea.org/publications/freepublications/publication/WEO2015SpecialReportonEnergyandClimateChange.pdf.

International Renewable Energy Agency (IRENA), 2015. Renewable Energy Zones for the African Clean Energy Corridor. https://www.irena.org/publications/2015/Oct/Renewable-Energy-Zones-for-the-Africa-Clean-Energy-Corridor.

International Renewable Energy Agency (IRENA), 2018. Renewable Power Generation Costs. https://www.irena.org/-/media/Files/IRENA/Agency/Publication/2019/May/IRENA_Renewable-Power-Generations-Costs-in-2018.pdf.

International Energy Agency, 2016. Renewables Information. Available on the Internet: http://www.iea.org/bookshop/668-Renewables_Information_2015.

Jankauskas, V., Rudzkis, P., Kanopka, A., 2014. Risk factors for stakeholders in renewable energy investments. Energetika 60 (2), 113–124.

Kahia, M., Kadria, M., Ben Aissa, M.S., Lanouar, C., 2017. Modelling the treatment effect of renewable energy. Journal of Clean Production 149 (15), 845–855. https://doi.org/10.1016/j.jclepro.2017.02.030.

Kamzon, M.A., Abderafi, S., Bounahmidi, T., 2016. Promising bioethanol processes for developing a biorefinery in the Moroccan sugar industry. International Journal of Hydrogen Energy 41, 20880–20896. https://doi.org/10.1016/j.ijhydene.2016.07.035.

Klevas, V., Murauskaitė, L., Kleviene, A., Perednis, E., 2013. Measures for increasing demand of solar energy. Renewable and Sustainable Energy Review 27, 55–64. https://doi.org/10.1016/j.rser.2013.06.050.

Kousksou, T., Allouhi, A., Belattar, M., Jamil, A., El Rhafiki, T., Zeraouli, Y., 2015a. Morocco's strategy for energy security and low-carbon growth. Energy 84, 98−105. https://doi.org/10.1016/j.energy.2015.02.048.

Kousksou, T., Allouhi, A., Belattar, M., Jamil, A., El Rhafiki, T., Arid, A., Zeraouli, Y., 2015b. Renewable energy potential and national policy directions for sustainable development in Morocco. Renewable and Sustainable Energy Reviews 47, 46−57. https://doi.org/10.1016/j.rser.2015.02.056.

Kraft, J., Kraft, A., 1978. On the relationship between energy and GNP. Journal of Energy and Development 3 (2), 401−403.

International Hydropower Association (IHA), 2018. Hydropower Status Report, 2018. https://www.hydropower.org/publications/2018-hydropower-status-report.

International Hydropower Association (IHA), 2019. Hydropower projects test draft climate resilience guide. https://www.hydropower.org/news/hydropower-projects-test-draft-climate-resilience-guide.

Luqman, M., Ahmad, N., Bakhsh, K., 2019. Nuclear energy, renewable energy and economic growth in Pakistan: evidence from non-linear autoregressive distributed lag model. Renewable Energy 139, 1299−1309. https://doi.org/10.1016/j.renene.2019.03.008.

L'Agence Marocaine pour l'Efficacité Énergétique (AMEE), 2019. Réeglement Thermique de Construction au Maroc. http://www.amee.ma/images/Text_Pic/Others/Reglement_thermique_de_construction_au_Maroc_-_Version_simplifiee.pdf.

Leidreiter, A., Boselli, F., 2015. 100% Renewable Energy: Boosting Development in Morocco. World Future Council. Available on the internet: http://www.energynet.co.uk/webfm_send/1606o.

Mahia, R., Arce, R., Medina, E., 2014. Assessing the future of a CSP industry in Morocco. Energy Policy 69, 586−597. https://doi.org/10.1016/j.enpol.2014.02.024.

Maji, I.K., Sulaiman, C., 2019. Renewable energy consumption and economic growth nexus: a fresh evidence from West Africa. Energy Reports 5, 384−392. https://doi.org/10.1016/j.egyr.2019.03.005.

Marano, V., Rizzoni, G., 2008. Energy and economic evaluation of PHEVs and their interaction with renewable energy sources and the power grid. In: IEEE International Conference on Vehicular Electronics and Safety (ICVES), pp. 84−89.

Medina, E., de Arcea, R., Mahía, R., 2015. Barriers to the investment in the Concentrated Solar Power sector in Morocco: a foresight approach using the Cross-Impact Analysis for a large number of events. Futures 71, 36−56. https://doi.org/10.1016/j.futures.2015.06.005.

Minitere de l'Energie, 2012. Loi no 47-09 relative a l'efficacite energetique. http://www.mem.gov.ma/SitePages/TestesReglementaires/Loi47-09.pdf.

Moore, S., 2017. Evaluating the energy security of electricity interdependence: perspectives from Morocco. Energy Research & Social Science 24, 21−29. https://doi.org/10.1016/j.erss.2016.12.008.

Moroccan Ministry of Energy, Mines, Water and Environment, 2016. Moroccan Solar Energy Program. Available on the Internet: http://www.mem.gov.ma/SitePages/GrandChantiersEn/DEREESolarEnergy.aspx.

Naimi, Y., Saghir, M., Cherqaoui, A., Chatre, B., 2017. Energetic recovery of biomass in the region of Rabat, Morocco. International Journal of Hydrogen Energy 42, 1396−1402. https://doi.org/10.1016/j.ijhydene.2016.07.055.

Narayan, P.K., Popp, S., 2012. The energy consumption-real GDP nexus revisited: Empirical evidence from 93 countries. Economic Modelling 29 (2), 303−308. https://doi.org/10.1016/j.econmod.2011.10.016.

Nfaoui, M., El-Hami, K., 2018. Extracting the maximum energy from solar panels. Energy Reports 4, 536−546. https://doi.org/10.1016/j.egyr.2018.05.002.

Norton Rose Fulbright, 2012. Renewable Energy in Morocco. http://www.nortonrosefulbright.com/knowledge/publications/66419/renewable-energy-in-morocco>.

Office National de l'Electricité et de l'Eau Potable (ONEE), Projet eolien integre 850 MW le marche sera finalement attribue au 1er trimestre 2016, Le Matin, 02/11/2015, 23 pg. Available on the Internet: http://www.one.org.ma.

Ozcan, B., Ozturk, I., 2019. Renewable energy consumption-economic growth nexus in emerging countries: a bootstrap panel causality test. Renewable and Sustainable Energy Reviews 104, 30−37. https://doi.org/10.1016/j.rser.2019.01.020.

Pantaleo, A.M., Camporeale, S.M., Sorrentino, A., Miliozzi, A., Shah, N., Markides, C.N., 2018. Hybrid solar-biomass combined Brayton/organic Rankine-cycle plants integrated with thermal storage: techno-economic feasibility in selected Mediterranean areas. Renewable Energy xxx, 1−19 (article(in press). https://doi.org/10.1016/j.renene.2018.08.022.

Pérez-Collazo, C., Jakobsen, M.M., Buckland, H., Fernández-Chozas, J., 2012. Synergies for a wave-wind energy concept. In: Working Paper, at DONG Energy at the 6th Annual INORE Symposium, Denmark, May,2012.

Power Engineering International (PEI), 2018. New Horizons for Renewables in Morocco and Africa. https://www.powerengineeringint.com/articles/print/volume-26/issue-6/features/new-horizons-for-renewables-in-morocco-and-africa.html.

Rodríguez, J.M., Martínez, G.S., Bennouna, El, BadrIkken, B., 2016. Techno-economic assessment of thermal energy storage solutions for a 1 MWe CSP-ORC power plant. Solar Energy 140, 206−218. https://doi.org/10.1016/j.solener.2016.11.007.

Samus, T., Lang, B., Rohn, H., 2013. Assessing the natural resource use and the resource efficiency potential of the Desertec concept. Solar Energy 87, 176−183. https://doi.org/10.1016/j.solener.2012.10.011.

Saidi, K., Mbarek, M.B., 2016. Nuclear energy, renewable energy, CO2 emissions, and economic growth for nine developed countries: evidence from panel Granger causality tests. Progress in Nuclear Energy 88, 364−374. https://doi.org/10.1016/j.pnucene.2016.01.018.

Sarkis, J., Tamarkin, M., 2008. Real options analysis for renewable energy technologies in a GHG emissions trading environment. In: Antes, R., Hansjürgens, B., Letmathe, P. (Eds.), Emissions Trading. Springer, New York, pp. 103−119.

Savino, M.M., Manzini, Della Selva, V., Accorsi, R., 2017. A new model for environmental and economic evaluation of renewable energy systems: the case of wind turbines. Applied Energy 189, 739−752. https://doi.org/10.1016/j.apenergy.2016.11.124.

Sebri, M., 2015. Use renewables to be cleaner: meta- analysis of the renewable energy consumtion - economic growth nexus. Renewable and Sustainable Energy Reviews 42, 657−665.

Sibley, M., Sibley, M., 2015. Hybrid transitions: combining biomass and solar energy for water heating in public bathhouses. Energy Procedia 83, 525−532. https://doi.org/10.1016/j.egypro.2015.12.172.

Sierra, J.P., Martín, C., Mosso, M., Mestres, M., Jebbad, R., 2016. Wave energy potential along the Atlantic coast of Morocco. Renewable Energy 96, 20−32. https://doi.org/10.1016/j.renene.2016.04.071.

Simelyte, A., Sevcenko, G., El Amrani El Idrissi, N., Monni, S., 2016. Promotion of renewable energy in Morocco. Entrepreneurship and Sustainability Issues 3 (4), 319−327. https://doi.org/10.9770/jesi.2016.3.4(2.

Šimelytė, A., Dudzevičiūtė, D., 2017. Consumption of renewable energy and economic growth. In: Contemporary Issues in Business, Management and Education'2017: 5th International Scientific Conference, 11−12 May 2017. Vilnius Gediminas Technical University (conference proceedings).

Solar energy, 2019. Invest in Morocco. Available on the Internet. http://www.invest.gov.ma/?Id=24&lang=en&RefCat=2&Ref=145.

Sovacool, B.K., 2011. The Routledge Handbook of Energy Security, first ed. Routledge, New York.

Tugcu, C.T., Topcu, M., 2018. Total, renewable and non-renewable energy consumption and economic growth: revisiting the issue with an asymmetric point of view. Energy 152, 64−74. https://doi.org/10.1016/j.energy.2018.03.128.

UNEP, 2012. Feed-inttariffs as Policy Instrument for Promotions Renewable Energies and Green Economies in Developing Countries. http://www.unep.org/pdf/UNEP_FIT_Report_2012F.pdf.

Wolsink, M., 2007. Wind power implementation: the nature of public attitudes: equity and fairness instead of backyard motives. Renewable and Sustainable Energy Review 11, 1188−1207. https://doi.org/10.1016/j.rser.2005.10.005.

World Council, 2015. 100% Renewable Energy: Boosting Development in Morocco. Available on the internet: http://www.worldfuturecouncil.org/inc/uploads/2016/01/WFC_2015_100_Renewable_Energy_boosting_Development_in_Morocco.pdf.

Wüstenhagen, R., Wolsink, M., Bürer, M.J., 2007. Social acceptance of renewable energy innovation: an introduction to the concept. Energy Policy 35, 2683−2691. https://doi.org/10.1016/j.enpol.2006.12.001.

Yildirim, E., Sarac, S., Aslan, A., 2012. Energy consumption and economic growth in the USA: evidence from renewable energy. Renewable and Sustainable Energy Reviews 16 (9), 6770−6774. https://doi.org/10.1016/j.rser.2012.09.004.

Further reading

Groupe Energies Renouvelables, Environnement et Solidarités, 2015. Morocco Biomass Sustainable Management and Dissemination of Efficient Appliances. Available on the Internet: https://www.geres.eu/en/wp-content/uploads/2012/06/projet-maroc-biomasse-en.pdf.

International Renewable Energy Agency (IRENA), 2014. Renewable Power Generation Costs in. Available on the Internet: http://www.irena.org/documentdownloads/publications/irena_re_power_costs_2014_report.pdf.

The International Renewable Energy Agency (IRENA), 2012. Renewable Power Generation Costs in 2012: An Review. Available on the Internet: http://costing.irena.org/media/2769/Overview_Renewable-Power-Generation-Costs-in-2012.pdf.

Social responsibility, social marketing role, and societal attitudes

Rasa Smaliukiene[1,2], *Salvatore Monni*[3]

[1]Vilnius Gediminas Technical University, Vilnius, Lithuania; [2]General Jonas Žemaitis Military Academy of Lithuania, Vilnius, Lithuania; [3]Roma Tre University, Rome, Italy

Introduction

Social responsibility of business and social marketing are two complementary forces that have a potential to transform energy supply and demand. This is particularly important as irresponsible energy production and consumption by businesses and private households is leading to degradation of the ecosystem and irreversible climate change. Unfortunately,

Energy Transformation towards Sustainability
https://doi.org/10.1016/B978-0-12-817688-7.00014-8

the present unsustainable trend in energy supply and consumption is exacerbating the crisis. Therefore, the main challenge today is finding a way to prevent the degradation at the same time ensuring socially beneficial economic growth. There are two dominating views toward the challenge of unsustainability in the field of energy. The first lays the responsibility with the energy companies while the other one attributes it to the endusers. As a result, responsibility and social marketing become extremely important.

In accordance with the first view, supply side of energy management and, in particular, corporate social responsibility, plays a vital role. More specifically, climate change depends mainly on how responsible energy producers are. Reflecting on the multiple ecological catastrophes in coal and oil mines as well as in the nuclear industry, suspicion and concerns regarding the fairness of the energy business have been raised in society. Additionally, energy suppliers are being blamed for important energy issues such as the exhaustion of traditional energy resources and rising energy prices that are causing energy poverty and other social problems. In response to the growing requirements to be ethical and responsible, energy companies are forced to take additional steps and act in a socially responsible way by delivering more sustainable business outcomes. What is more, the catastrophic oil spill in the Gulf of Mexico by British Petroleum (BP) in 2010 marked the beginning of a new era of social responsibility in the energy business. Nowadays, not only global giants like Exxon Mobil, Shell, or BP are implementing strategies of corporate social responsibility and becoming investors in renewable (Hopkins, 2009; Parrett, 2018) but also local energy suppliers are advancing toward cooperation with local communities to sustain the environment (Strachan et al., 2015). Notwithstanding the fact that companies accept the inevitability of social responsibility, socially responsible actions are highly challenging due to the specific characteristics of the energy sector that we are discussing below.

As for the second view, climate change is understood primarily as the outcome of irresponsible demand and consumption of energy. For that reason, attitudinal changes of society toward energy consumption are vital in transforming the behavior of households and communities. With regard to this approach, governmental and nongovernmental institutions are using social marketing to speed up the change for energy conservation. In this chapter, we will analyze how social marketing can change the demand side of energy management. Taking into consideration that "households are accountable for nearly three-quarters of global carbon emissions"(Druckman and Jackson, 2016), it is essential to apply social marketing that transforms destructive behavior resulting in these emissions. Social marketing is particularly relevant as it is capable of changing the behavior of energy users as their needs and wants become grounded in all activities of social initiatives. Data on household energy consumption indicates that rational initiatives such as building renovation, smart technology in buildings, and improved metering of energy consumption do not lead to desired results and the progress in energy conservation is very slow. This knowledge leads us to an understanding that energy consumption has to be based on specific energy users' needs and wants instead of rational economic values. Therefore, a user-centric approach rooted in social marketing is analyzed in the chapter.

This chapter presents conceptual and methodological insights into the complementarity of corporate social responsibility and social marketing. Specifically, it analyzes the role of the former in transforming the supply side energy management; as for the demand side, we are retooling social marketing to promote sustainable energy use. In the chapter we occasionally use empirical results and cases to illustrate how the conceptual approach is being implemented in practice.

Supply side: the role of corporate social responsibility

It is not an understatement to say that energy companies have become active in addressing social and environmental issues. This raises a question: What does "socially responsible" actually mean when it comes tothe energy business? Two arguments must be taken into consideration. The first one is based on dominating literature and refers to the so-called "social license to operate," which means that companies in order to be effective have to meet the expectations of the society at large and of local communities in particular. The second argument promotes the natural link between business social responsibility and corporate strategy, i.e., companies have to maximize their long-term profitability by incorporating social responsibility into their core business decisions and operations. The contradiction and complementarity of these two arguments stimulate the ongoing discussion regarding social responsibility among scholars and practitioners in the energy sector.

Social responsibility as "social license to operate"

Energy companies are perceived as social agents that force positive social changes, i.e., produce power for economic and societal development, invest in renewables and transform the energy sector, and help endusers change their behavior for energy conservation.

Ever since the 1970s the emphasis on business responsibility has been seen in the literature. Churchill's (1974) theories of social accounting along with Carroll's (1979) pyramid of corporate social responsibility with four types of responsibilities have laid down the foundations for today's practice of business social responsibility. Accordingly, socially responsible practice is perceived as a win-win strategy (Falck and Heblich, 2007) where a value-driven business creates value for the society and performs well financially. In line with this approach, social responsibilities become a business concept that represents a range of potential business returns from the government, partners, users, or the society at large. The most common explanation of business involvement in social responsibility is related to the positive business image and long-term profitability (Chen et al., 2018; Khojastehpour and Johns, 2014). Nonetheless, a business can perform effectively only if it has a "social license to operate" (Dunfee and Donaldson, 2011; Demuijnck and Fasterling, 2016). The idea behind the license is that in order for a business to be effective it needs tooperate not only in line with legal requirements but also act positively toward the well-being of the society.

Furthermore, it should be pointed out that "social license to operate" is commonly used in the context of confrontation or disapproval between business activities and societal norms. It also alerts that ignoring the latter and harming social well-being could harm business interests, too (Demuijnck and Fasterling, 2016; Morrison, 2014). Accordingly, transition to socially responsible business is slow and based on the hard lessons from irresponsible business practice. The case of Royal Dutch Shell PLC is probably the best illustration of this transition.

Case of Royal Dutch Shell PLC for societal license to operate

Today Shell is one of the biggest investors in new energies with plans to invest $2billion a year in renewable energy sources (Lempriere, 2018). It also joins forces with other companies to increase its commitment to serving new and underserved markets with hydrogen energy (Grayson and Hodges, 2017) and offshore wind energy at home in the Netherlands (Williams, 2019). Such environmentally responsible business decisions are complemented with social responsibility, which is already integrated into daily practice of the company together with social accountability. Namely, its network of around 100 community liaison officers integrates communities' feedback in their projects or assets and it enables the company to track their social performance (Shell Sustainability, 2018).

Current practice by Shell is based on the hard lessons the company learned through its long history of transnational business activities in oil-rich regions around the world. Probably the best known and most widely analyzed case is related to Ogoni in Nigeria, where Shell subsidiaries were accused of gross violations of human rights and blamed for environmental irresponsibility resulting in oil spills during their explorations in the 1990s. According to a comprehensive analysis provided by Yusuf and Omoteso (2016), the company had long legal proceedings at national as well as international courts. Since 1996 a number of cases were brought against Shell in the United States, which were followed by the Dutch Parliament public hearing on its oil giant's operations. In the end, the District Court of The Hague ruled the practice to hold subsidiaries for the harmful practices in foreign countries unfair. Finally, "the company was under pressure to respond to the concerns being expressed by many stakeholder groups including its stakeholders"(Blowfield, 2014) and a number of human rights organizations were involved in this process and acted against Shell. The company only partially admitted its liability out of court under the Nigerian law in 2011 and offered the sum of $83 million to those who had been affected by the oil spills during their explorations. The case damaged Shell's reputation and challenged its social license to operate. It comes as no surprise that now one of three strategic ambitions of Shell is oriented toward"strong license to operate" (our Strategy, 2019).

The concept of social license to operate is based on the stakeholder theory, which suggests that it is beneficial for a business to operate in a socially responsible way and contribute to the social well-being; otherwise, nonfinancial stakeholders might not support the company, which then may lose its image and customers. Hence, support from the society is central to the stakeholder theory and it accentuates that a business can continue to exist only if its core values resonate with the ones of the society where it operates (Blowfield, 2014).

However, the application of the theory in developing a socially responsible energy business is controversial. With regard to Blowfield (2014), such a business has to balance profit maximization and stakeholders' needs, making the definition highly challenging for multinational energy corporations as stakeholders' needs are incompatible in many cases. Thus, energy companies are forced to prioritize whose needs have to be met and whose needs are less important.

One of the most comprehensive publications in the area of social responsibility in the energy sector is a book by Yakovleva and Crowther (2005) that illustrates the consequences of imbalanced prioritization of stakeholders' needs. Based on the cases in the Russian Federation, specifically inthe regions of East Siberia and the country's Far East, Yakovleva and

Crowther (2005) assesses the role of the mining industries within main areas of corporate social responsibility. Research results showed that the mining industry takes full responsibility for its financial stakeholders and extends its economic responsibility toward the society it operates in, i.e., it makes a positive economic impact on the region and the society by contributing to the region's budget, employment, and social welfare at large. However, its overall social impact is negative as social responsibility is not integrated into main decisions and operations of the business. The mining industry makes a severe negative environmental impact as it pollutes river flows, disturbs forest and land resources, and uses nuclear exposures that create a danger to human health. Additionally, the industry's dominating effect in the region gives decision-makers the power to distribute the financial benefit in a socially unjust way, creating social inequality and social exclusion in the region. Due to harshness and the range of impacts caused, the mining industry is pressured by international organizations, national governments, employees, and local communities to improve their environmental and social performances (Yakovleva and Crowther, 2005). The case is a good illustration that the energy sector can be an issue not only for environmental but also for social reasons in the society if social responsibility is not of utmost importance for the business.

As we can see from the cases in oil and mining industries, governments and international organizations are big players in stimulating social responsibility in the energy business. Unfortunately, the power of both governmental and nongovernmental organizations is limited because of their conceptual attitude toward business social responsibility. Despite the fact that these organizations are seeking to stimulate social responsibility and encourage companies to integrate social responsible solutions into their decisions and operations, voluntary social responsibility remains in their rhetoric (Smaliukienė, 2005). For example, even though the International Labor Organization (ILO) sees social responsibility of any business as the central part of economic and social advancement, it promotes companies' volunteering contribution tothe society. Thus, social responsibility becomes a voluntary activity more akin to charity than to an integrated business solution resulting in the controversial practice of the business. Hence, social responsibility is perceived as an obligation of doing business (Dobrea and Găman, 2011) instead of being a natural part of it.

Integrating social responsibility into corporate strategy

Looking at the patterns inthe energy business in many countries, there is a huge shift in addressing the gap between social responsibility and corporate strategy. Indeed, there is a clear understanding among scholars and practitioners that the code of ethics, sponsorships and donations, and social reporting are too limited and too disconnected from the strategy (Dobrea and Găman, 2011; Galbreath, 2009; Smaliukienė et al., 2017). Thus, it remains a question as to why there are still obstacles when integrating social responsibility into a business strategy. Engert (2016) provide probably the most reasonable explanation. Based on their content analysis of over 100 scientific journal articles, they conclude that social responsibility is rooted in sustainability-related topics instead of being integrated into strategic management research streams and strategic management decision practice. This means that social responsibility remains an obligation rather than a natural part of any business.

To overcome this main challenge, Grayson and Hodges (2017) step further and develop a framework of corporate social opportunity. It is composed of seven steps that transform a business from a "have-to-do"-based approach to "want-to-do" business mentality and integrates social responsibility into the core of corporate strategy (Fig. 14.1).

According to Grayson and Hodges (2017), the first thing a company has to do is identify the triggers that impact its business (step 1). They are usually prompted by stakeholders. In the case of the energy business, the most challenging thing is the intersection of regulation, societal expectations, and commerce. Thus, environmental issues are only one of the possible triggers for the energy business as social issues are equally important. For example, energy poverty and energy inaccessibility are social concerns as inefficient energy management leaves low-income users suffering from cold in their houses even in robust economies (Reyes et al., 2019); or communities are left without access to affordable energy sources and stay disconnected from the knowledge economy in developing economies (Ahmed, 2010). Once the trigger is identified, a company then identifies potential business strategies (step 2). It could be a new business model that transforms the entire value chain of the energy supply, or it could be a new market for underserved energy users. When a potential business strategy is selected, an organization must convert it into a business case (step 3). The best practice is to start with corporate goals and other organizational considerations that integrate socially responsible decisions into daily operations (Khalili, 2011) (step 4). The case of the company First Solar is a good example how social responsibility is integrated into a business strategy by identifying social issues and converting them into a business case.

FIGURE 14.1   Integrating social responsibility into a business strategy. *Based on Grayson, D., Hodges, A., 2017. Corporate Social Opportunity!: Seven Steps to Make Corporate Social Responsibility Work for Your Business, Taylor & Francis.*

Case of First Solar: a link between social responsibility and corporate strategy

In 2008 First Solar was the first solar panel company that managed to decrease its manufacturing cost to $1 per watt (Kinnear, 2008). It was a considerable achievement in fighting energy accessibility in developing economies. First Solar identified the social issue, shaped their innovation strategy, and reallocated their resources in order to exploit a business trigger. As a result, their priority was to provide energy for the markets that had a compelling need for alternative energy sources (First Solar Sustainability, 2019), thus making First Solar one of the best performing renewable energy producers in the world (Thomson Reuters top, 2019). $1 per watt as a business trigger marked its first business development stage.

The second business development stage of First Solar was related to the environmental issue: one of the biggest issues in the photovoltaic (PV) solar energy business is recycling as "growing PV panel waste presents a new environmental challenge" (Weckend et al., 2016). The company jumped ahead of this challenge and introduced their recycling service 8 years before the directive for electronic waste was established in the European Union. The initiative was based on the goal to minimize the impact of environmental sensitivity and maximizere source recovery. As a result, PV recycling service became a complementary cost-effective part of the main business at First Solar as up to 90% of semiconductor material and glass are recycled using state-of-the-art recycling facilities and their expertise in PV manufacturing (Recycling, 2019). Recycling solves more than just environmental issues. The company's recycling plants in the United States and Germany contribute considerably to job creation (Field, 2018), thus solving social unemployment problems. In recent years, the company has made a further step forward and introduced mobile recycling modules that mitigate unnecessary shipping and createa new potential revenue source at the same time, creating new jobs in developing countries (Corporate, 2019).

As a consequence of this long-term investment and focus on social and environmental triggers, the company has become one of the industry leaders in the energy sector not only for its economic performance or investor confidence but also for social responsibility. According to Thomson Reuters top (2019), the company is ranked higher than average in the sector for protecting public health and respecting business ethics as well as human rights, for focusing on the effectiveness toward job satisfaction and work conditions, as well as integrating economic, social, and environmental decisions into daily operations.

As it can be seen from the case of First Solar, if socially responsible goals are set reasonably, putting a strategy into action is a rather natural process as organizational values, leadership, and systems contribute to it. According to Grayson and Hodges (2017), a very important step in the framework of "want-to-do" social responsibility is related to resource integration and gathering (step 5). A company must assess the resources needed for strategy implementation as well as identify resource gaps and find potential sources through cooperation and partnership. In the energy sector, we can list numerous national and international initiatives when energy companies share their resources with other ones for socially responsible business initiatives (see Monni et al., 2017; Mezher et al., 2010). Moreover, as stakeholders' engagement (step 6) is crucial, a company has to assess its strategy's impact on stakeholders and vice versa. Notwithstanding the fact that internal and external stakeholders may have very different expectations, they should be closely and formally involved in business activities and their supervision (Hopkins, 2009; Smaliukiene, 2007). Finally, the last step in the proposed framework is measuring and reporting that show how effective achieved goals are. Evaluation includes subjective measures and facilitates the discussion with stakeholders involved in the strategy implementation (Khalili, 2011; Tvaronavičienė, 2018), but financial business performance is of particular importance. The integration of social responsibility into corporate strategy enables a business to look for advanced solutions that would be both cost-effective and valuable for the society's well-being.

Demand side: the power of social marketing in promoting sustainable energy use

The primary aim of social marketing is to change a destructive behavior and convert it into a constructive one. Thus, social marketing is capable to change the demand side while transforming behavior of energy users. Numerous research has already demonstrated the value of social marketing in promoting energy efficiency and in changing users' behavior (Anda and Temmen, 2014; Gordon et al., 2018). Still, household energy efficiency could contribute up to 25% of carbon emissions reductions (UNEP, 2017), but for this purpose extensive communication with the users is necessary. As it was stated in the UN Emissions Gap Report 2018 (United Nations Environment Program, 2018), there is a need to increase public awareness and the public needs to be informed about the greater impact of their behavior on cleaner air and human health. Hence, social intervention that changes understanding and behavior of a population requires more attention.

The irrational energy user

Taking users' needs and wants into consideration, many scholars, who perceive the public as being rational, emphasize the economic benefit of energy efficiency. This perception, however, is not enough. On the one hand, economic benefit is a very important stimulus for behavioral changes. When social marketing campaigns take place, they highlight the economic value of participating in a specific activity, for instance, when using energy star–qualified bulbs, installing PV panels in residential buildings, or using other greener forms of energy for lighting and heating. The benefit (Evans et al., 2014) or valueinbehavior in using energy efficiently (Butler et al., 2016) is based on the understanding that the public is rational and each household will act in its economic self-interest. Consequently, economic values have become the core message in social marketing campaigns to achieve energy efficiency.

On the other hand, there are a number of examples of inefficient residential energy efficiency campaigns that ignored the value of social interaction and relied solely on information delivery regarding the economic value of the new behavior. The cases (see McKenzie-Mohr, 2000) provide a range of evidence on nonsignificant impact of energy efficiency campaigns. In economic literature this phenomenon is called the energy efficiency paradox, when users neglect cost-effective opportunities and do not take logical measures at current energy prices to decrease their spending on energy (Ramos et al., 2015). It is clear that information about economic benefit is not enough to change users' behavior. Therefore, a more sophisticated benefit-focused social marketing approach is needed as different benefits are of different importance for individual users. The perception of what benefit is can vary. It could be willingness to cut energy bills or minimizing energy poverty, increasing energy security, or fighting climate change. Taking the spectrum of needs into consideration, we review the principles and tools of social marketing in promoting sustainable energy use.

Retooling social marketing to promote sustainable energy use

Behavioral change toward sustainable energy use requires retooling of social marketing, as when social marketing is advocating energy efficiency, it is usually associated with reducing consumption and decreasing demand for unsustainable energy sources. However, it is important to realize that the intention to "reduce" and "decrease" contradicts the common culture of consumerism in the globalized world; therefore, the tools for this purpose have to be revised.

The potential of social marketing in promoting sustainable energy use lies in the domain of traditional marketing, but instead of selling goods and services it changes the behavior to increase the well-being of households and communities. Probably the most adopted approach to using social marketing tools for sustainable energy use has been developed by McKenzie-Mohr (2000). According to him, social marketing can be very effective if it is communitybased instead of being an information-intensive campaign. McKenzie-Mohr's framework of the former enhances energy efficiency in several steps. First McKenzie-Mohr (2000) suggests selecting the behavior we need to change. Second, following the exchange theory, he proposes identifying barriers and benefits of the new behavior and designing a strategy to remove the obstacles to reach the goal. Finally, strategy is piloted in a small segment of society. McKenzie-Mohr's (2000) framework is simple and practice-oriented. For this reason it is used for residential energy efficiency programs by the US Department of Energy (2019). However, it is worth mentioning that McKenzie-Mohr's (2000) approach is oriented exclusively toward small communities; therefore, the vital three-stage marketing research process (segmentation, targeting, and positioning) is excluded.

To overcome this limitation, the principles of social marketing have to be integrated. According to Kotler and Lee (2016), thus social marketing is different from the commercial one, the former has to follow the principles of the latter in changing the behavior for societal gain. They provide six main principles for any social marketing (Kotler and Lee, 2016):

— Userorientation. Orienting all marketing activities toward the needs and wants of individuals to change their behavior.
— Exchange is the main theoretical concept. The users must perceive the value of changing their behavior.
— Marketing research is carried out from the very beginning until the very end. An effective strategy is developed only if specific needs and wants of a target audience are understood and reflected in the entire process of behavioral change.
— Marketing decisions have to be different for different target audiences due to their specific needs and wants, thus they are divided into segments.

- Activities are included in a marketing mix. The strategy is implemented with an integrative approach and activities are not limited to persuasive communication only.
- Results are measured and used for improvement. Social marketing is continuously improving its performance based on the feedback.

Very similar principles are presented by Peattie and Peattie (2009). Additionally, the authors emphasize not only the behavioral changebut also the behavioral maintenance when social marketing goes beyond decreasing consumption. They stress the importance of adoption and maintenance of significantly different lifestyle.

Following these viewpoints, the redesigned social marketing process consists of five stages: (1) selecting the behavior, (2) user orientation, (3) exchange, (4) marketing mix: elements of intervention, and (5) measuring behavior change for energy transformation (Fig. 14.2). The process integrates three vital elements of marketing research as user orientation is composed of a three-stage marketing research process including user segmentation, targeting, and positioning.

Step 1: selecting behavior for sustainable energy use

Behavioral is the final goal of any energy efficiency campaign and the starting point when considering social marketing. According to Rangan and Karim (1991), social marketing is about "changing attitudes, beliefs, and behaviours of individuals or organizations for a social benefit < … >and the social change is the primary purpose of the campaign." Even though the change is the backbone of social marketing, it is essential to note that it is neither a donation nor a sacrifice, it is rather a conscious participation in the process of exchanging costs and benefits.

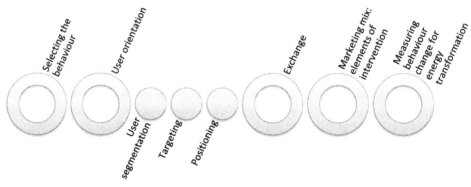

FIGURE 14.2   Elements of social marketing for sustainable energy use. *Based on Dibb, S., 2014. Up, up and away: social marketing breaks free. Journal of Marketing Management 30, 1159–1185. https://doi.org/10.1080/0267257X.2014. 943264, Kotler, P., Lee, N., 2016. Social Marketing: Changing Behaviors for Good/Nancy R. Lee, University of Washington and Social Marketing Services, Inc, Philip Kotler, Kellogg School of Management, SAGE, Los Angeles; Bird, S., 2010. Benchmark Criteria for Social Marketing: Bristol Social Marketing. Centre Spotlight on Social Marketing #2; Peattie, K., Peattie, S., 2009. Social marketing: a pathway to consumption reduction? Journal of Business Research 62, 260–268. https://doi.org/10. 1016/j.jbusres.2008.01.033.*

There are many examples of deliberate and targeted energy efficiency marketing campaigns organized by municipalities with measurable objectives of behavioral change. Still, the most recent and prominent example of behavioral changes in energy use in a city took place in Amsterdam, the Netherlands.

Case of behavioral changes in energy use in Amsterdam

Although Amsterdam's strategy on circular economy is not finished, the city's Circular Innovation Program already won the World Smart City Award in 2017 (Benchmarking Study, 2018). This award was given for the platform that accelerates the city's transition. Implementing a variety of projects and intensive marketing campaigns of a smart city with a circular economy, Amsterdam's municipality aims to redesign 20 product or material chains (Amsterdam's Circular, 2018). It is expected that the majority of new material chains involve technological as well as behavioral results that would lead to added monetary value in millions of euro, growth of employment rate, as well as material savings and reduction in CO_2 emissions (Circular Amsterdam, 2018).

The cooperation between policy content experts and communication and marketing companies is the key of the marketing approach of the Amsterdam's Circular Innovation Program (Benchmarking Study, 2018). As a result, there is a comprehensive approach toward main marketing elements such as product, place, promotion, and personnel. There are three impulses when it comes to marketing of this program: "linking projects […] and events to city marketing objectives; and starting new marketing projects and include those which already exist with an integral approach" (Benchmarking Study, 2018). The end goal of the city is to implement a circular economy that "requires rethinking market strategies and models that encourage competitiveness in different sectors and the responsible consumption of natural resources"(Circular Economy, 2018). This shift would change production processes and consumer behavior as the program not only stimulates energy savings and investments into solar energy but also tries to transform the mind-set of the residents. New solutions for energy saving (including food and water cycles) and new forms of renewable energy (using innovative collection and sorting of waste, etc.) are based on behavioral changes of Amsterdam residents (Amsterdam's Circular, 2018; Circular Economy, 2018; Amsterdam Smart City, 2019). What is important, values and behavior of the local community were perceived as vital by program implementers.

As can be seen from these and other cases, projects on energy efficiency integrate research, best practices, and theories of social marketing to understand attitudes as well as the social context in which the demanded behavioral change has to occur. In such projects, destructive behavior is changed into the constructive.

Step 2: user orientation

The second step in our social marketing in fostering demand for sustainable energy—user orientation—deals with three stages of marketing research process, i.e., segmentation, targeting, and positioning. This essential process helps operationalize the concept of user-oriented and puts marketing theory into practice. Although these three stages have been developed (and are actively used) as concepts of commercial marketing with the intention to sell the goods, nowadays they have become a vital part of behavioral change interventions for social purposes

(Dibb, 2014). Their application for effective energy use is rather straightforward; however, it is not as wide as in business. Let's discuss the meaning and application of these three concepts.

Segmentation

Energy users' segmentation divides a large population into groups according to their shared values, wants, and needs. According to segmentation theory, people in the same group are likely to respond to behavioral interventions similarly. Typically, any population is segmented according to demographic characteristics (such as age, gender, ethnicity, etc.); however, as technologies of the Internetera shape everyday behavior, energy users' segmentation is based more on attitudes and lifestyles than on wants and needs. As a result, segmentation of energy users identifies one or more segments in the target audience (Thøgersen, 2017) as there is an in-depth understanding that it is impossible to be effective across all the population. It, therefore, has to be segmented into groups and only a few segments can be targeted with social marketing mix.

As already mentioned, social marketing adopts the methods of commercial marketing, yet its purpose is very different. In business, the same segments are targeted with a variety of accompanying products they might prefer to use. In contrast, social marketing targets behavior only with one goal and this goal is usually associated with the decrease in consumption. There are a few segmentation approaches developed to understand how a population can be segmented according to its attitude toward the environment. As an example, Table 14.1 presents segmentations of UK and US markets. According to these segmentation examples, energy users can be divided into three large groups based on their attitude toward the environment—environmentalists, the environmentally concerned, and the disinterested. How large these groups are and how many segments compose each group depends on values of the society at large. As we can sees fromthe UK and US segmentation results, UK society has more segments that are environmentalist and environmentally concerned. Meanwhile US society's segmentation identifies more unique segments that are indicated as disinterested in the environmental impact of their consumption.

Target audience

One or a few target audiences are selected after a population is divided into groups according to demographic, value-based, lifestyle, and behavioral criteria. This step requires consideration regarding the potential efficiency of each segment. According to the mainstream marketing authors, any target audience has to meet several criteria. These criteria vary from author to author and organizations have to choose the most important ones according to their marketing objectives and measurement benchmarks (Dietrich et al., 2016; Sarstedt and Mooi, 2014). In spite of differing views on targeting, there is a common agreement in mainstream as well as in social marketing literature about three most important criteria: target audience has to be large enough to make marketing program effective in scale, each segment of the target audience needs different benefits, and target audience is accessible with marketing messages.

To continue the examination of UK and US segmentation examples, segmentation divides population into nearly equal and substantial groups according to their attitude toward

TABLE 14.1   Segmentation of UK and US populations according to attitude toward environment and climate change.

Segments of the UK population	Segments of the US population	Segment description
Positive greens Waste watchers	Liberal greens	Environmentalists: are very worried about environmental issues and feel interconnected with the nature, try to conserve whenever they can
Concerned consumers Sideline supporters	Outdoor greens Religious greens	Environmentalists: are very worried about environmental issues, environment-friendly behavior makes them feel better
Cautious participants Long-term restricted	Middle-of-the-roaders	Environmentally concerned: are generally concerned about the environment, but behave environmentfriendly only because of constraints
Stalled starters Honestly disengaged	Homebodies Disengaged Outdoor browns Religious browns Conservative browns	Disinterested: they tend toward apathy when it comes to environmental issues, environmental issues do not resonate with them

Based on the data from Public Opinion and the Environment: The Nine Types of Americans. Yale School of Forestry & Environmental Studies, 2015. and Defra, January 2008. A Framework for Pro-environmental Behaviours.

environmental issues and willingness to act on behalf of the environment (Fig. 14.3). The UK's case in particular illustrates good practice in targeting an audience as each segment not only differs in terms of needs but they also were reached with effectively selected marketing messages. According to Giorgi et al. (2016) some audiences received "only information, whereas others received a mixture of information and activities, depending on the target and existing behaviors and attitudes." Hence, we see not only segmentation but also targeting, which is based on segmentation results.

Positioning against competing alternatives

Thermal comfort, car dependency, and other lifestyle norms compete against behavioral interventions that would lead to sustainable energy use. It is the issue that social marketing is trying to solve by using positioning, i.e., an act that distinguishes the offer from the competing alternatives, makes it even more attractive, and provides inspiration and parameters as to "how […] the desirable behavior [has] to be seen by the target audience" (Kotler and Lee, 2016). Asocial marketing positioning statement shows how to overcome the barriers for the new behavior. The most powerful positioning is based on the message "energy-saving" (Ben and Steemers, 2018), but the message itself has to integrate different values for different segments of any target audience based on their needs and wants. Additionally, Giorgi et al. (2016) suggest that the positioning statement has to provide real examples that show how

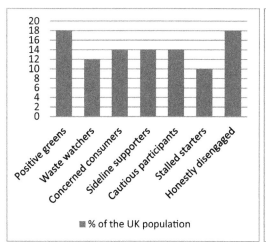 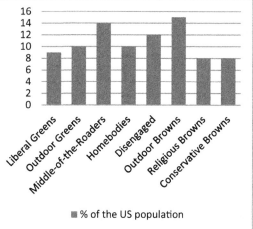

FIGURE 14.3 The distribution of UK and US populations according to the attitude toward environment and climate change. *Based on the data from Public Opinion and the Environment: The Nine Types of Americans. Yale School of Forestry & Environmental Studies, 2015. and Defra, January 2008. A Framework for Pro-environmental Behaviours.*

others are doing. Once this statement is developed, specific strategies as to how to position new demanded behavior are developed and implemented in the stage of the marketing mix.

Step 3: exchange

As was already discussed, voluntary exchange is a mainstay of social marketing. According to the exchange theory, social marketing has to offer users benefits in exchange for their behavioral change. Giorgi et al. (2016) point out that they agree to change their behavior toward more sustainable energy use in exchange for lower cost, convenience, and lifestyle choice. Respectively, marketers have to consider the alternatives as to what will motivate users to change their behavior and what should be offered as a value in exchange. The only concern is that the meaning of value is different for different segments. The exchange in energy consumption can be motivated by self-interest, social norms, or concern for the common good. While environmentalists "can leave comfort and cleanliness behind in the pursuit of a contemporary natural purity" (RCZM, 2019), the disinterested change their behavior solely because of cost saving. This, consequently, leads to different proposals for behavioral change. They can be very simple or very complex depending on the segment and its willingness to change, i.e., adjust the temperature, use more efficient vehicles, avoid unnecessary flights, manage energy better, recycle more, waste less food, etc. (Jonkhof and-van der Kooij, 2019).

Giorgi et al. (2016) provide a comprehensive list of proposals how to offer value to a different target audience based on their attitudes and preferences. For environmentalists, they suggest specific measures that would help incorporate changes into their lifestyles, while for the disinterested cost saving has to be the key entry point to stimulate their behavioral change. Despite different attitudes and preferences (Ramos et al., 2015), research results show that all segments are more willing to participate in exchange when its value is clear.

Step 4: marketing mix

The marketing mix is the core concept adopted from commercial marketing for the behavioral intervention. While commercial marketing mix is created of 4Ps (product, price, place, and promotion) or 7Ps (people, product, price, promotion, place, process, and physical evidence) (Lovelock et al., 2015), social marketing mix contains eight elements (product/service, price, place, promotion, public, partnership, policy, purse strings) (Peattie and Peattie, 2016). This way, social marketing solves more challenging tasks than any business.

Despite following the eight elements of social marketing mix, the application of marketing tools and techniques remains problematic in promoting sustainable energy. Accordingly, Gordon et al. (Galbreath, 2009) suggest moving away from the traditional marketing mix and propose to abolish the elements that come from commercial marketing. It is a reasonable proposal as even the first element—product—is difficult to interpret. Social marketing, as it was mentioned previously, is oriented toward a new effective behavior that changes the lifestyles of individuals or communities; therefore, when using this term one can mean a conscious energy usage at home or house renovation as well as a new tax reform that introduces a tax for carbon emissions. Peattie and Peattie (2016) suggest using the concept "social marketing proposition" instead. Additionally, they require a clear narrative on how the behavioral change would benefit the users. One of the most interesting cases of such social marketing proposition is presented by a fossil fuel subsidy reform in Iran. The government's subsidy reform on energy was "carefully prepared by clear government communication through various channels, such as websites and hotlines to answer questions about the reform" (United Nations Environment Program, 2018). The "proposition" in this case was "country's economic competitiveness by creating more jobs and using its oil resources more efficiently" (Atansah et al., 2019).

The link between "price" and "cost" is no less challenging. Some scholars propose to change the element "price" into "cost" (Dibb, 2014) as the latter can deal with both monetary costs as well as the costs of inconvenience. Other elements of social marketing mix emphasize the interrelationship and impact of a community and society on a person's behavioral change. Hence, Gordon et al. (Galbreath, 2009) convert social marketing mix into a new model with five interrelated elements: promotion, nudge, rewards and exchanges, service and support, and relationships and communities. The model enhances the understanding of communities and other stakeholders and their impact on behavioral change. Moreover, it resonates with other contemporary approaches in marketing and behavior such as the theory of value cocreation and value coproduction (Vargo et al., 2008; Osborne, 2017; Smaliukiene et al., 2014). Most importantly, new approaches on marketing mix stress the importance of users' motivation as it directs their behavior.

Step 5: measuring behavioral change for sustainable energy use

Evaluation of behavioral changes can generate strong implications about the impact of social marketing and confront criticism regarding the value of the intervention. However, many social marketing programs are either evaluated poorly or not at all (Grier and Bryant, 2005). An example of existing good practice is the case of the city of Macau, China, where social marketing results were quantified.

Case of behavioral changes in energy use in the city of Macau

Macau was challenged by a steady increase in energy consumption over the last 25 years and decided to focus on building-use energy, especially air-conditioning and lighting systems. Energy-saving publicity campaigns were conducted in schools and for the general public as well as for the business sector. The campaign increased public awareness and knowledge, which resulted in the population acting more responsibly. As a result, energy-saving behavior became very common in the daily life of city residents and businesses. The measurable impact was most important as behavioral changes affected energy consumption; energy intensity per million electricity meter operators decreased four times and the energy consumption per capita decreased by more than 10%.

Adapted from Song, Q., Li, J., Duan, H., Yu, D., Wang, Z., 2017. Towards to sustainable energy-efficient city: a case study of Macau. Renewable and Sustainable Energy Reviews 75, 504–514. https://doi.org/10.1016/j.rser.2016.11.018.

Since social marketing is often a continuing activity that runs over long periods of time, it is not easy to do so. That is why impact evaluation looks at the effect rather than the outcome of each program. Alternative evaluation, on the other hand, provides important insights while observing behavioral changes instead of measuring energy saved. While impact evaluation deals with user-specific information that is collected through surveys, interviews, consumer panels, opinion polls, feedback from program participants, etc. (Global CCS Institute, 2020), actual evaluation or the cost-effectiveness of the programs are very difficult as social marketing aims to change the behavior of energy users.

Concluding remarks

While analyzing the academic discourse, we found that social responsibility and social marketing are powerful sources for transforming energy production and its use toward sustainability. The former in the context of energy transformation represents the supply side of energy management, while the latter is related to the demand side. Thus, both of them are two complementary discourses of the same issue.

Business social responsibility is associated with the always increasing pressure on controlling climate change while the energy business is liable for the supply side of sustainable energy. In other words, energy business is directly or indirectly responsible for the social issues arising from extreme weather, food security, water supply, etc. Therefore, as it is not enough for the energy business to see social responsibility as only a social license to operate, an integrative approach that would integrate social responsibility into corporate strategy is needed. More specifically, it is no longer sufficient to refer only to the level of acceptance of stakeholders. This is especially important when taking the diversity of stakeholders from multinational energy corporations into consideration. Working only under a social license to operate is becoming a highly challenging task since energy companies have to decide which stakeholders are more important as well as address their needs in order of priority. Consequently, wrong prioritization of the needs may result in the loss of the social license when nonfinancial stakeholders turn away from the company.

Accordingly, in this analysis we propose integrating social responsibility into corporate strategy instead of focusing only on the social license. Based on the analyzed cases, it is evident that the framework of corporate social opportunity is a realistic way to do exactly that. More specifically, regardless of the size of the energy company, it has to move from a "have-to-do"-based approach to "want-to-do" business mentality and integrate social responsibility into the core of its corporate strategy, which can happen only when social challenges are perceived as business triggers and are transformed into the corporate strategy. For example, it could be a new business model that transforms the entire value chain of the energy supply, or it could be a new market for underserved energy users. Whatever social challenge a company is dealing with, what is paramount is that its corporate strategy is at the same time oriented toward the society's well-being and business prosperity. This claim is not just a theoretical concept as business can balance its dual objectives in a highly efficient way and supply energy from highly sustainable sources.

The demand side of energy sustainability is greatly affected by the behavior of household energy users. By examining the impact of social marketing on energy sustainability, we show its power in changing their destructive behavior. Our discourse and case studies also illustrate how social marketing can change the demand side of energy. By examining the demand side of energy it is important to realize that the behavior of household users is irrational and their energy consumption is often driven by their lifestyle and values rather than by economic-rational motives. Thermal comfort, car dependency, and other lifestyle norms compete against behavioral interventions that would lead to sustainable energy use. Taking this irrationality and complementing contemporary theoretical advantages into consideration, we suggest retooling social marketing for sustainable energy use. The new tools are integrated into the framework that transforms the selected destructive behavior into a sustainable one. First, we suggest using the user orientation concept that divides the society into three groups based on their attitude toward environmental issues, i.e., environmentalist, the environmentally concerned, and the disinterested. Then, by applying the exchange theory, we point out what would motivate users to change their behavior and what should be offered as a value in exchange. When considering the latter, segmentation results are likely to have the strongest impact on this decision. Finally, we propose to reconsider social marketing mix and reframe it in accordance with energy-user behavior matters.

The analysis provided some interesting insights related to the link between social sustainability and social marketing in developing energy sustainability. Although these two social practices represent and shape supply and demand, both of them tackle the same social issues. In both cases the issues are transformed into the strategy that provides the base for targeted actions. As for the results, they are measured from the perspective of societal well-being; in addition, they have to be economically feasible to eliminate the short-term impact.

Our academic discourse provides a more complete picture of two practically interlinked but theoretically separated discourses. Knowing how social responsibility is manifested as the supply side of sustainable energy and how social marketing can shape the demanded behavior of household energy users, further research has to continue to broaden the scope of these two fields of research while at the same time looking for more links between them.

References

Ahmed, M., 2010. Economic dimensions of sustainable development, the fight against poverty and educational responses. International Review of Education 56, 235–253. https://doi.org/10.1007/s11159-010-9166-8.

Amsterdam Smart City, Circular City. https://amsterdamsmartcity.com/themes/circular-city.

Amsterdam's Circular Innovation Programme Changing Behavior – "Google" paieška. https://www.google.com/search?q=Amsterdam%E2%80%99s+Circular+Innovation+Programme+changing+behavior&ei=sW0WXJ6FEbHqrgSsw4DABw&start=10&sa=N&ved=0ahUKEwjeoZna06TfAhUxtYsKHawhAHgQ8NMDCJMB&biw=1680&bih=882.

Anda, M., Temmen, J., 2014. Smart metering for residential energy efficiency: the use of community based social marketing for behavioural change and smart grid introduction. Renewable Energy 67, 119–127. https://doi.org/10.1016/j.renene.2013.11.020.

P.Atansah, M.Khandan, T.Moss, A.Mukherjee, J.Richmond, When Do Subsidy Reforms Stick? Lessons from Iran, Nigeria, and India. https://www.cgdev.org/publication/when-do-subsidy-reforms-stick-lessons-iran-nigeria-and-india.

Ben, H., Steemers, K., 2018. Household archetypes and behavioural patterns in UK domestic energy use. Energy Efficiency 11, 761–771. https://doi.org/10.1007/s12053-017-9609-1.

Benchmarking Study: Amsterdam –Branding at Its Best. A Well Co-ordinated, Managed and Marketed Place: Future Place Leadership. https://futureplaceleadership.com/wp-content/uploads/2018/04/Case-Amsterdam-by-Future-Place-Leadership.pdf.

Bird, S., 2010. Benchmark Criteria for Social Marketing: Bristol Social Marketing. In: Centre Spotlight on Social Marketing #2.

Blowfield, M., 2014. Corporate Responsibility a Critical Introduction, third ed. Oxford University Press, Oxford, New York.

Butler, K., Gordon, R., Roggeveen, K., Waitt, G., Cooper, P., 2016. Social marketing and value in behaviour? Journal of Social Marketing 6, 144–168. https://doi.org/10.1108/JSOCM-07-2015-0045.

Carroll, A.B., 1979. A three-dimensional conceptual model of corporate performance. Advances in Magnetic Resonance 4, 497–505. https://doi.org/10.5465/amr.1979.4498296.

Chen, Y.-C., Hung, M., Wang, Y., 2018. The effect of mandatory CSR disclosure on firm profitability and social externalities: evidence from China. Journal of Accounting and Economics 65, 169–190. https://doi.org/10.1016/j.jacceco.2017.11.009.

Churchill, N.C., 1974. Toward a theory for social accounting. Sloan Management Review 15.

Circular Amsterdam: A Vision and Action Agenda for the City and Metropolitan Area, 2018. https://www.oecd.org/governance/observatory-public-sector-innovation/innovations/page/circularamsterdamavisionandactiona-gendaforthecityandmetropolitanarea.htm#tab_description.

Circular Economy in Cities: White Paper, 2018. World Economic Forum.

Corporate. http://www.firstsolar.com/en/PV-Plants/Corporate.

Defra, A., January 2008. Framework for Pro-environmental Behaviours.

Demuijnck, G., Fasterling, B., 2016. The social license to operate. Journal of Business Ethics 136, 675–685. https://doi.org/10.1007/s10551-015-2976-7.

Dibb, S., 2014. Up, up and away: social marketing breaks free. Journal of Marketing Management 30, 1159–1185. https://doi.org/10.1080/0267257X.2014.943264.

Dietrich, T., Rundle-Thiele, S., Kubacki, K., 2016. Segmentation in Social Marketing: Process, Methods and Application. Springer Singapore.

Dobrea, R., Găman, A., 2011. Aspects of the correlation between corporate social responsibility and competitiveness of organization. Revista: Economia, Seria Management 14, 236.

Druckman, A., Jackson, T., 2016. Understanding households as drivers of carbon emissions. In: Clift, R., Druckman, A. (Eds.), Taking Stock of Industrial Ecology. Springer Open, Cham, pp. 181–203.

Dunfee, T.W., Donaldson, T., 2011. Social contract approaches to business ethics: bridging the "is-ought" gap. In: Frédérick, R. (Ed.), A Companion to Business Ethics. Blackwell Publishers Ltd; Wiley-VCH, Malden, MA, pp. 38–55.

Engert, S., 2016. Exploring the integration of corporate sustainability into strategic management: a literature review. Journal of Cleaner Production 112, 2833–2850.

Evans, W.D., Pattanayak, S.K., Young, S., Buszin, J., Rai, S., Bihm, J.W., 2014. Social marketing of water and sanitation products: a systematic review of peer-reviewed literature. Social Science & Medicine 110, 18–25. https://doi.org/10.1016/j.socscimed.2014.03.011.

Falck, O., Heblich, S., 2007. Corporate social responsibility: doing well by doing good. Business Horizons 50, 247–254. https://doi.org/10.1016/j.bushor.2006.12.002.

Field, K., 2018. First Solar Breaks Down Its Plans for Solar Module Recycling #SPI2018. https://cleantechnica.com/2018/12/04/first-solar-breaks-down-its-plans-for-solar-module-recycling-spi2018/.

First Solar Sustainability Report. http://www.firstsolar.com/-/media/First-Solar/Sustainability-Documents/First-Solar_SustainabilityReport.ashx.

Galbreath, J., 2009. Building corporate social responsibility into strategy. European Business Review 21, 109–127. https://doi.org/10.1108/09555340910940123.

Giorgi, S., Fell, D., Austin, A., Wilkins, C., 2016. Public Understanding of the Links between Climate Change and (I) Food and (II) Energy Use (EV0402): Final Report.

Global CCS Institute, 2020. Case Studies on Innovative Communication Campaign Packages on Energy Efficiency. Global CCS Institute. https://hub.globalccsinstitute.com/publications/energy-efficiency-recipe-success/case-studies-innovative-communication-campaign-packages-energy-efficiency.

Gordon, R., Butler, K., Cooper, P., Waitt, G., Magee, C., 2018. Look before you LIEEP. Journal of Social Marketing 8, 99–119. https://doi.org/10.1108/JSOCM-04-2016-0017.

Gordon, R., Dibb, S., Magee, C., Cooper, P., Waitt, G., 2018. Empirically testing the concept of value-in-behavior and its relevance for social marketing. Journal of Business Research 82, 56–67. https://doi.org/10.1016/j.jbusres.2017.08.035.

Grayson, D., Hodges, A., 2017. Corporate Social Opportunity!: Seven Steps to Make Corporate Social Responsibility Work for Your Business. Taylor & Francis.

Grier, S., Bryant, C.A., 2005. Social marketing in public health. Annual Review of Public Health 26, 319–339. https://doi.org/10.1146/annurev.publhealth.26.021304.144610.

Hopkins, M., 2009. Corporate Social Responsibility and International Development: Is Business the Solution? Routledge, London.

E.Jonkhof, E.van der Kooij, Towards the Amsterdam Circular Economy. https://assets.amsterdam.nl/publish/pages/580742/towards_the_amsterdam_circular_economy_web.pdf (accessed 13 January 2019).

Khalili, N.R., 2011. Practical Sustainability: From Grounded Theory to Emerging Strategies. Palgrave Macmillan, Basingstoke.

Khojastehpour, M., Johns, R., 2014. The effect of environmental CSR issues on corporate/brand reputation and corporate profitability. European Business Review 26, 330–339. https://doi.org/10.1108/EBR-03-2014-0029.

Kinnear, J., 2008. Information Review of First Solar | Solar Power Authority. https://www.solarpowerauthority.com/first-solar/.

Kotler, P., Lee, N., 2016. Social Marketing: Changing Behaviors for Good/Nancy R. Lee. University of Washington and Social Marketing Services, Inc, Philip Kotler, Kellogg School of Management, SAGE, Los Angeles.

Lempriere, M., 2018. Shell Moves into Renewables: Big Splash or a Dip in the Water? https://www.power-technology.com/features/shell-moves-renewables-big-splash-dip-water/.

Lovelock, C.H., Patterson, P., Wirtz, J., 2015. Services Marketing: An Asia-Pacific and Australian Perspective, sixth ed. Pearson Australia, Frenchs Forest, N.S.W.

McKenzie-Mohr, D., 2000. Fostering sustainable behavior through community-based social marketing. American Psychologist 55, 531–537.

Mezher, T., Tabbara, S., Al-Hosany, N., 2010. An overview of CSR in the renewable energy sector: Examples from the Masdar Initiative in Abu Dhabi. Management of Environmental Quality: An International Journal 21 (6), 744–760.

Monni, S., Palumbo, F., Tvaronavičienė, M., 2017. Cluster performance: an attempt to evaluate the Lithuanian case. Entrepreneurship and Sustainability Issues 5, 43–57.

Morrison, J., 2014. The Social License: How to Keep Your Organization Legitimate. Palgrave Macmillan, Basingstoke.

Osborne, S.P., 2017. From public service-dominant logic to public service logic: are public service organizations capable of co-production and value co-creation? Public Management Review 20, 225–231. https://doi.org/10.1080/14719037.2017.1350461.

Our strategy. https://www.shell.com/investors/shell-and-our-strategy/our-strategy.html.

Parrett, D., 2018. The Biggest Investors in Renewable Energy Will Shock You. https://moneymorning.com/2018/10/10/chart-the-biggest-investors-in-renewable-energy-will-shock-you/.

Peattie, K., Peattie, S., 2009. Social marketing: a pathway to consumption reduction? Journal of Business Research 62, 260−268. https://doi.org/10.1016/j.jbusres.2008.01.033.

Peattie, S., Peattie, K., 2016. Ready to fly solo?: reducing social marketing's dependence on commercial marketing theory. Marketing Theory 3, 365−385. https://doi.org/10.1177/147059310333006.

Public Opinion and the Environment: The Nine Types of Americans, 2015. Yale School of Forestry & Environmental Studies.

Ramos, A., Gago, A., Labandeira, X., Linares, P., 2015. The role of information for energy efficiency in the residential sector. Energy Economics 52, S17−S29. https://doi.org/10.1016/j.eneco.2015.08.022.

Rangan, V.K., Karim, S., 1991. Focusing the Concept of Social Marketing. Harvard Business Review.

RCZM, Energy use behaviour change. http://www.rczm.co.uk/post-5.html.

Recycling. http://www.firstsolar.com/en-EMEA/Modules/Recycling.

Reyes, R., Schueftan, A., Ruiz, C., González, A.D., 2019. Controlling air pollution in a context of high energy poverty levels in southern Chile: clean air but colder houses? Energy Policy 124, 301−311. https://doi.org/10.1016/j.enpol.2018.10.022.

Sarstedt, M., Mooi, E. (Eds.), 2014. A Concise Guide to Market Research: The Process, Data, and Methods Using IBM SPSS Statistics, second ed. Springer Berlin Heidelberg; Imprint; Springer, Berlin, Heidelberg.

Shell Sustainability Report 2017, 2018. https://reports.shell.com/sustainability-report/2017/.

Smaliukiene, R., 2007. Stakeholders' impact on the environmental responsibility: model design and testing. Journal of Business Economics and Management 8, 213−223. https://doi.org/10.1080/16111699.2007.9636171.

Smaliukiene, R., Chi-Shiun, L., Sizovaite, I., 2014. Consumer value co-creation in online business: the case of global travel services. Journal of Business Economics and Management 16, 325−339. https://doi.org/10.3846/16111699.2014.985251.

Smaliukienė, R., 2005. Public−private partnership and its influence to corporate social responsibility. Public Policy and Administration 1, 69−76.

Smaliukienė, R., Bekešienė, S., Chlivickas, E., Magyla, M., 2017. Explicating the role of trust in knowledge sharing: a structural equation model test. Journal of Business Economics and Management 18, 758−778. https://doi.org/10.3846/16111699.2017.1317019.

Song, Q., Li, J., Duan, H., Yu, D., Wang, Z., 2017. Towards to sustainable energy-efficient city: a case study of Macau. Renewable and Sustainable Energy Reviews 75, 504−514. https://doi.org/10.1016/j.rser.2016.11.018.

Strachan, P.A., Cowell, R., Ellis, G., Sherry-Brennan, F., Toke, D., 2015. Promoting community renewable energy in a corporate energy world. Sustainable Development 1. https://doi.org/10.1002/sd.1576.

Thøgersen, J., 2017. Housing-related lifestyle and energy saving: a multi-level approach. Energy Policy 102, 73−87. https://doi.org/10.1016/j.enpol.2016.12.015.

Thomson Reuters Top 100 Global Energy Leaders Report. https://www.thomsonreuters.com/content/dam/ewp-m/documents/thomsonreuters/en/pdf/reports/thomson-reuters-top-100-global-energy-leaders-report.pdf.

Tvaronavičienė, M., 2018. Towardssustainable and secure development: energy efficiency peculiarities in transport sector. Journal of Security & Sustainability Issues 7.

UNEP, 2017. TheEmissions GapReport 2017: A UN Environment Synthesis Report. United Nations Environment Programme (UNEP), Nairobi, Kenya.

United Nations Environment Programme, 2018. Emissions Gap Report. http://wedocs.unep.org/bitstream/handle/20.500.11822/26895/EGR2018_FullReport_EN.pdf.

US Department of Energy, Office of Energy Efficiency & Renewable Energy, Community-Based Social Marketing Toolkit. https://www.energy.gov/eere/better-buildings-residential-network/downloads/community-based-social-marketing-toolkit.

Vargo, S.L., Maglio, P.P., Akaka, M.A., 2008. On value and value co-creation: a service systems and service logic perspective. European Management Journal 26, 145−152.

Weckend, S., Wade, A., Heath, G., 2016. End-of-life Management: Solar Photovoltaic Panels. https://www.irena.org/-/media/Files/IRENA/Agency/Publication/2016/IRENA_IEAPVPS_End-of-Life_Solar_PV_Panels_2016.pdf.

D.Williams, Shell Pushing for More Dutch Wind Power. https://www.powerengineeringint.com/articles/2017/10/shell-pushing-for-more-dutch-wind-power.html.

Yakovleva, N., Crowther, P.D., 2005. Corporate Social Responsibility in the Mining Industries. Taylor and Francis, Florence.

Yusuf, H.O., Omoteso, K., 2016. Combating environmental irresponsibility of transnational corporations in Africa: an empirical analysis. Local Environment 21, 1372−1386. https://doi.org/10.1080/13549839.2015.1119812.

Subject Index

Author Index

Printed in the United States
By Bookmasters